AF248542

196
Topics in Current Chemistry

Springer

Berlin
Heidelberg
New York
Barcelona
Budapest
Hong Kong
London
Milan
Paris
Singapore
Tokyo

Carbon Rich Compounds I

Volume Editor: A. de Meijere

With contributions by
A. de Meijere, R. Haag, S. Hagen, H. Hopf, B. König,
D. Kuck, T. J. Seiders, J. S. Siegel, Y. Sritana-Anant

Springer

This series presents critical reviews of the present position and future trends in modern chemical research. It is addressed to all research and industrial chemists who wish to keep abreast of advances in the topics covered.

As a rule, contributions are specially commissioned. The editors and publishers will, however, always be pleased to receive suggestions and supplementary information. Papers are accepted for "Topics in Current Chemistry" in English.

In references Topics in Current Chemistry is abbreviated Top. Curr. Chem. and is cited as a journal.

Springer WWW home page: http://www.springer.de
Visit the TCC home page at http://www.springer.de/

ISSN 0340-1022
ISBN 3-540-64110-6
Springer-Verlag Berlin Heidelberg New York

Library of Congress Catalog Card Number 74-644622

Cover design: Friedhelm Steinen-Broo, Barcelona; MEDIO, Berlin
Typesetting: Fotosatz-Service Köhler OHG, 97084 Würzburg

SPIN: 10649212   66/3020 – 5 4 3 2 1 0 – Printed on acid-free paper

Preface

Not too long ago, graphite and diamond were the only two known modifications of carbon. That changed dramatically with the discovery of C_{60} in 1985 and all the higher fullerenes soon thereafter. Nevertheless, this breakthrough did not stand alone in paving the way to the new era of chemical and physical research into carbon rich compounds that we are now enjoying. The last 20 years have witnessed the development of a powerful repertoire of new carbon-carbon bond forming processes, especially metal-catalyzed and metal-mediated ones. These, together with other useful organic synthetic methodologies, including important protection and deprotection procedures, have made the synthesis of new targets possible that were inconceivable previously.

Almost 100 years of aromatic and polycyclic aromatic hydrocarbon chemistry were compiled in two volumes of the then famous book by Erich Clar in 1964. But what has polycyclic aromatic hydrocarbon chemistry undergone during the last thirty years? There are completely new ways of assembling six carbon atoms in a cycle, and there are surprisingly efficient and selective modes of constructing oligosubstituted benzene derivatives and impressive aggregates of benzene rings with other substructures. Nowadays there seems to be no limits to achieving what the fantasy of the chemist can come up with. Combinations of aromatic moieties with aliphatic subunits or with five-, four-, and three-membered rings, rod-shaped oligoacetylenes with bulky end groups, almost everything seems to be possible.

High-temperature pyrolysis techniques are now applied with surprising selectivity, leading to all sorts of bowl-shaped molecules, which are substructures of C_{60} and higher fullerenes. Combinations of saturated and unsaturated, acyclic and cyclic fragments with acetylenic subunits have led to two-dimensional and three-dimensional arrays and all-carbon networks.

In short, the time has undoubtedly come to present recent development in the field of carbon rich compounds compiled in the form of several up-to-date volumes with articles written by active practitioners of these arts. The term "carbon rich" is used here to refer to everything that has a carbon to hydrogen ratio of $1:(\leq 1)$, but it can also include formally saturated hydrocarbons of the general formula C_nH_n which frequently can serve as precursors to compounds with a higher C:H ratio.

Göttingen, May 1998   Armin de Meijere

Contents

Contents of Volume 194
Microsystem Technology
in Chemistry and Life Sciences

Volume Editors: A. Manz, H. Becker

Contents of Volume 195

Biosynthesis
Polyketides and Vitamins

Volume Editors: F. J. Leeper, J. C. Vederas

Design of Novel Aromatics using the Loschmidt Replacement on Graphs

Yongsak Sritana-Anant · T. Jon Seiders · Jay S. Siegel

Department of Chemistry, University of California, San Diego, La Jolla, California 92093–0358.
E-mail: jss@chem.ucsd.edu

Graphs are seen as an excellent place to start the design of a non-natural product for total chemical synthesis. In the design process, the vertices of the graph are replaced by a chemical "radical" to create a molecular graph. Classical examples of this are platonic caged hydrocarbons that arise from CH replacements for the vertices of a platonic polyhedron. In the present context of carbon-rich aromatic structures, we demonstrate how benzene-ring replacement can lead to larger targets as well as a greater variety of theoretically interesting molecules. In honor of the first chemist to think of benzene as a group element, we call this replacement process the Loschmidt replacement.

Keywords: Cyclophane, Synthesis, Benzenoid Hydrocarbons, Arenes, Polyphenyls.

1
Introduction

Mathematical models such as platonic solids, topological structures, and molecular graphs have served to motivate challenging non-natural product total

Topics in Current Chemistry, Vol. 196
© Springer Verlag Berlin Heidelberg 1998

syntheses. In each case, the relationship between the mathematical model and the chemical constitution is abstract and to some extent subjective, which leads to a greater emphasis on the molecular design. The chemist must decide which features of the mathematical model need to be represented and which chemical features are allowed to manifest that representation. For example, preparation of a spherical molecule is impossible beyond the single atom, although many chemists would consider the synthesis of buckminsterfullerene (C_{60}) as a reasonable surrogate. As such, the highly symmetrical reticulation of 60 carbon atoms on a sphere becomes the sphere. In this example, a shape is mapped to a graph which is mapped to a molecule. Dodecahedrane might also be accepted as a spherical molecule. It has the same symmetry as C_{60} and can also be represented as a graph inscribed on the surface of a sphere. Both examples use a dodecahedral graph as their inspiration point, but each substitutes a distinct chemical entity to represent the points of the graph. For dodecahedrane the point is replaced by CH, and for C_{60} each point is expanded to a benzene ring. The point-CH replacement has been commonly used in caged hydrocarbon chemistry to lead to marvelous structures and syntheses. What follows is a discussion of how the point-benzene replacement can equally enrich the design and synthesis of new carbon rich structures.

Historically, chemists have marveled at the stability of the benzene nucleus. A myriad of reactions can be done without altering the core of six carbons. As such, benzene could loosely be called a group element. Indeed, this is the position taken by Loschmidt in his 1860 book "Konstitutionsformeln der Organischen Chemie in Graphischer Darstellung" [1]. Therein, he represents all elements as annotated circles and ascribes elemental character to the ring of six carbons in benzene. Arguably then, it was Loschmidt who first used the point-benzene replacement as a way to focus his thoughts about aromatic chemistry. In deference to his seminal contribution in this area, we will name the point-benzene replacement as the Loschmidt replacement.

The goal of this exercise is to stimulate the imagination and thereby generate new chemistry. Toward that goal, we shall discuss selected aromatic molecules and their syntheses within the context of the Loschmidt replacement. We will also set down some designs that can be generated from this analysis and suggest some target families for future pursuit. As stated at the outset, the nature of this area is somewhat fuzzy, but this should be seen as a benefit for expanding the abstraction and discussion of our ideas of chemical structure. This taxonomy should not be taken as archival, because of the danger that the discussion will spiral downward toward arguments as to what class a specific molecule belongs rather than how a specific graph can be transformed into new chemistry. Therefore, we will loosely group structures by category, but take poetic license to impose our own ad hoc classification if ambiguity arises.

Some examples of how graphs and molecules might be transformed follow, but greater details will be given in a later section. Simple linear graphs will correspond to the polyphenyls or linked phenyl chains. For example, two points connected by a single line would represent biphenyl as well as bibenzyl. At this level of discussion no distinction is made between ortho-, meta-, or para-terphenyl; all three are mapped onto a linear chain of three points. Fused rings are

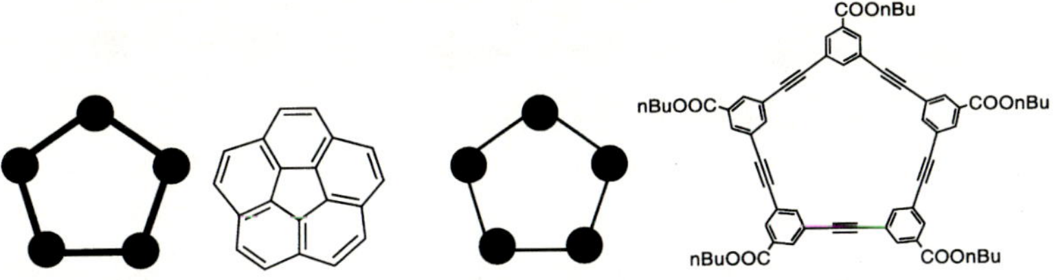

Fig. 1. Graph representations of corannulene and Moore's cyclic phenylacetylene pentamer

represented as a weighted graph (i.e., a graph with weighted edges). For example, naphthalene maps onto a graph with a single line of order two between two points; minus the weighting factor this approach to graphing polynuclear aromatic hydrocarbons (PAHs) is present in the concept of the "inner dual" [2, 3]. Dihydroanthracene would be depicted as two points with two independent lines; i.e., a cyclic graph with two points. Anthracene and phenanthrene on the other hand would be represented as linear graphs of three points in which the connections between adjacent points were of order two. Thus, corannulene [4–6] would be a pentagon with each connection of order two, whereas the cyclic phenylacetylene pentamer of Moore et al. [7] would be a simple pentagon (Fig. 1).

In general, graphs can be linear, branched, cyclic and polycyclic. To a first approximation, this classification acts as a reasonable outline for discussion. Starting from the simple linear series and progressing through the topologically constrained polycyclic graphs, we will sample some classical and contemporary aromatic constructs with an eye on how they might have been designed from a Loschmidt replacement strategy, and a rudimentary knowledge of graph theory.

2
Nature of the Graph

The concept of the molecular graph is rarely taught in basic chemistry classes but is so intuitive that it is almost second nature to every chemist [3]. In graph theory, a graph is simply a collection of points and their connections. Compare that to the chemist's definition of a molecule as a collection of atoms and bonds; atoms are often regarded as "point-atoms" and bonds are defined as that which connects atoms in a stable array. For an assembly of "point-atoms," given that we agree on the bonding, we could all write down the "molecular graph" without any problem. Assuming the two-electron covalent bonds of Lewis, the molecular graph looks very much like a Lewis dot structure. Complications can arise if different degrees of interaction are considered important, for example, hydrogen bonds, dative bonds, or even polar and van der Waals interactions. In this area, it is the chemical definition of a connection that alters the nature of the graph. For the extreme case in which all pairwise atomic interactions are considered, we arrive at what has been defined as the "complete molecular graph" [8] in which each pair of atoms in a molecule are considered to have a "weighted" connection. Within the workings of a computer, as in an ab initio computation,

Fig. 2. Simple, weighted and multiple connections

the complete molecular graph can be quite useful, but the complexity of such graphs is often unnecessary for common chemical parlance and therefore the valence bond limit is a useful convention.

In the above context, the perspective was one of how do we describe a molecule in terms of a graph; however, it is the reverse mapping which leads to new molecular designs. So let us review quickly some basic features of graphs, specifically connected graphs, as molecular prototypes.

Points in a graph that are directly connected are called neighbors, and the number of neighbors for a given point define that point's degree or valence, much the same as the valency concept in chemistry. Connections between two points in a graph can be simple (a single line of unit weight), weighted, or multiple (two or more independent connections) (Fig. 2). Weighted and multiple connections may seem chemically confusing or even redundant at first, but keep in mind that our discussion is going to evolve beyond simple point-atoms and their neighboring connections. A special class of connections, called loops, connect a single point with itself; such connections are part of "general" but not simple graphs.

The fundamental structures of a graph can be in the form of trees and cycles. Trees have termini (points of valence 1), and can take on forms of chains, branches, and stars. Pure cyclic or polycyclic graphs have points with a minimum valence of 2. Complex graphs can be derived from hybrids of these tree and cycle structural types. The valence of the points in a graph can be used to categorize a graph further. For example, if all points have the same valence, the graph is called "regular" and if all valences, for a graph of n points, equal n-1 then the graph is complete, as is the case in the complete molecular graph described above. Coloring the points of a graph such that no two neighbors have the same color leads to what are called partite graphs and when the number of colors is restricted to two they are called bigraphs (bipartite graphs).

Relationships between graphs are important, just as relationships between molecules. Two basic relationships are isomorphs and homeomorphs. Isomorphs are chemically closest to stereoisomers. For two graphs to be isomorphic they must have the same number and type of points and connections, and differ only in their arrangement in space (what mathematicians call their representation). The chemical counterpart of homeomorphs is rooted in our use of the functional group or chemical family categorizations. For example, the molecular graphs of all monocyclic diketones are homeomorphic. Rigorously, two graphs are homeomorphic if they can be rendered isomorphic by addition or deletion of points with valency of two. Additionally, it is often desirable to group together graphs that share a fundamental connectivity pattern called a sub-graph. A sub-graph of a larger graph is simply a subset of points and connections all of which are contained in the larger graph. Two homeomorphs always have at least one identical subgraph. In the realm of molecular graphs, the mathematical correspondent can be restricted to what are called spatial graphs [9].

Graphs have long been related to polyhedra. For example, graphs with points, connections, and cycles adhering to Euler's polyhedral formula ($P - Cn + Cy = 2$) are called polyhedral graphs. Specifically, the complete graph of four points (K_4) is a tetrahedral graph. Highly symmetric polyhedral graphs which correspond to one of the five Platonic solids form a special sub-class called Platonic graphs.

Graphs can also be related to topologies [10, 11]. For example, graphs which can be represented in two dimensions without crossing lines are called planar graphs. The complete graph of five points (K_5) and the complete bigraph of six points ($K_{3,3}$) are the simplest graphs for which no planar representation is possible and as such are topologically different from planar graphs and called nonplanar graphs. All other nonplanar graphs have K_5 or $K_{3,3}$ as subgraphs. Non-crossing representations of K_5 and $K_{3,3}$ can be "embedded" onto a Möbius strip. Cyclic graphs can have representations that show crossings when presented in two dimensions, for example, knots and links. These graphs are not only different topologically from the planar graphs, but also from the non-planar graphs related to K_5 and $K_{3,3}$. In the case of the linked or knotted cyclic graphs, there exists an isomorphic graph that has a planar representation, but for K_5 and $K_{3,3}$ no planar isomorph exists.

In these few paragraphs, we can see a wide variety of fundamental graphs and their related shapes and topologies. It remains for us to delve into how these graphs can be used to design new synthesis targets and enhance our appreciation of classical molecules.

3
From Graph to Molecule by Loschmidt Replacement

Within the constraints of the Loschmidt replacement, points on a graph become transformed into benzene nuclei and the connections between points are "wild card" variables. Molecular progeny will stem from each of the basic graph types: linear, branched, cyclic, polycyclic, Platonic, and topological.

A simple two dot picture can be transformed into a chain terminated by phenyl rings (e.g., biphenyl, tolan, or any α,ω-diphenylalkane). Branching graphs transform into structures from the simple 1,3,5-triphenylbenzene to hyperbranched polymers and dendrimers.

Loops correspond to compounds like ansa cyclophanes, but equally well they might induce the design of novel annelated benzenes. Simple and polycycles manifest the familiar classes of cyclophanes and cryptophanes commonly synthesized by practitioners of "supramolecular" chemistry [12, 13]. The Platonic graphs are a subset of the polycyclic, but, because of their potential for high symmetry and structural simplicity, they form an especially provocative group.

The topological graphs present a complexity because the spatial orientation of their connections becomes an integral part of their design. Adding a restriction to the convention that each connection is constricted to an area such that no two connections cross, makes possible several topologically distinct designs. A simple example comes from the bigraph (bipartite graph) $K_{3,3}$, for which there

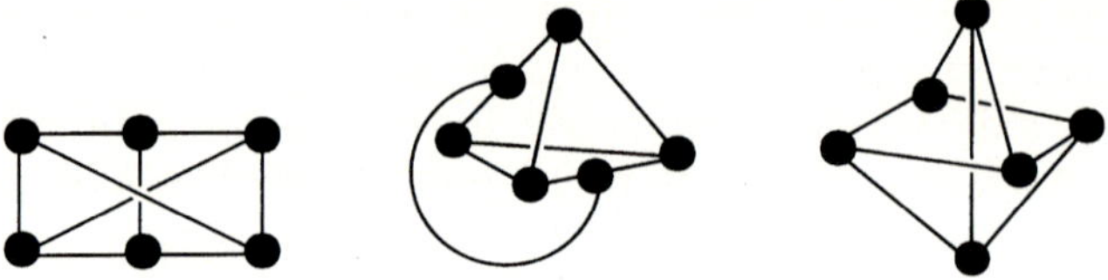

Fig. 3. Three representations of $K_{3,3}$: C_2, C_s, and D_{2d} symmetry

are several topological representations (e.g., Möbius C_2, looped C_s, and netted D_{2d} symmetric) (Fig. 3). Even though all points in the graph represent benzene rings the bigraph nature is required. Membership in one of the two families of three points is a consequence of the adjacency enforced by the nonplanar connectivity; a point in one family of three points is adjacent to all members of the other family but to none of its own family.

The transformation between molecules and graphs within our convention does not provide a seed for every benzene-containing molecule. One problem that arises is the fact that in chemistry not all connectors are restricted to connecting only two points together. Take for example triaryl-X compounds like triphenylmethane or triphenylphosphine. In order to include such structures into the plan, alternative point replacements are necessary. The multipartite idea is one approach where this might be useful. Point replacements coming from benzene would form one family (or color) of vertices and points coming from "connectors" of more than two benzene points would form another. In such a scenario triphenylmethane could then stem from a bigraph in which the benzenes would be represented by solid points and the CH that connects to each of them would be an open point (Fig. 4). Extensions, in general, will be discussed further in Sect. 9.

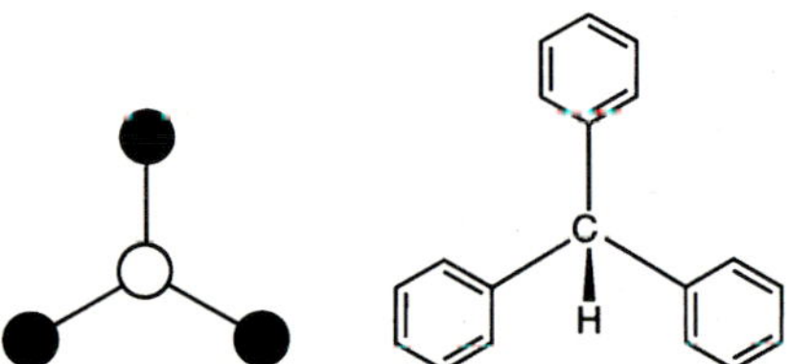

Fig. 4. Graph to molecule relationship using a bigraph

Just as the Loschmidt replacement gives us the freedom to generate many different molecules from one fundamental graph, it is also useful to recognize that a single structure may be generated by several different graphs. This fuzziness can arise in situations where a six-membered ring is created by the connections between several other six-membered rings. The context of the generation will be of great importance in such cases. For example, triphenylene can be generated from a simple triangular graph or from a radial spoked graph in which each spoke connection is of weight 2 (Fig. 5). Despite the redundancy

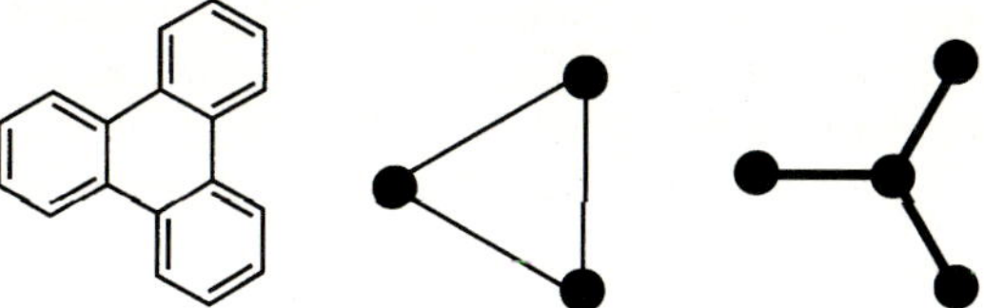

Fig. 5. Two graphs representing triphenylene

of mappings between graphs and molecules the process is not random, and changing the generator graph frees one to think of a single structure in multiple contexual ways.

One point to keep in mind is that the goal of this discussion is to stimulate the creation of new and perhaps wonderfully complex molecular designs from relatively simple graphs. As such, a balance between the number and kind of extensions, and the variety of molecules generated, should be maintained.

4
Linear and Branched Structures

In this and the following sections, we want to exemplify the immense variety of structures that can be related to these simple graphs. Selections from the classical chemical literature serve as a foundation, but we have tried to place an emphasis on newer syntheses (1990–1997) that show extreme extensions of a simple or common motif.

The simplest graph is a point, which we have already correlated to benzene (**1**) as by Loschmidt. [1] In principle, any PAH could be mapped to the point as well, but for our exposition we restrict ourselves to the six-member all sp^2-carbon ring. Connections between two points can have two weighted variations in our discussion. A weight of one, symbolized by a light line, represents connections between two independent six-membered rings as in biphenyl (**2**) or tolan (**3**). A weight of two, symbolized by a bold line, represents fused connections as in naphthalene (**4**). From these two weighted connections various chains can be constructed.

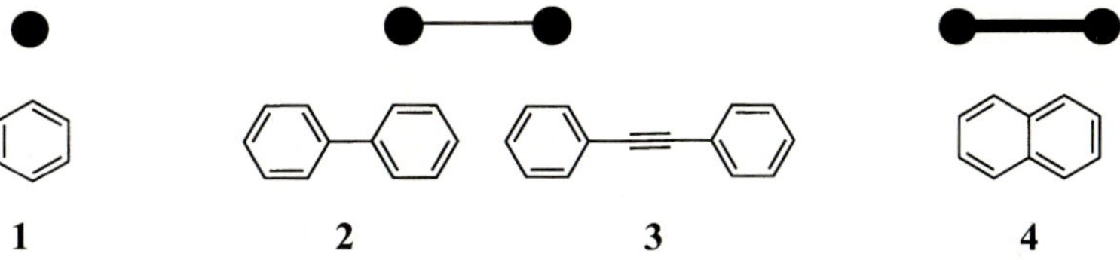

1   2   3   4

The "infinite" linear chain of light connections has become popular with the heightened interest in the materials properties of polyphenylene, polyphenylenevinylene, and polyphenylacetylenes. From the topological point of view, whether the linkage is ortho-, meta-, or para- is immaterial, so the same graph leads to the "rigid-rod" polymers of Tour et al. (**5**) [14] or Weiss et al. (**6,7**) [15], the zig-zag systems of Grubbs and Kratz (**8**) [16], or the helical folding polymers of Moore et al. (**9**) [17]. The beautiful enantiomerically pure helical binaphthol polymers of Pu et al. (**10**) [18] stem from a light main-chain graph with bold spurs.

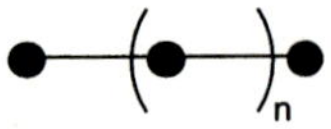

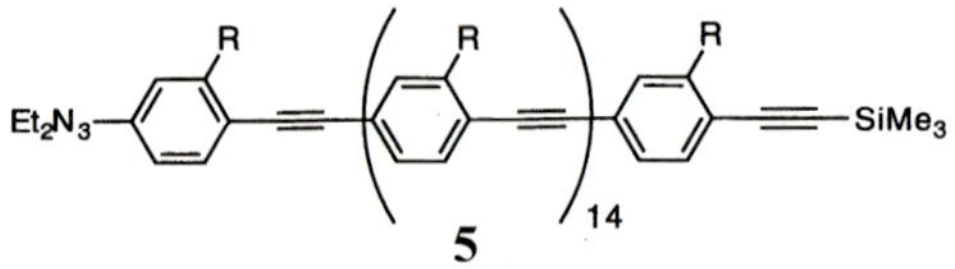

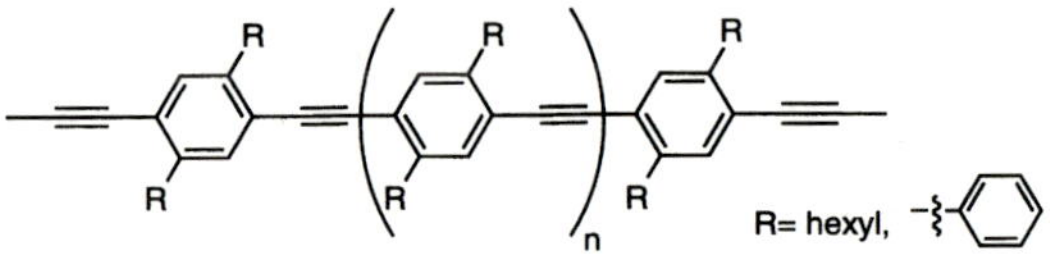

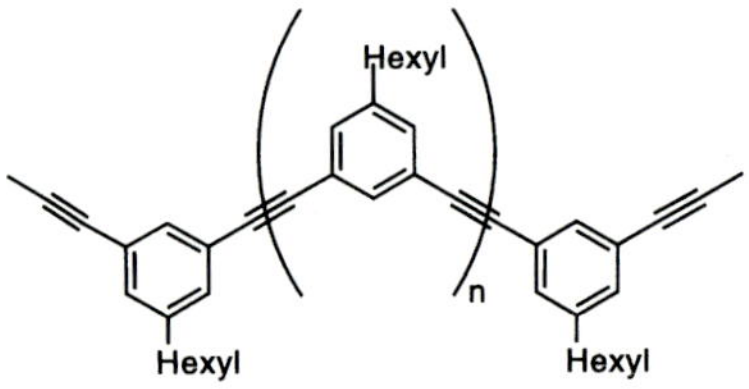

Long bold chain graphs can be seen in the classical helicene creations of Newman and Lednicer (**11**) [19], and Martin and Baes (**12**) [20], modernized by Katz et al. [21], and in the recent phenacene syntheses of Mallory et al. (**13**) [22]. Bold main chains with light spurs are the theme of many new polyphenyl arenes

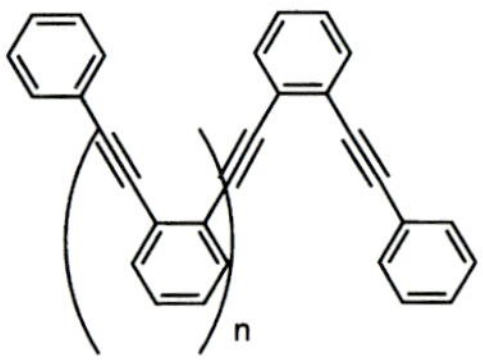

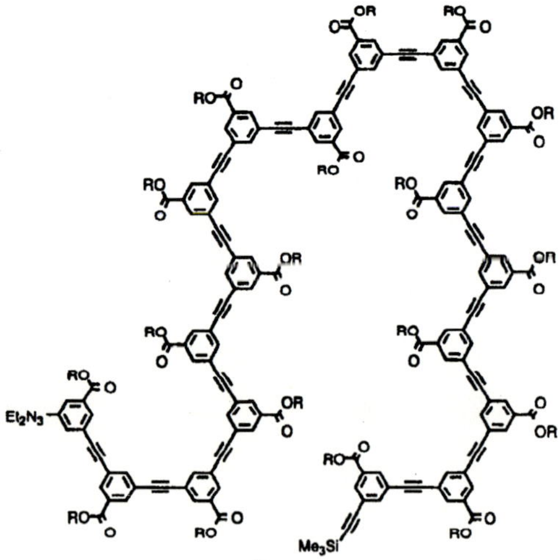

R = OC_6H_{13}

10

11

12

Pentyl   Pentyl   Pentyl   Pentyl

13

Ph Ph
Ph   Ph
Ph   Ph
Ph Ph

14

Ph Ph Ph
Ph   Ph
Ph   Ph
Ph Ph Ph

15

Ph Ph Ph
Ph
Ph
Ph Ph Ph

16

Ph Ph   Ph Ph
Ph   Ph Ph   Ph
Ph   Ph
Ph Ph   Ph Ph

17

coming from Pascal (**14–17**). Indeed, Pascal has mixed and matched light and bold connection to make his various "twistoflex" [23, 24] and "albatross" [25] structures.

Starting from a central point and bursting out in a star topology has been the theme of dendrimer chemistry in general. Star structures, such as those of

MacNicol et al. (**18**) [26], have shown interesting inclusion phenomena of which more remains to be explored. Heck coupling has made possible many carbon-rich systems such as the hexaphenylethynylbenzene executed by Heck et al. (**19**) [27], and the hexaphenylethenylbenzene of de Meijere et al. (**20**) [28, 29]. This idea has been extended beautifully by Müllen et al. (**21**) [30] and Moore and Xu (**22**) [31] in their starburst phenylenes.

"Starphene" (**23**) was coined by Clar and Mullen for structures that expanded on the terphenylene motif [32], an all bold connected star graph. Pascal has pushed the envelope on the parent system by preparing the highly distorted perchlorotriphenylene (**24**) [33], despite "common opinion" that this molecule could not exist.

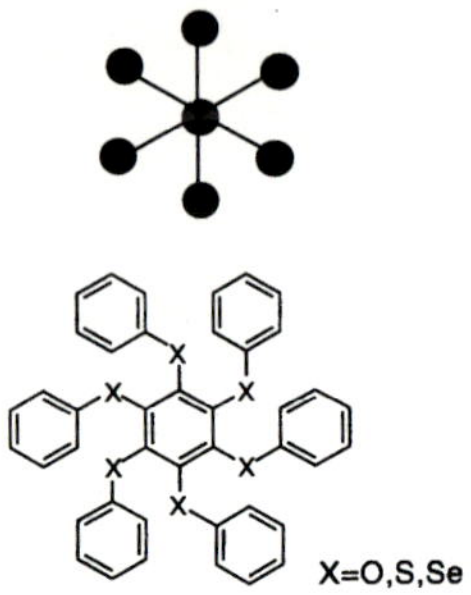

18 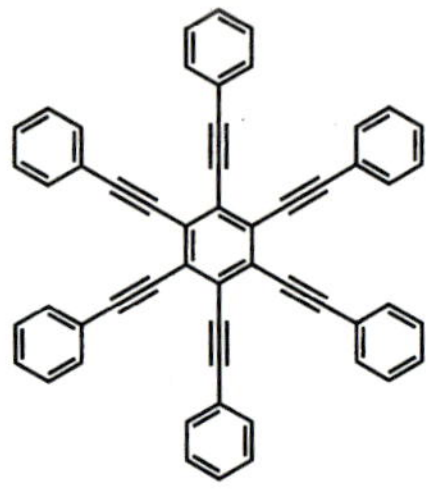   **19** 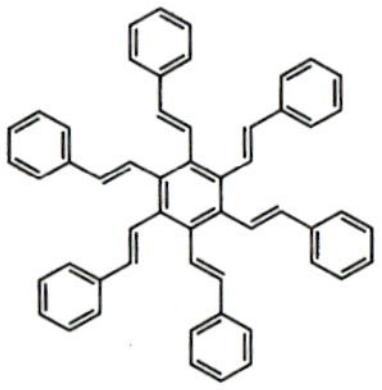   **20**

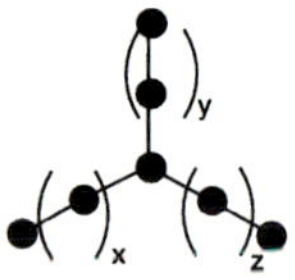

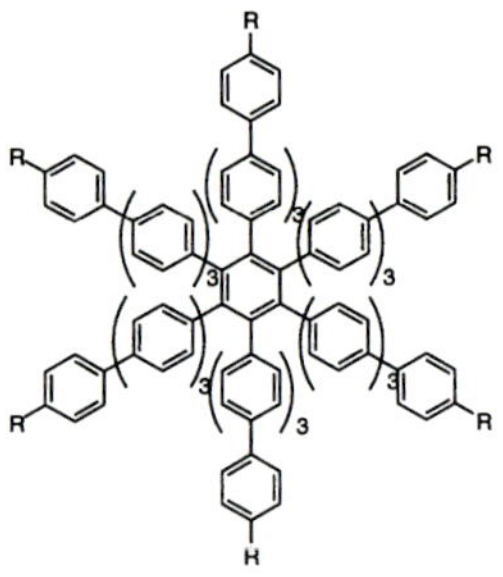

21 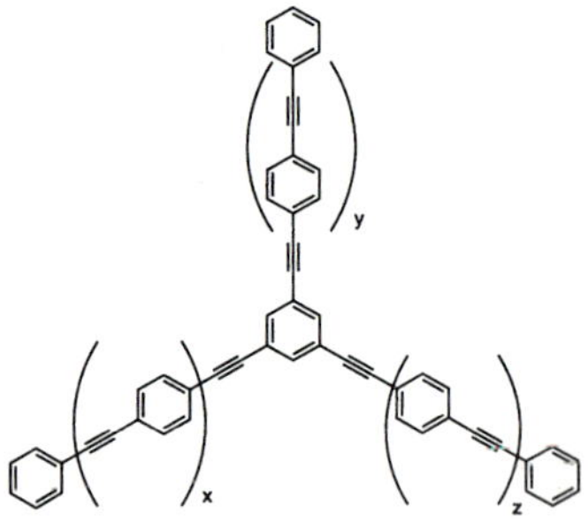   **22**

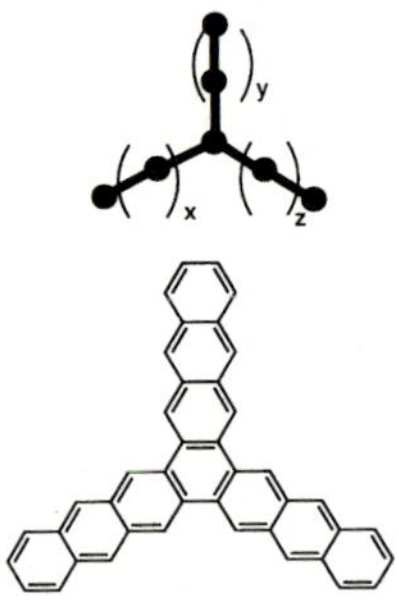

23 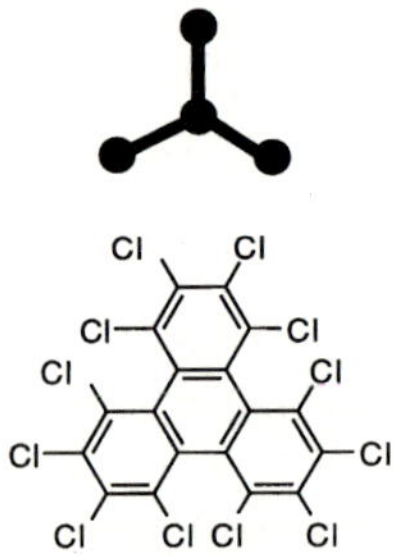   **24**

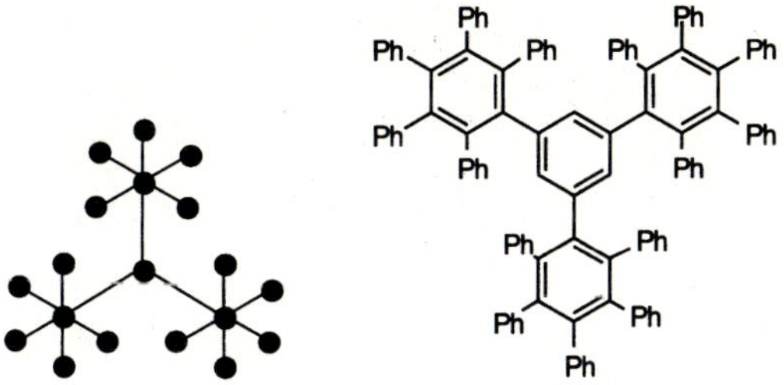

25

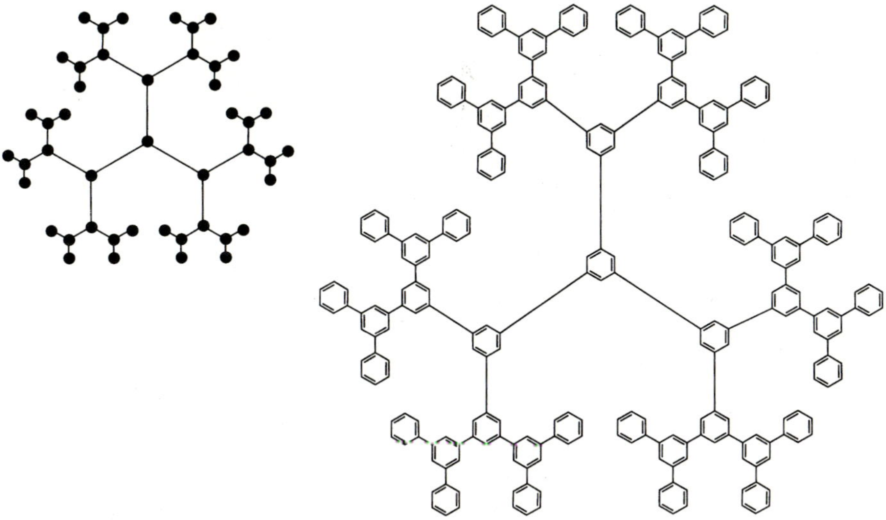

26

Full fledged dendritic graphs are well represented. Their genera range from the compact tris(pentaphenylphenyl)benzene (**25**) of Pascal et al. [25] to the dendritic fifth-generation phenylene of Miller et al. [34]; the latter adds the special flavor of a derivative with a perfluorinated last generation (**26**), which alters the chemical properties dramatically. The Godzillas of this family are produced by Fréchet et al.'s convergent dendrimer synthesis (**27**) [35], and Moore et al.'s orthogonal coupling chemistry (**28**) [36, 37]. The dendritic class is based primarily on light connected graphs and still has not seen the full exploitation of the mixed bold/light motif.

Such is a sampling of what can be dreamed up within the restrictions of acyclic graphs with only six-membered rings for points and two types of connection. In the next section we elaborate our graphs with cycles to explore more intricate systems.

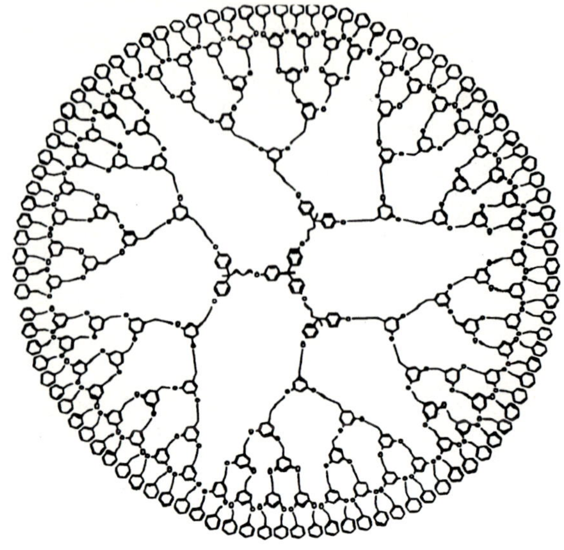

27

28

5
Monocyclic Structures

Cyclic structures begin with monocycles and monocycles begin from loops. In a loop a single point is connected back to itself. The chemical relevance of this graph is readily seen in the classical cyclophane work of Bickelhaupt et al. (**29a** (n = 4), **29b** (n = 5)) [38, 39] and Tochtermann et al. (**29c** (n = 7)) [40]. Modern twists on this theme of simple benzene derived cyclophanes have come through the adamantyl bridged molecule of Vögtle et al. (**30**) [41]. Clearly, tetralin would also fall into this class, as would the highly strained cyclopropabenzene, first synthesized by Anet and Anet [42].

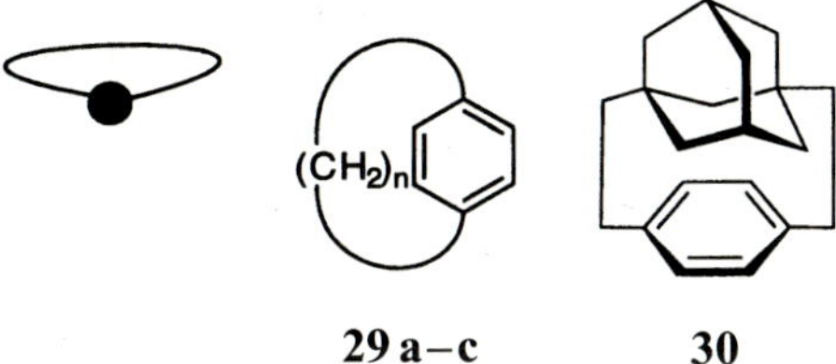

Given benzene's six possible connections it is reasonable to have multiple loops from the same point. In this line of cyclophane chemistry come the chiral "betweenabenzene" structures of Nakazaki et al. (**31**) [43]. Here let us point out that the spatial restrictions placed on the molecule are not reflected in the graph; thus, the strained achiral [*a*]-cyclopropa-[*c*]-cyclobutabenzene (**32**) [44] of Vollhardt and Saward maps to the same doubly-looped graph. The spatial restriction of graph connections is going to play a bigger role when we speak of graph/molecules with topological isomers.

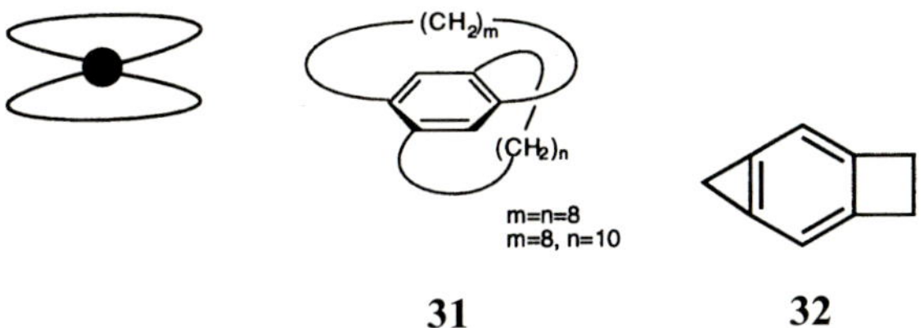

The extension to three loops has led to the highly strained trisannelated benzene of Billups et al. (**33**) [45], as well as the "cyclohexatriene" of Siegel et al. (**34**) [46, 47]. Still unrepresented are analogs of the betweenabenzenes with all three pairwise paraconnections hooked up.

Loops can occur from any point in a graph, as seen in the naphthalenophanes of Bickelhaupt et al. (**35**) [48] and the anthracenophanes of Tobe et al. (**36**) [49]. Such motifs also play well in the design of molecular receptors, such as multiple crown ethers organized around a central aromatic template.

The first true cycle comes about when two points are linked by two connections. This is not to be confused with a single bold line (weight 2) between two six-membered rings; in our convention naphthalene is linear. [2.2]Paracyclophane (**37**), first prepared and identified by Brown and Farthing [50, 51] would be a classic example of the two-point cycle. More recently, Hopf et al. (**38**) [52] and Fallis and Romero (**39**) [53] have incorporated alkyne functionality into the classic design, which has gussied up the synthesis as well as the resultant chemistry. A nice mix of bold and light connections in the simple two-point cycle is reflected in Dougherty et al.'s "jump rope" naphthalenes (**40**) [54]; these compounds show the potential importance of entropic control over conformational dynamics.

Three-point cycles abound. The [2.2.2]paracyclophane of Cram and Dewhirst (**41**) [55], tribenzo[12]annulene originated by Campbell, Staab and coworkers (**42a**, x = y = z = 1) [56, 57] and updated by Youngs et al. [58], and the cyclotriveratrylenes (**43**) [59] all represent high profile examples of these structures. Expanded versions of the annulenes have since come from Vollhardt et al. (**42b**, x = 2, y = z = 1) [60] and Haley et al. (**42c**, x = y = 2, z = 1) [61], and an all-hydrocarbon version of cyclotriveratrylene (**44**) [62] has come from Magnus et al. An interesting cognate of the [n.n.n]paracyclophane was made by Gleiter et al. and uses a diazacyclodecadiyne (**45**) [63] as the connector. Combinations of cyclic and dendritic motifs can lead to structures where the core is a cavity and the shell is sterically closest packed. Collet and Malthête have started down that path with their grafted cyclotriveratrylene (**46**) [64].

41

42 a–c

43

44

4 45

∠46

The abundance of four-point cycles could allow things to get out of hand, so we severely constrain our discussion to hydrocarbons. The elegant cyclotetraphenylenemethylene of Miyahara et al. (**47**) [65] is a classic in this area. Calixarenes are typified by calix[4]arene (**48**) and comprise so many beautiful variations as to constitue a class of their own which has been amply reviewed [66, 67].

More recently, Biali and Goren have tackled the problem of deoxygenating calixarenes and made viable a number of important hydrocarbon derivatives like **49** [68]. Introduction of acetylene spacers has become the rage, especially given the great synthetic advances made in palladium catalyzed carbon-carbon bond forming reactions. Vögtle et al. (**50** (n=1,2)) [69], Moore et al. (**51**) [7], Oda et al. (**52**) [70], Haley et al. (**53** (x=1 y=z=0 and x=y=z=0)) [61], and Vollhardt et al. (**53** (x=z=0=1)) [60] have each presented important additions to this area. Herges et al. have contributed a mixed four- (**54**) [71] and five-point [72] cycle shaped like a carbon tube.

The classic cyclic hexaphenyl (**55**) [73] of Wittig and Rumpler and spherand of Cram et al. (**56**) [74] reveal that the expanded cycles are well within the chemist's grasp. Higher order cycles (e.g., 5–10) with meta-related acetylene spacers (**57**) have been prepared with remarkable ease by Moore et al. [7]; ortho derivatives with up to 50 benzene rings have been claimed by Youngs et al. [75, 76]. Oda et al. have recently generalized an approach to the vinyl analogs (n=4,6, and 8) (**58,59**) [77]. In the polyphenylenemethylene series the para-sub-

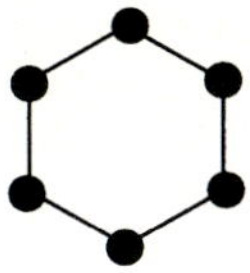

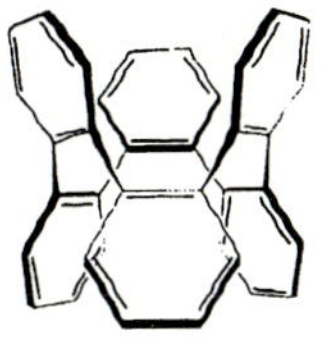

55

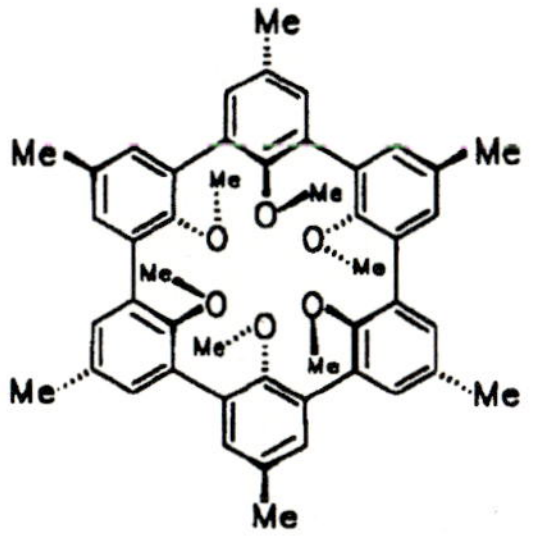

56

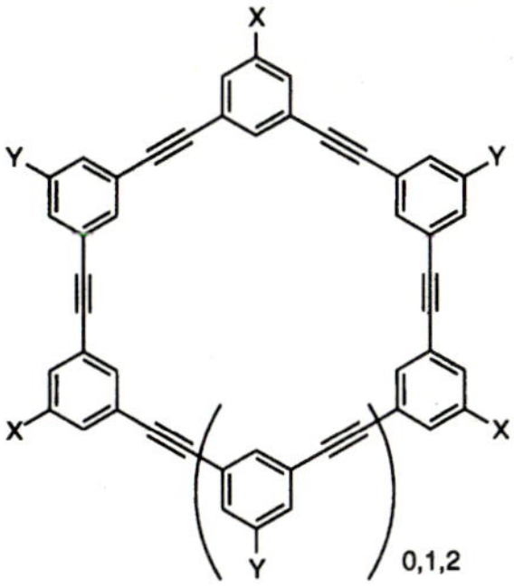

57

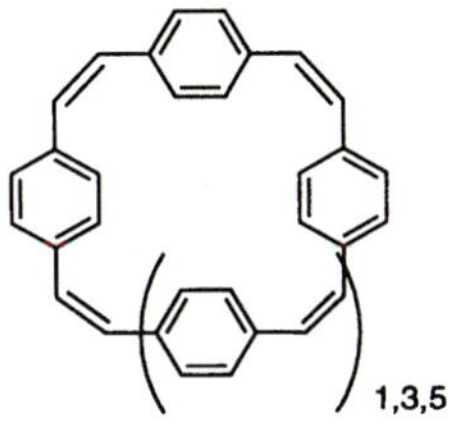

58 1,3,5

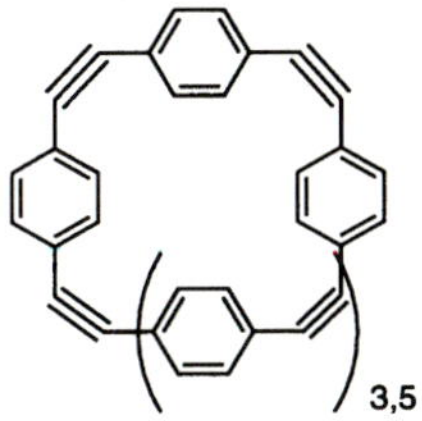

59 3,5

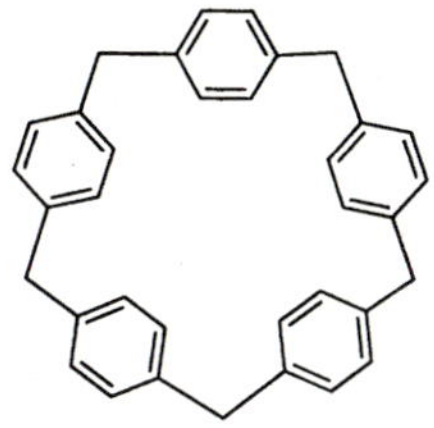

60

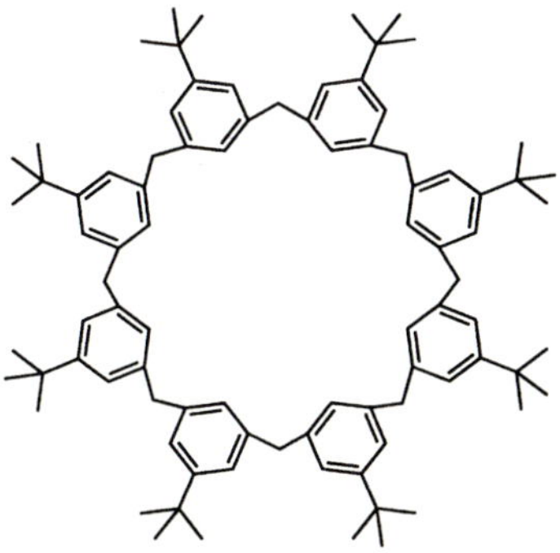

61

stituted pentamer of Gribble and Nutaitis (**60**) [78] and the deoxycalix[8]arene (**61**) [68] of Biali and Goren stand out as signal examples.

The homogeneous bold connected cycles comprise the family named circulenes by Wynberg and Dopper [79]. Corannulene (**62**) [4–6], coronene (**63**) [80], and pleidannulene (**64**) [81] represent the [5]-, [6]-, and [7]circulenes, respectively. A related 12-membered cycle, but not a circulene, can be seen in the structure of Kekulene (**65**) [82, 83].

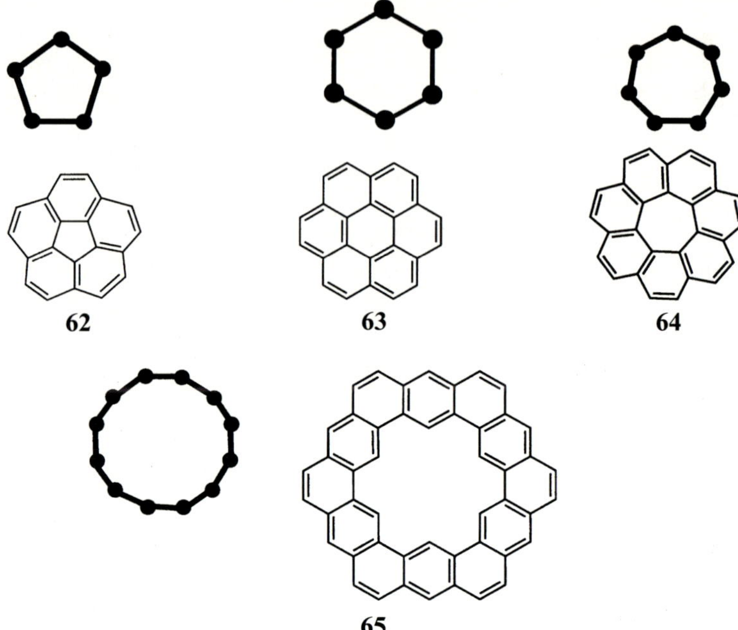

Mixed cycles of bold and light connections have stimulated a number of interesting molecular architectures. The phenanthrene dimer and related compounds of Wennerström and Thulin (**66, 67**) [84, 85] and the alkyne-linked binaphthol trimer of Diederich et al. (**68**) [86] highlight this motif.

6
Polycyclic Structures

The simplest polycycles stem from two triply-bridged points. These systems are exemplified by Trost et al.'s 4,8-dihydrodibenzo[*cd,gh*]pentalene (**69**) [87, 88], a precursor for a purturbed [12]annulene dianion, and Mislow et al.'s double-bridged biphenyl derivatives (generally shown as **70**) [89], where X is methylene, carbonyl, or various heteroatoms. Other longer bridged biphenyls include the triple-bridged cyclophanes **71** made by Hubert and Dale [90], their unsymmetrical relatives by Cram and Reeves [91], and the recent polyalkynyl cyclophane (**72**) made by Rubin et al. [92], a proposed fullerene precursor.

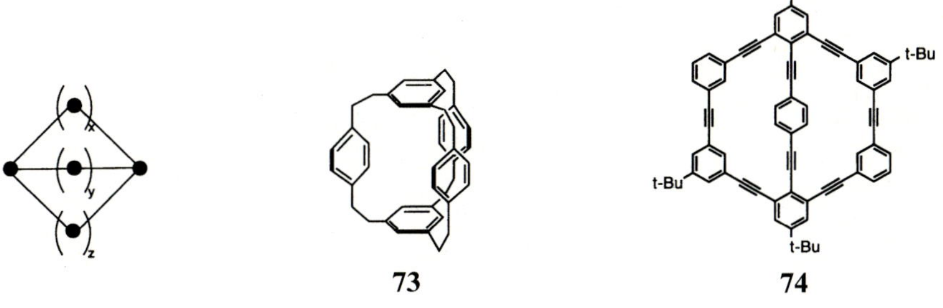

Homeomorphs of the two-point bicycle show up in an interesting collection of structures, ranging from the simple cage-type bicyclophane (**73**) [93, 94] to the extended homeomorphs by Lehn et al. [95], Vögtle et al. [96], and Moore et al. [97]. Other molecular representations are Moore and Bedard's flat molecular turnstile (**74**) [98], Hart and Vinod's cuppedophanes (**75**) [99 – 102], and Okazaki et al.'s bowl-shaped bicyclic cyclophane (**76**) [103, 104].

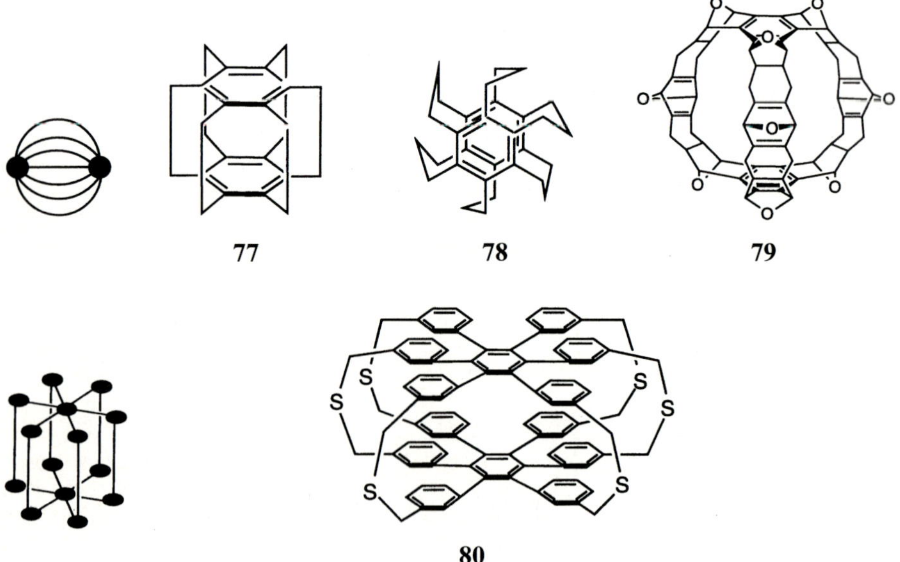

75   **76**

Numerous examples of multibridged cyclophanes have appeared in the literature. Of great interest are those with maximum connections between units. The elegant "superphane" (77), first synthesized by Boekelheide et al. [105–107], is the archetype of this family. Other variations are Shinmyozu et al.'s molecular pinwheel (78), the three-atom bridged analog of superphane [108, 109], and Stoddart et al.'s trinacene 79 [110], a triply-stripped bridged cage from a "substrate-directed" synthesis [111]. A larger homeomorphic version of superphane (80) has been synthesized by Vögtle in an impressive one-step synthesis [112].

Chained polycycles were extensively explored by Misumi et al. in their multilayered cyclophanes. The system with up to six layers (81 (n=4)) has been synthesized and studied [113–115] mainly with regard to its transannular charge transfer interactions. Vögtle et al.'s molecular ribbons (82) with n up to 7 [116, 117] form an attractive variation in this category. The diversity of chained polycycles includes Vögtle et al.'s triple-layered cyclophane (83) and its isomers [96], the interesting double helical octaphenylene (84) by Rajca et al. [118], and polycycles (85) and (86, R = $C_{12}H_{25}$) made by de Meijere et al. [119] and Schlüter et al. [120], respectively. The two latter examples demonstrate how mixing different connections and branching generates complex graphs and molecular structures.

77   **78**   **79**

80

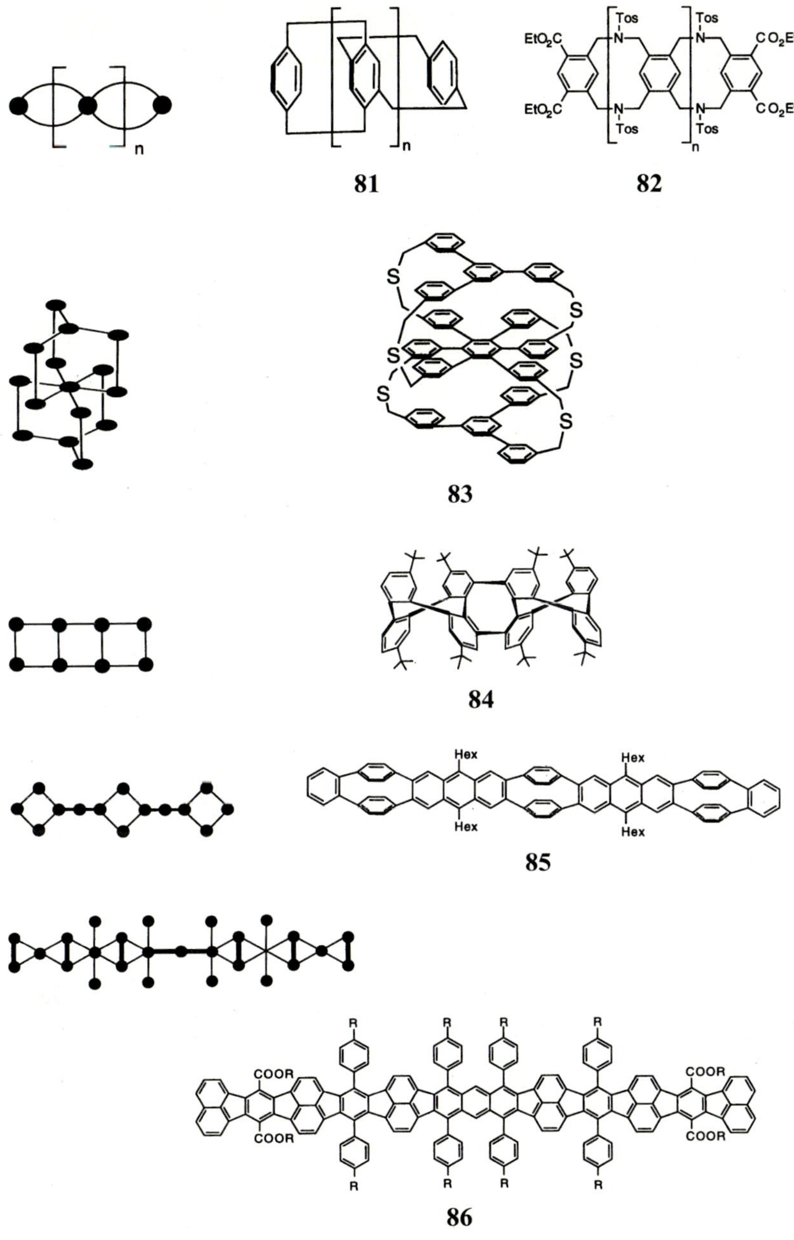

The polycyclic version of the star graph is represented by Diercks and Vollhardt's triangular [4]phenylene (**87**), or "starphenylene" [121]. Some earlier examples are trifoliaphane (**88**) by Hopf and Psiorz [122], its meta-linked isomer by Vögtle and Kissener [123], and decacyclene (**89**) for which the crystal structure was only recently reported [124].

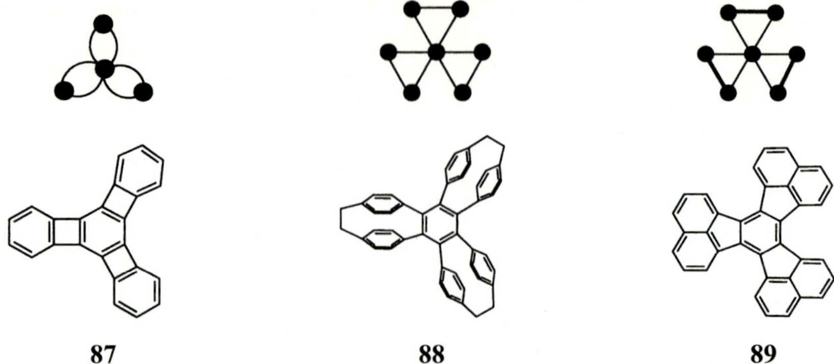

87   88   89

The "cycle of cycles" is a class of graphs in which several polycycles connected together form a macrocycle. The simplest structure of the family is Boekelheide et al.'s deltaphane (**90**) [125], representing the triangular polycycle. Other three-point analogs include Vögtle's deformed biphenylene (**91**) [126], precursors to molecular belts, such as Stoddart et al.'s [14]cyclacene derivative (**92**) [127], and Cory and McPhail's [8]cyclacene trisquinone (**93**, R = C_6H_{13}) [128]. Connecting

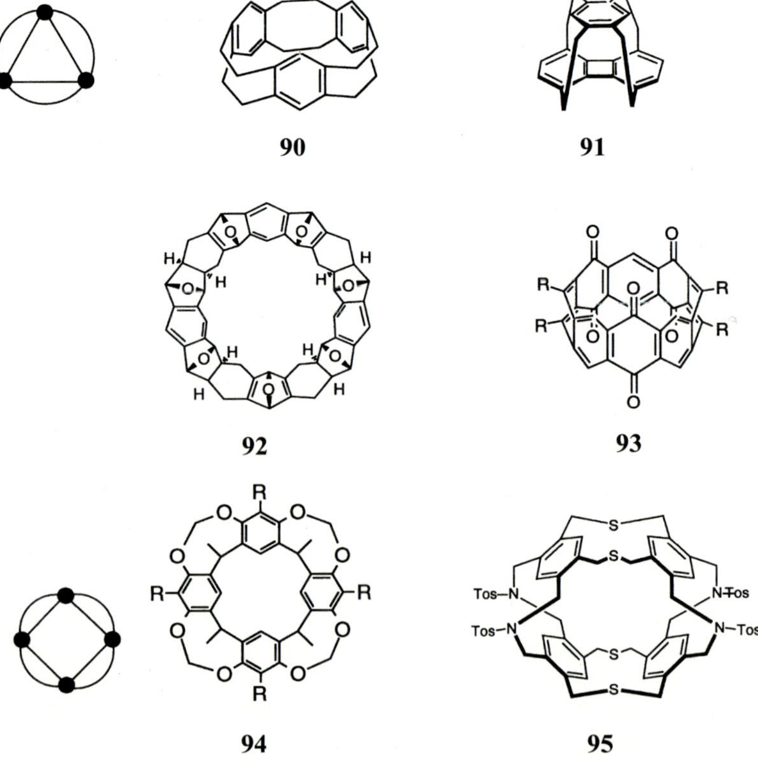

90   91

92   93

94   95

all oxygen atoms in exo-calix[4]arene creates a four point macropolycycle, cavitand (**94**) [129, 130]. Higher symmetric structures are illustrated by Vögtle et al.'s short molecular tube (**95**) [131], Stoddart et al.'s [12]collarene (**96**) [132], and its relative (**97**) proposed by Klärner et al. [133]; the last two cases represent macropolycycles with six points.

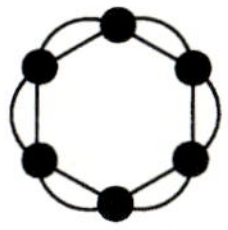

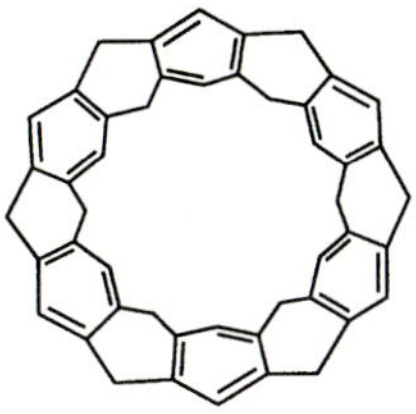

96

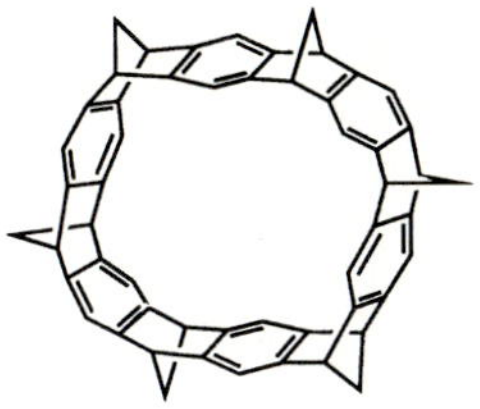

97

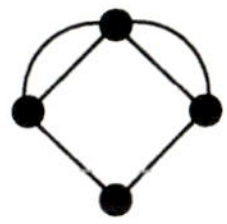

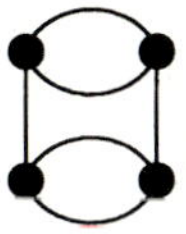

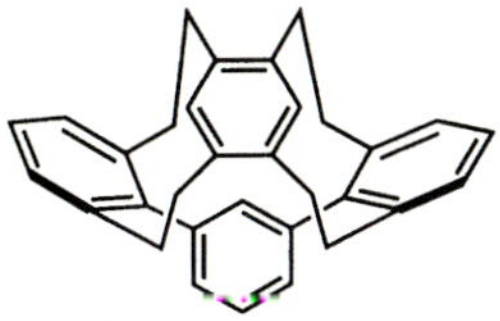

98

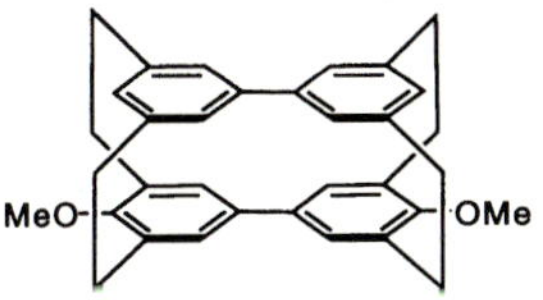

99

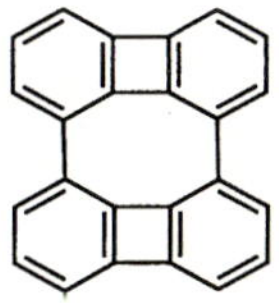

100

Once again, mixing different connections into earlier basic graphs will create even more diverse structures. For example, adding two paths onto a square graph generates Hart and Vinod's cappedophane (**98**) [99–102], and "cylinder-graph" molecules such as Tani et al.'s parallel biphenylophane (**99**) [134] and Rajca et al.'s biphenylene dimer (**100**) [135]. Dougherty et al.'s cation binder (**101**) [136] is an example of a cylinder-graph homeomorph. Higher oligomers of this cylindrical motif appear in the larger macropolycycles such as Cram et al.'s saddle-shaped macrocycle (**102**) [137]. Continuing on this line, Reinhoudt et al.'s calix[4]arene dimer (**103**, R=t-butyl) [138, 139] and trimer (**104**, R=t-butyl) will represent a square based cylinder and triangular oligomer, respectively. Diederich and Peterson's steroid receptor (**105**) [140] combines various-elements into a sophisticated structure depictable as a hexagonal dimer graph.

PAHs are attractive structures for chemists working with graph theory. Pyrene (**106**) and coronene (**63**) [80] are among the simplest polycycles with the graphic presentations or "inner duals" [2, 3] as shown. Coronene is a special case that could also be a monocycle of six points depending whether the central hexagon is counted as a unit. The dual concept, as described in Sect. 1, has been

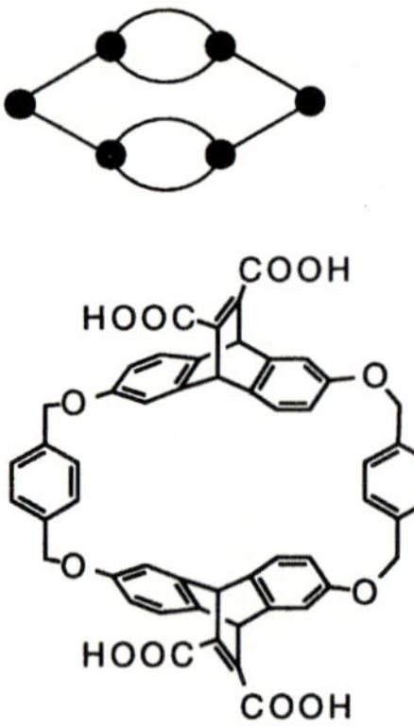

101

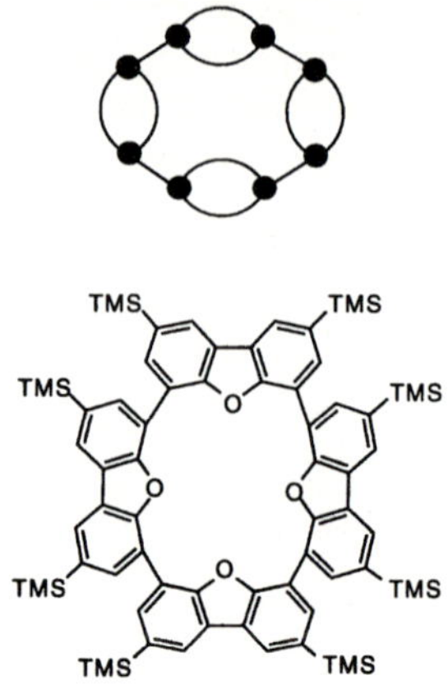

102

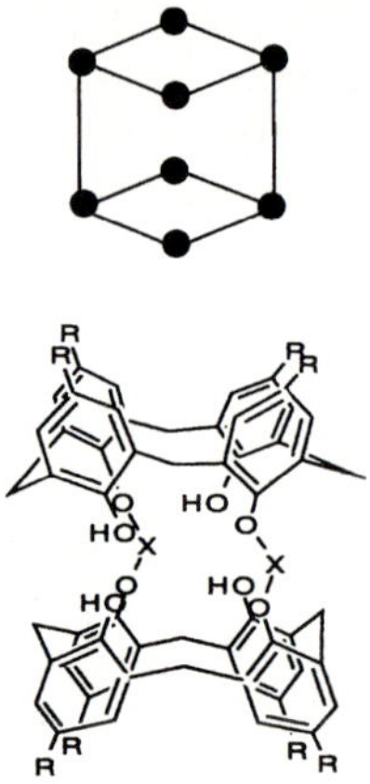

103

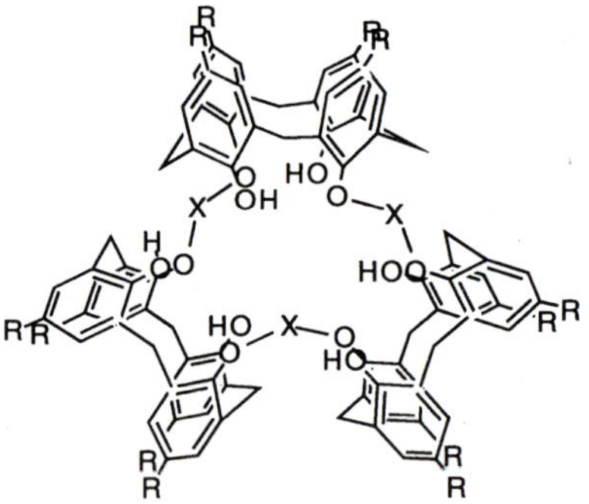

104

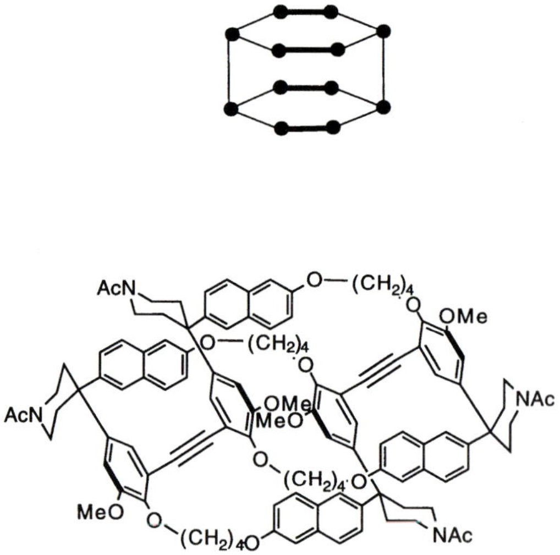

105

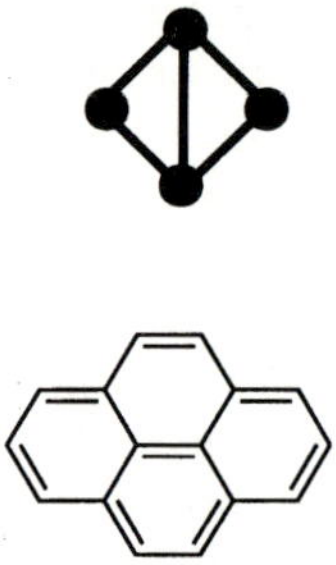

106

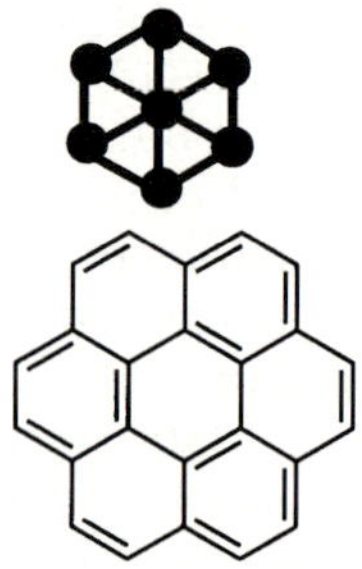

63

used and expanded to simplify the notation of various other classical benzenoid derivatives [2, 3, 141, 142]. In this context, we will limit our coverage to some recent examples of these exceptionally large PAHs.

Large polybenzenoids have been extensively studied by Müllen et al. [143–145]. Among various sizes and shapes of these "superacenes", as he named them, are **107, 108,** and the 50-benzene-unit PAH (**109**). These compounds are generally constructed from their dendritic polyphenylene precursors in good yields, albeit difficult to characterize conclusively. Even larger sheets have been proposed and where the limits lie is still unknown.

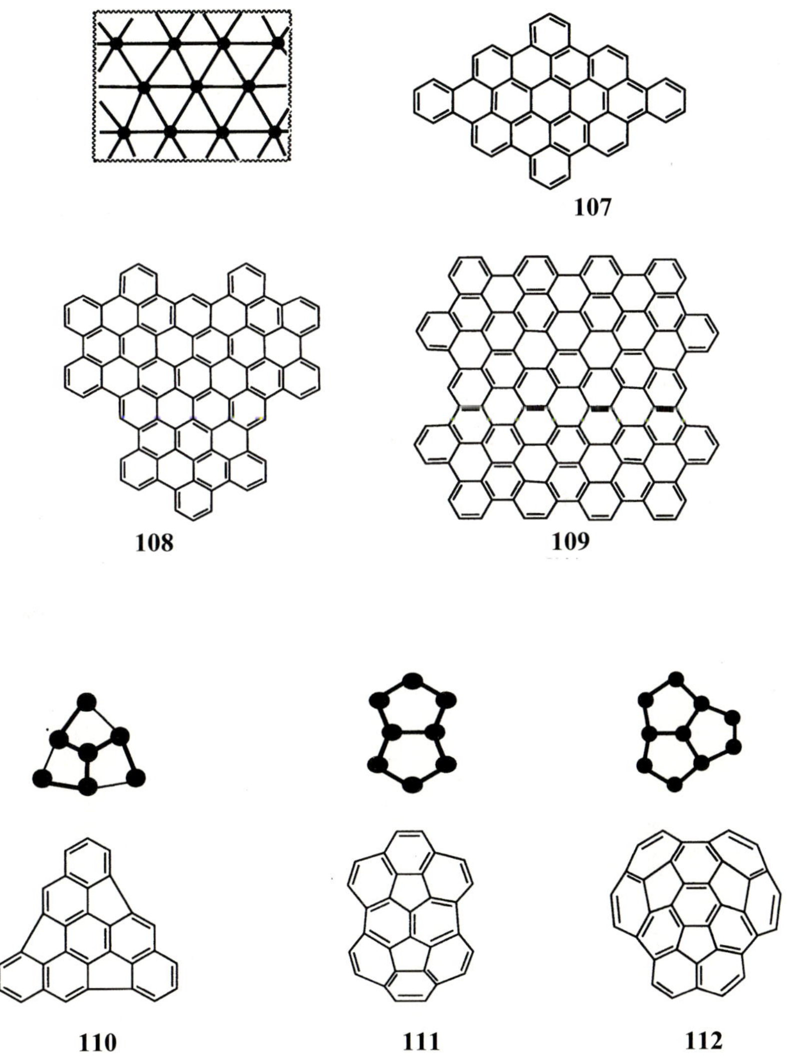

107

108

109

110

111

112

Incorporation of five-membered rings into the polyhexagon sheet creates curvature in the PAH, as seen in corannulene. Larger polyaromatic $C_{30}H_{12}$ bowls, (**110** and **111**), have been recently synthesized by Rabideau et al. [146, 147] and Scott et al. [148]. The largest molecule yet synthesized in this family, $C_{36}H_{12}$ (**112**), was achieved by Scott et al. [149]. Cyclophanes from curved PAHs are a natural extension. Two examples shown here are Bodwell et al.'s pyrenophane (**113**) [150] and Siegel et al.'s corannulene cyclophane (**114**) [151].

A few other variations are included here to demonstrate the diversity of this class with regard to molecular shape. Vögtle et al.'s multiple ansa compound (**115**, $R = CH_2CH_2OCH_2CH_3$), an unexpected product obtained from an attempt to synthesize biscalixarenes with aromatic linkers, has two bridges threaded through the macrocycle formed by joining the two calixarenes subunits [152]. Siegel et al.'s

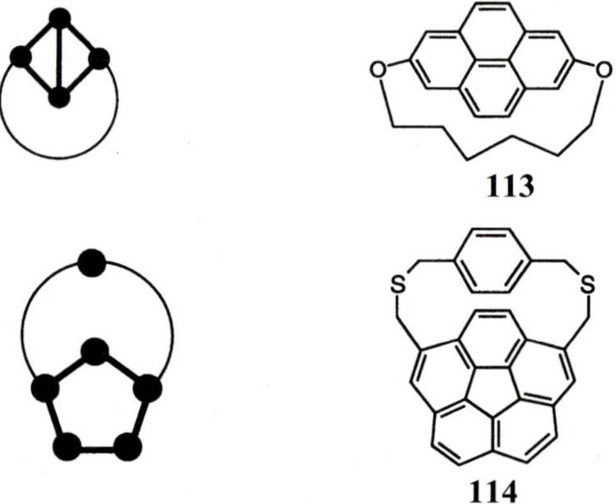

113

114

D_{2h} isomer of Kuratowski's cyclophane (**116**) [153, 154], where branching is introduced within the bridges, shows a cylindrical graph with an internal cross diameter connection. Okazaki et al.'s lantern-shaped molecule (**117**) [155, 156] approximates a pyramidal shape, a high symmetric structure related to the Platonic tetrahedron. Platonic and other highly symmetrical graphs are compiled in the next section.

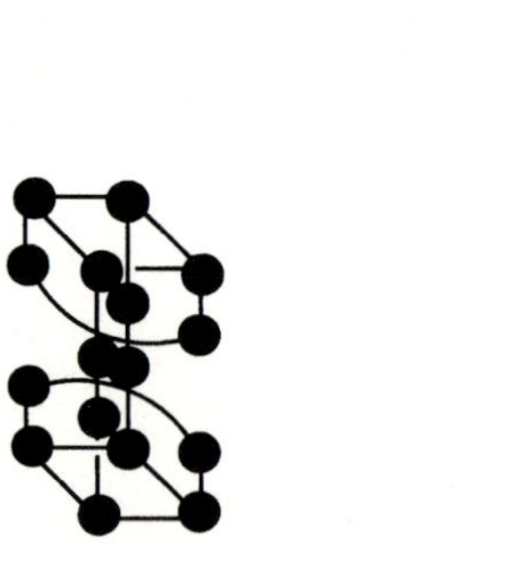

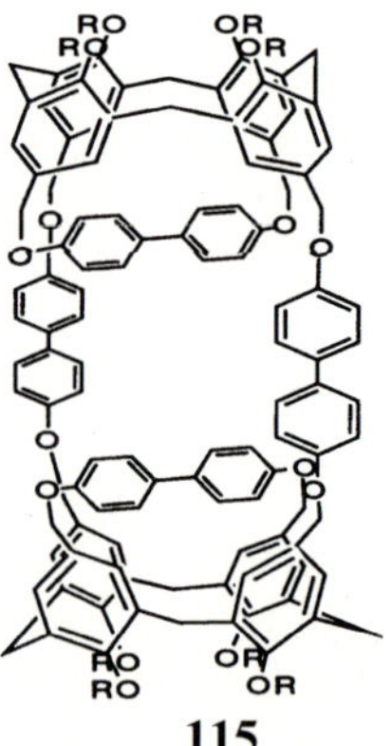

115

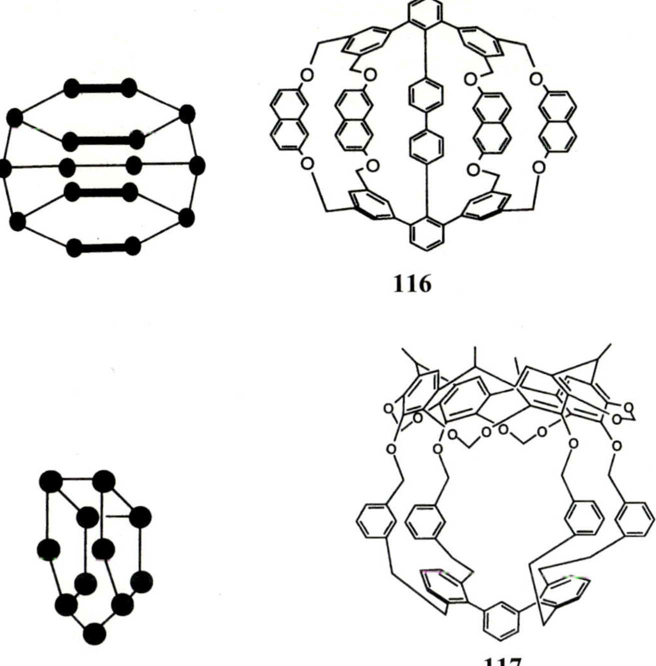

116

117

7
Platonic Structures

Platonic graphs are polycyclic graphs with highly symmetric geometries represented by platonic solids. In the carbon atom-point graph system, incredible successes in synthesizing these molecules have been demonstrated for quite some
time [157]. For macromolecules, however, the challenge of new designs and syntheses still remains. On the basis of our analysis, various high symmetry shapes
such as cage, cup, bowl, tube, etc. intrinsically carry or relate to one of the platonic
graphs.

After successes in the syntheses of several bicyclophanes, Wennerström and
Norinder [158] proposed a highly symmetric cage cyclophane with the tetrahedral arrangement of four benzene rings. Later, Vögtle et al. synthesized a
tetrahedral analog with saturated bridges and named it "spheriphane" (**118**) [159].

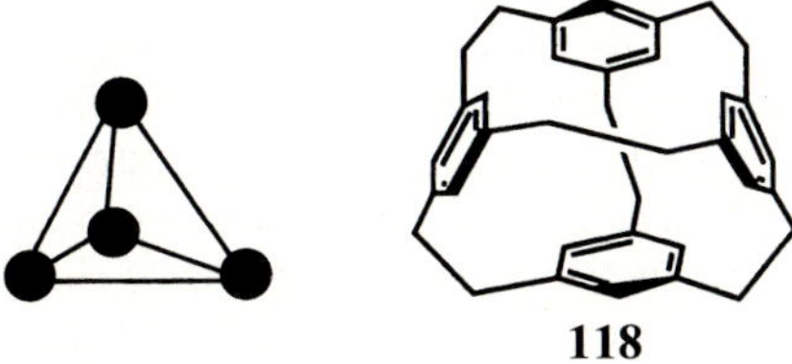

118

Other molecules that contain four key aromatic units arranged in a tetrahedral array, do not necessarily have T_d symmetry. Distortion creates two groups of lower symmetric structures, D_{2d} and C_{3v}. Two examples of what stem from D_{2d} symmetry are the adamantane-like macrotricycle (119, R=C_6H_{13}), synthesized by Moore and Wu [160], and Tani et al.'s cross-oriented biphenylophane (120) [134] obtained with its parallel polycyclic isomer (99).

119   **120**

Another group of distorted tetrahedra contains triangular pyramidal graphs with C_{3v} symmetry. Vögtle and Wambach were the first to synthesize compound 121 and several of its analogs in which the "floor plate" benzene is the bottom of the basket or the peak of the pyramid [161]. Still et al. designed and synthesized 122 (R = CH_2Ph), a similar "cup" shape molecule with different linked bridges between peripheral benzene rings [162]. When replacing the three benzene rings along the triangle base by naphthalenes, the cup is widened and increases the binding properties toward neutral guests [163].

121   **122**

Addition of vertices to the edges of a tetrahedron creates homeomorphs of tetrahedral geometry. When one vertex is added onto each edge of one of the triangular faces, a triply bridged hexagonal pyramid is obtained. Examples are bowl or basket-shaped molecules structurally quite similar to the earlier C_{3v} distorted tetrahedral structures. Pascal et al.'s molecular bowl (123) [164], with sulfur as a one-atom linker between aromatic rings, can be obtained in a few synthetic steps through consecutive high dilution cyclizations. Lehn et al.'s basket molecule (124) has been tested as a receptor of the speleand type that binds quaternary ammonium cations [165]. Shinkai et al. obtained a capped calix[6]arene (**125**, R=CH$_3$) with a basket shape that also binds an ammonium cation [67, 166].

Another interesting variation of pyramidal shape molecules arises from the group of concave hydrocarbons (such as **126**) [167, 168] synthesized by Vögtle et al. In this structure, a new point is added to three convergent edges. The addition of aromatic rings rigidifies the molecule and allows its cavity to retain its size and shape suitable for inclusion of an appropriate guest.

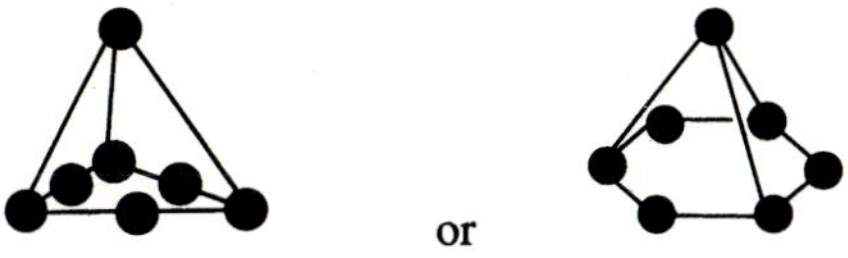

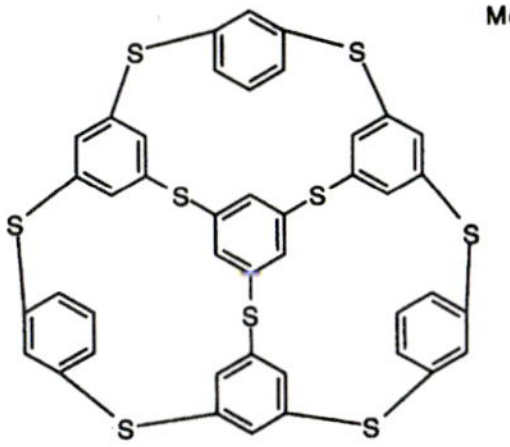

123 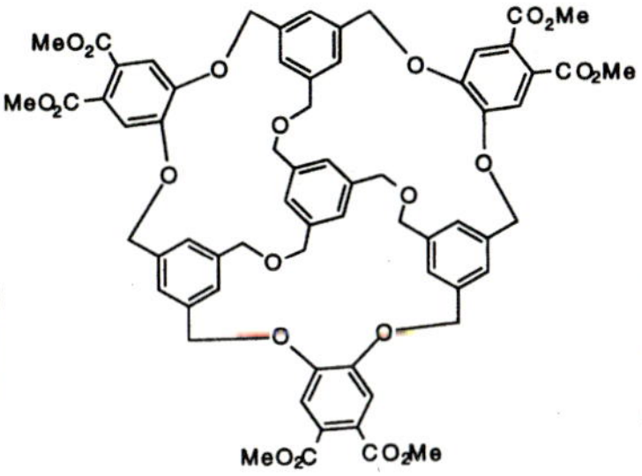   **124** 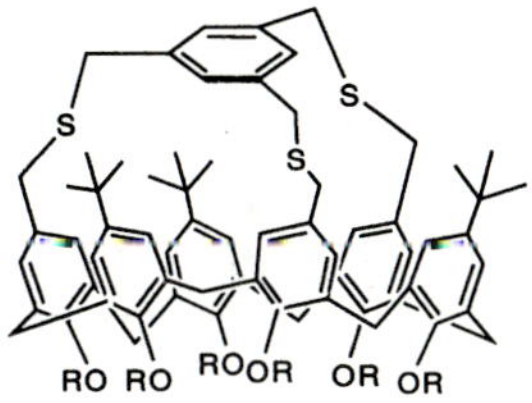   **125**

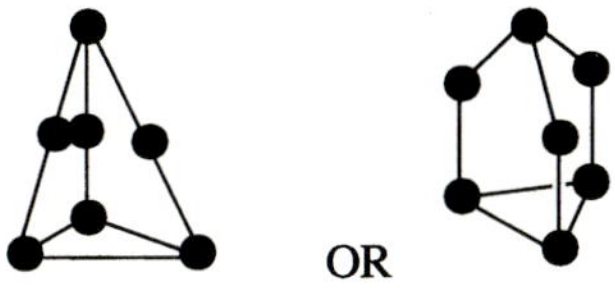

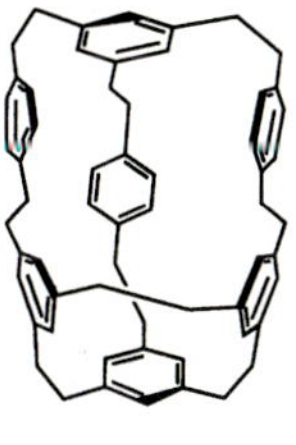

126

A molecule representing a cube, with benzene corners and methylene edges (C$_{60}$H$_{48}$), has been proposed and computationally studied [169]. Attempts to synthesize such a molecule through several strategies have not yet been successful [170]. Distortion by elongation of the cube creates a square-prismatic structure. Prisms are not rigorously platonic solids, but they are highly symmetric and share an esthetic appeal with the platonic structures.

The triangular prisms are largely exploited by Collet et al. as a class of macrocycles called "cryptophanes" (**127**) [59, 171]. These compounds are constructed from joining two cyclotriveratrylene subunits. Several possible connections between the two units create an interesting series of stereoisomers [172–174]. Cram et al. also synthesized, a similar kind of this class of compounds (**128**) [175], using biacetylene units as bridges. Cram et al. prepared an unsymmetrical prism (**129**) [176] by joining one cyclotriveratrylene unit and another triaromatic unit with longer 3-atom bridges.

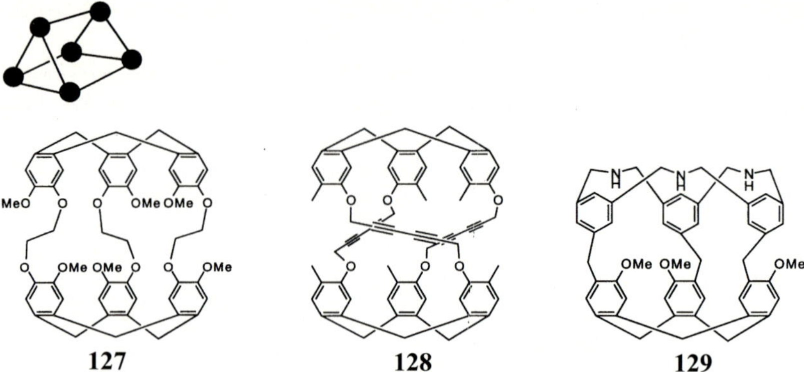

127   **128**   **129**

Square-prismatic graphs or "capsule"-shaped molecules have been developed from calixarenes. Böhmer et al. were first to synthesize the mono-, doubly-linked and quadruply-linked double calixarenes such as **130** (X=(CH$_2$)$_{10}$, R=H) [177] using all methylene chains as the bridges between two calixarene units. Later Shinkai et al. made a close analog of **130** (X=CH$_2$OCH$_2$CH$_2$OCH$_2$, R=CH$_3$) [67, 178] by, instead, using ethylene glycol units to bridge the two calixarenes. Other variations in this family include Blanda and Griswold's octasulfide [179] and similar skeletons with other organic bridging groups [180]. An unsymmetric square prism has been made by Reinhoudt et al., who has connected one calix[4]arene and one cavitand to yield a hybrid square prism (**131**, R=C$_3$H$_7$, R′=C$_{11}$H$_{23}$) [181, 182].

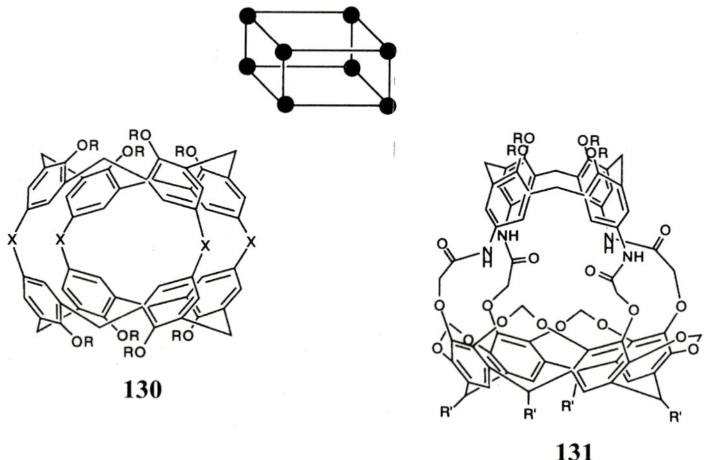

130   **131**

Alternatives to the square prisms are known from Cram's container molecules called carcerands [183, 184]. Molecules, such as **132** [185, 186], are made by quadruply linking two cavitand units, thus creating a void inside capable of incorporating a small organic molecule which is then called a "carceplex." The encapsulated guests vary from simple solvent molecules to highly reactive intermediates such as cyclobutadiene and benzyne [187, 188]. In a strict view carcerands have a more complicated embedded-graphs than just a simple square prism. All vertices of the square units are linked together by double bridges as its parent cavitands (**93**).

For other higher order prisms, Shinkai et al. successfully synthesized a compound represented by the hexagonal prismatic graph by joining two calix[6] arene units as the hexasulfide (**133**, R = CH$_3$) [189]. An interesting macrocycle, formed during Reinhoudt's synthesis of **131**, is obtained from connecting two units of each starting monocycle. This compound (**134**, R = C$_3$H$_7$, R′ = C$_{11}$H$_{23}$), named holand (a molecule with a large hole) [181], has expanded the graphic connections along the side of the prism. When viewed sideways, the molecule can be considered as a symmetric octagonal prism.

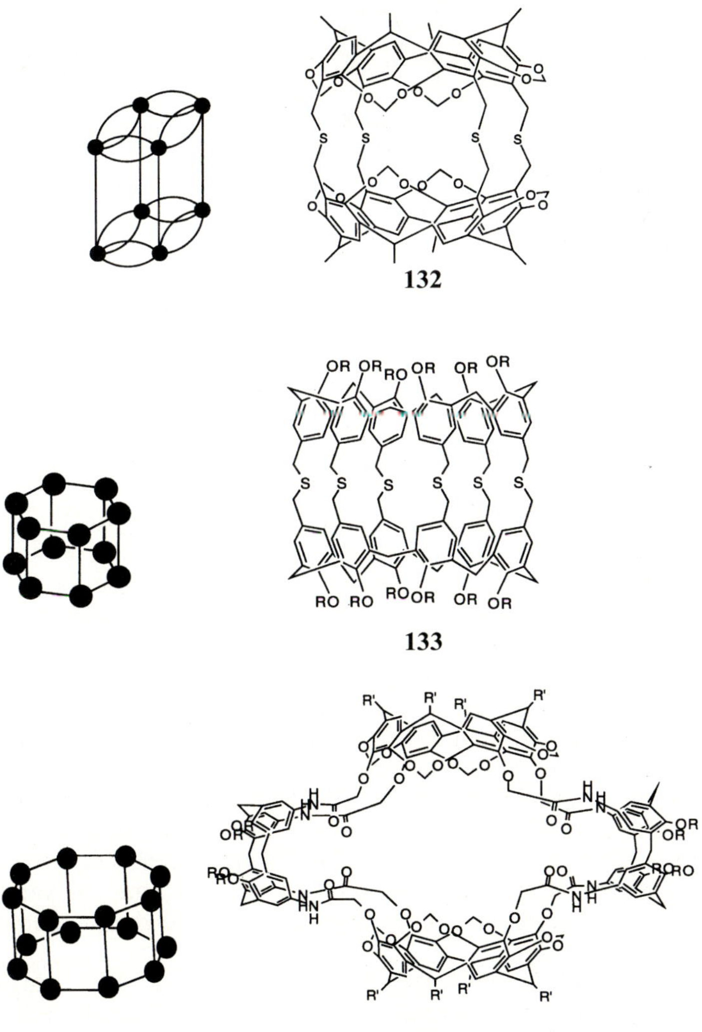

132

133

134

A striking highly symmetric allotrope of "carbon" known as C_{60} or buckminsterfullerene (**135**) [190, 191] was discovered in 1985. The molecule consists of 20 benzene units fused together in a soccer ball-like shape. By our definition, these 20 units are arranged as a perfect dodecahedron geometry with the double-weighted lines linked between vertices. Other higher fullerenes, with five-fold symmetry as in C_{70} (**136**) [190, 191], are an extension of this dodecahedral graph. Chiral fullerenes, such as C_{76}, provide a novel cage graph that creates a chiral reticulation on the sphere [192–194]. A polyaromatic example representing the icosahedron is to our knowledge, unknown.

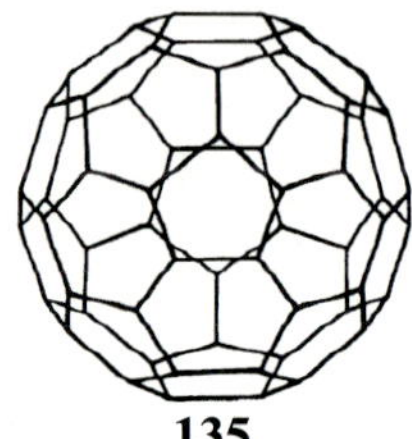

135

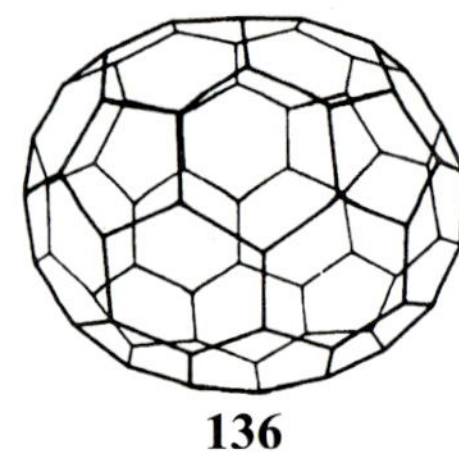

136

8
Topological Structures

Topological graphs are groups of cyclic graphs that have distinct character only by their connectivity and require no Euclidean molecular rigidity to remain different. The theoretical background in this subject has already appeared in several excellent reviews [10, 195,196]. To avoid redundancy and keep this review within an appropriate length, readers are referred to these references for further information. Only a brief explanation through examples will be presented here.

Several kinds of topological constructions have been suggested in a theoretical paper by van Gulick [11]. On the basis of this proposal and the successfully synthesized molecules in the literature, we decided to describe three basic groups: the simple catenanes or interlocking rings, knots and higher links, and polycycles possessing nonplanar graphs.

The first catenane synthesis was reported back in 1960 by Wasserman [197]. During the earlier attempts, aromatic moieties were involuntarily excluded from the linked main cycles, although their important role in directing the link became apparent in the Schill-Lüttringhaus synthesis [198]. Catenanes with aromatic backbones were first introduced by Sauvage et al. using metal complex directed synthesis [199–201]. The simplest [2]catenane (**137**) [202, 203] and [3]catenane (**138**) [204] have been synthesized using copper(I) diphenanthroline complex directing the crosslink between two cycles. Numerous variations of this strategy have been used including the triply linked bicyclophane or tris[2]catenane (**139**) [205].

Another successful strategy is that developed by Stoddart et al. using electron donor-acceptor pairs [206]. Several catenanes have been synthesized ranging from the simple [2]catenane (**140**) [207] to [5]catenane (**141**) which is named

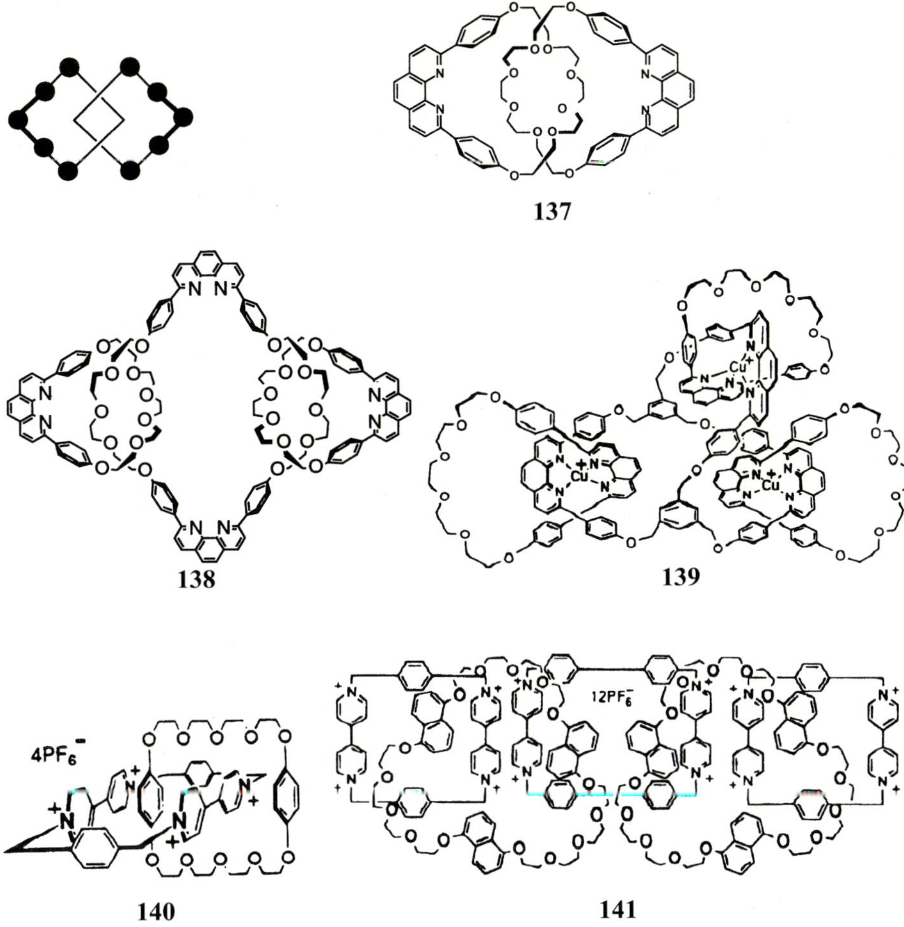

137

138

139

140

141

"olympiadane" [208, 209]. The links can also be directed by hydrogen bonds between two complementary functional groups within each cycle. The [2]catenanes of this type are **142** (R=H), accomplished by Hunter [210], and its methoxy derivative (R=OMe) by Vögtle et al. [211, 212]. For knots and higher links, only one example each has been achieved. Sauvage et al. have further applied their copper-phenanthroline template to the synthesis of the trefoil knot (**143**) [213, 214] and doubly interlocked [2]catenane (**144**) [215].

The first organic molecules carrying nonplanar K_5 subgraph were attained by Simmons et al. [216, 217] and Paquette et al. [218, 219] as a trioxa derivative of a centrohexaalicyclic compound. Later Kuck and Schuster added the aromatic flavor into a K_5 hydrocarbon which he called centrohexaindane [220, 221]. These molecules, however, construct the K_5 graph all from five carbon atoms. The aromatic rings in Kuck and Schuster's compound are incorporated as part of the bridges between units. In the case of the nonplanar $K_{3,3}$ graph, Walba et al.'s polyether Möbius strip would be the front runner [222]. The first benzene-point

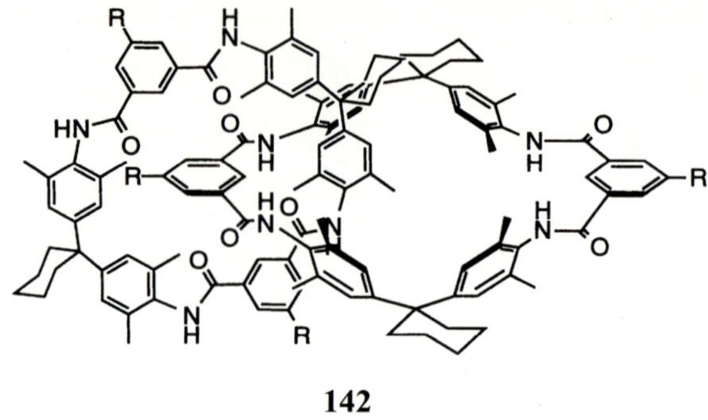

142

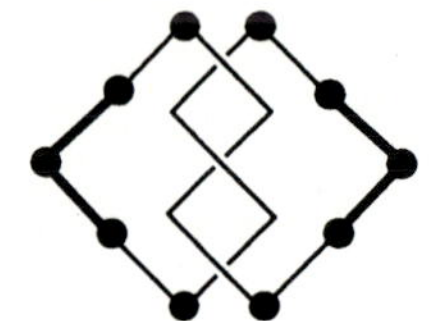

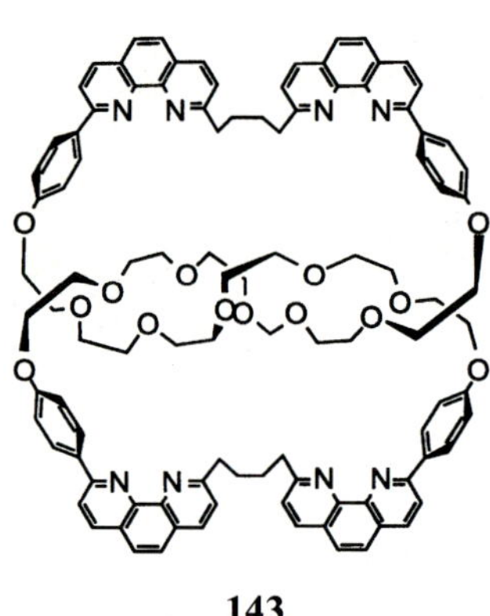

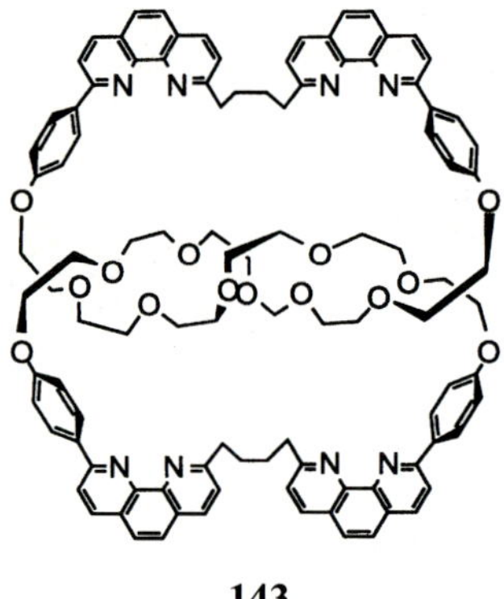

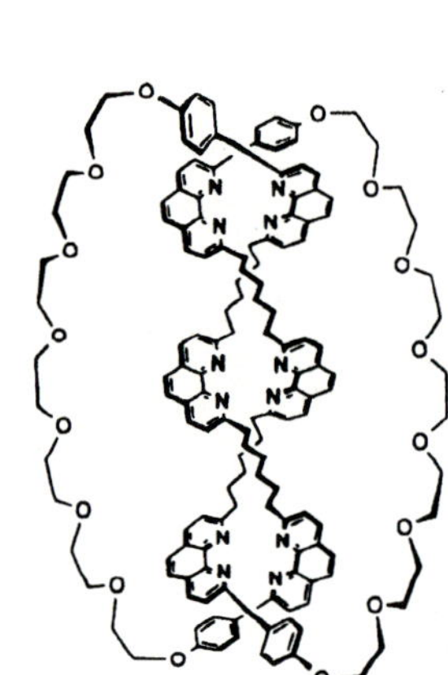

143

144

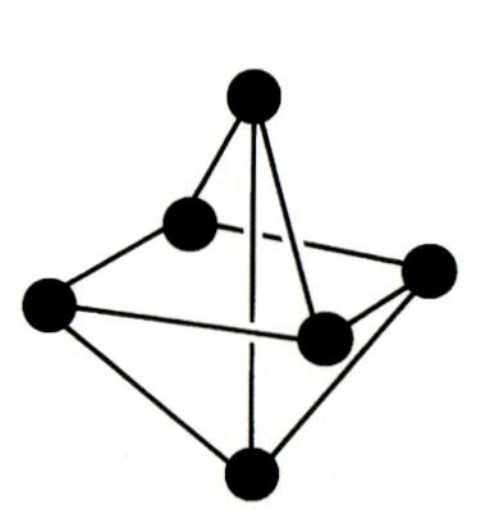

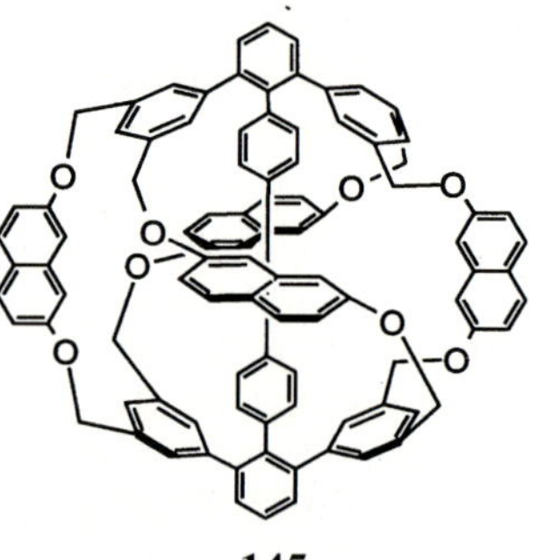

145

replacement on the $K_{3,3}$ graph was achieved recently by Siegel et al. [153, 154]. The molecule **145** is named Kuratowski's cyclophane, after Kuratowski who first introduced the nonplanar graph theorem.

9
Extending the Point Replacement Idea: Bigraphs

Our discussion up to this point has been restricted to graphs where every point can be related to a benzene ring and all other atoms are lumped into "the connections." In simple graph theory, connections occur only between two points; i.e., there is no equivalent of a three-center two-electron bond. This restriction has meant that all of our branched and polycyclic graphs have had to diverge at benzene rings, a condition which excludes many molecular structures from our discussion. How then might we extend the Loschmidt replacement convention but still retain the relative simplicity of the design rules? One possibility would be to implement the idea of multipartite graphs to distinguish points which map to Loschmidt replacements from those that symbolize "branching connectors" for lack of a better term. For example, the related structures of 1,3,5-triphenylbenzene and triphenylmethane would map to similar graphs with the distinction that the triphenylmethane graph would have an open point at the center to distinguish its non benzenoid character (Fig. 6). The bipartite nature of these graphs comes from different replacement rules: closed points are replaced by benzene, open points are replaced by "branching connectors" like $CH(R)_3$.

Very quickly the number of describable and designable structures expands. It would be injudicious for us to attempt to cover the myriad of possible structures, as the design rules should be transparent and easily applied by anyone who has followed our discussion up to this point. For esthetic reasons, however, we would like to point out some signal examples of molecules that fall into this expanded regime.

The simplest cyclic bigraph in our context would be a bicycle comprising one closed and one open point. A beautiful illustration of this graph comes from Pascal et al.'s in-cyclophane (**146**) [223] where the CH representing the open point jams its hydrogen into the face of the benzene ring, its closed-point counterpart.

Returning to the triaryl-X template, we can see that tethering the closed circles in a cycle provides the basis for a trioxatricornan structure (**147**). Capping the trioxatricornan structure with a benzene ring creates a cyclophane

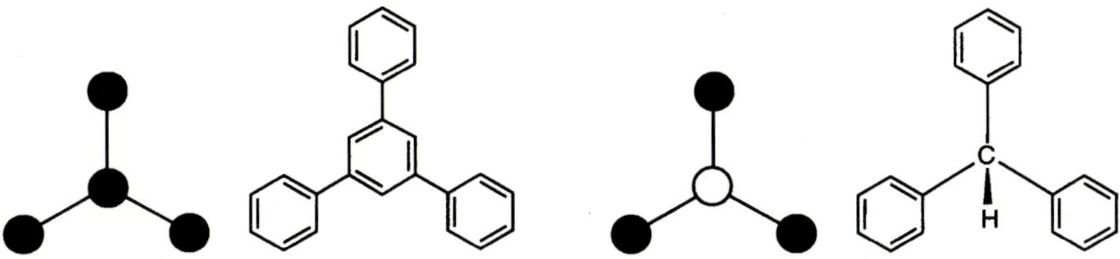

Fig. 6. Comparison of graphs for triphenylbenzene and triphenylmethane

(148) that maps to a trigonal bipyramid with an open circle at one apical point [224, 225].

Cyclophanes with novel inclusion properties have been designed from the dimerization of two triaryl-X fragments. This prismatic graph cognates from the labs of Breslow et al. (**149** (R=Me)) [226], Vögtle and Berscheid (**149**, R=H) [227], and Whitlock et al. (**150**) [228, 229] show not only the power of the bigraph design but also how easy it is to mix in additional structural features to the graph replacement idea.

146

147

148

149

150

Triptycene (**151**) is a simple example of a star topology for a bigraph. Harold Hart has built on this motif in his design of heptaiptycene (**152**) tritriptycene (**153**) and supertriptycene [230]. These rigid hyperbranched structures are among the first dendritic structures and show remarkable inclusion properties. Depicting these structures as their graph reductions evokes additional mole-

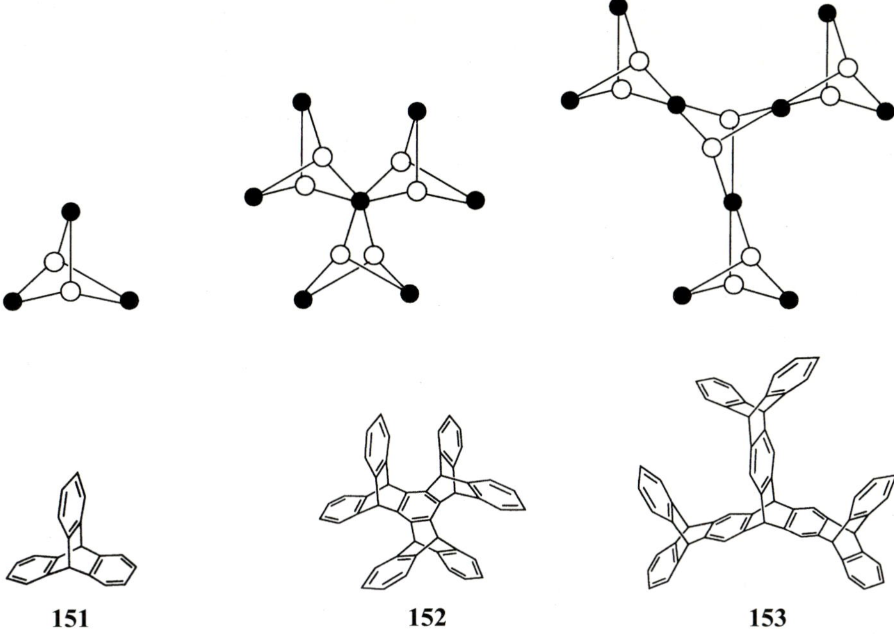

151 **152** **153**

cular designs that may have similar physical properties. Indeed, throughout this discussion it has been our hope that we are not merely categorizing known structures, but stimulating the reader to expand on these designs.

10
Conclusions

The graph replacement idea is a simple one: given a convention for points and connections the abstract graph can be converted into a class of molecules. We have attempted to illustrate how a little knowledge of graph theory and a convention like the Loschmidt replacement can create a different view of carbon-rich aromatic compounds. With this new perspective, we hope to have given the reader a set of design principles that will stimulate the directed total synthesis of more exciting and elaborate aromatic structures. Graph abstractions, which we find interesting and for which no molecular realization yet exists, include a Loschmidt cube, Borromean rings, dendridic PAHs, cascade "cycles-of-cycles" and Császár's polyhedron (K_6) (Fig. 7) [231]. No doubt each of these will succumb to the pursuits of synthetic chemists and with each total synthesis our appreciation of molecular form and function will be enriched.

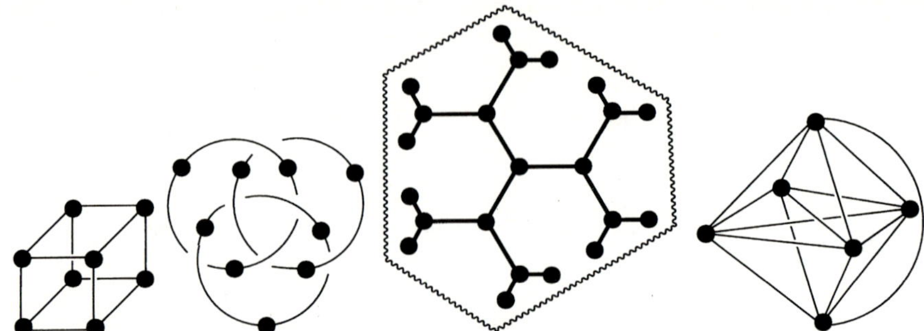

Fig. 7. Design graphs for a Loschmidt cube, Borromean rings, dendritic PAH's, and Császár's polyhedron (K_6)

Acknowledgement. This work was supported by the US National Science Foundation.

11
References

1. Loschmidt J (1860) Konstitutionsformeln der Organischen Chemie in Graphischer Darstellung. Wilhelm Engelmann, Leipzig
2. Fleischener HJ, Geller DP, Harary F (1974) J Indian Math Soc 38:215
3. Trinajstić N (1992) Chemical graph theory. CRC Press, Boca Raton
4. Barth WE, Lawton RG (1966) J Am Chem Soc 88:380
5. Scott LT, Hashemi MM, Meyer DT, Warren HB (1991) J Am Chem Soc 113:7082
6. Borchardt A, Kilway KV, Baldridge KK, Siegel JS (1992) J Am Chem Soc 114:1921
7. Zhang J, Pesak DJ, Ludwick JL, Moore JS (1994) J Am Chem Soc 116:4227–4239
8. Mislow K (1977) Bull Soc Chim Belg 86:595
9. Liang C, Mislow K (1997) Croat Chem Acta 70:735–744
10. Walba DM (1985) Tetrahedron 41:3161–3212
11. van Gulick N (1993) New J Chem 17:619–625
12. Lehn J-M (1995) Supramolecular Chemistry. VCH, Weinheim
13. Vögtle F (1993) Supramolecular Chemistry. John Wiley & Sons, New York
14. Schumm JS, Pearson DL, Tour JM (1994) Angew Chem Int Ed Engl 33:1360
15. Weiss K, Michel A, Auth E, Bunz UHF, Mangel T, Müllen K (1997) Angew Chem Int Ed Engl 36:506
16. Grubbs RH, Kratz D (1993) Chem Ber 126:149
17. Nelson JC, Saven JG, Moore JS, Wolynes PG (1997) Science 277:1793
18. Huang WS, Hu QS, Zheng XF, Anderson J, Pu L (1997) J Am Chem Soc 119:4313
19. Newman MS, Lednicer D (1956) J Am Chem Soc 78:4765
20. Martin RH, Baes M (1975) Tetrahedron 31:2135
21. Katz TJ, Liu L, Willmore ND, Fox JM, Rheingold AL, Shi S, Nuckolls C, Rickman BH (1997) J Am Chem Soc 119:10,054
22. Mallory FB, Butler KE, Evans AC, Brondyke EJ, Mallory CW, Yang C, Ellenstein A (1997) J Am Chem Soc 119:2119
23. Qiao X, Padula MA, Ho DM, Vogelaar NJ, Schutt CE, Pascal RA Jr, (1996) J Am Chem Soc 118:741–745
24. Qiao X, Ho DM, Pascal RA Jr (1997) Angew Chem Int Ed Engl 36:1531
25. Tong L, Ho DM, Vogelaar NJ, Schutt CE, Pascal RA Jr (1997) J Am Chem Soc 119:7291
26. MacNicol DD, Mallinson PR, Murphy A, Sym GJ (1982) Tetrahedron Lett 23:4131–4134
27. Tao W, Nebitt S, Heck RF (1990) J Org Chem 55:63

28. de Meijere A, Meyer FE (1994) Angew Chem Int Ed Engl 33:2379
29. Prinz P, Lansky A, Haumann T, Boese R, Noltemeyer M, Knieriem B, de Meijere A (1997) Angew Chem Int Ed Engl 36:1289
30. Keegstra MA, De Feyter S, Schryver FC, Müllen K (1996) Angew Chem Int Ed Engl 35:774
31. Xu Z, Moore JS (1993) Angew Chem Int Ed Engl 32:1354
32. Clar E, Mullen A (1968) Tetrahedron 24:6719
33. Shibata K, Kulkarni AA, Ho DM, Pascal RA Jr (1995) J Am Chem Soc 60:428
34. Miller TM, Neenan TX, Zayas R, Bair HE (1992) J Am Chem Soc 114:1018
35. Wooley KL, Hawker CJ, Fréchet JMJ (1991) J Am Chem Soc 113:4252
36. Kawaguchi T, Walker KL, Wilkins CL, Moore JS (1995) J Am Chem Soc 117:2159–2165
37. Devadoss C, Bharathi P, Moore JS (1996) J Am Chem Soc 118:9635
38. Kostermans GBM, Bobeldijk M, de Wolf WH, Bickelhaupt F (1987) J Am Chem Soc 109:2471–2475
39. Jenneskens LW, de Kanter FJJ, Kraakman PA, Turkenburg LAM, Koolhaas WE, de Wolf WH, Bickelhaupt F (1985) J Am Chem Soc 107:3716–3717
40. Tochtermann W, Olsson G, Mannschreck A, Stuhler G, Snatzke G (1990) Chem Ber 123:1437
41. Lemmerz R, Nieger M, Vögtle F (1993) J Chem Soc Chem Commun : 1168–1170
42. Anet R, Anet FAL (1964) J Am Chem Soc 86:525
43. Nakazaki M, Yamamoto K, Ito M, Tanaka S (1977) J Org Chem 42:3468
44. Saward CJ, Vollhardt KPC (1975) Tetrahedron Lett 51:4539
45. Billups WE, Arney BE, Jr, Lin LJ (1984) J Org Chem 49:3436–3437
46. Frank NL, Baldridge KK, Siegel JS (1995) J Am Chem Soc 117:2102–2103
47. Bürgi HB, Baldridge KK, Hardcastle K, Frank NL, Siegel JS, Ziller J (1995) Angew Chem Int Ed Engl 34:1454
48. van Es DS, de Kanter FJJ, de Wolf WH, Bickelhaupt F (1995) Angew Chem Int Ed Engl 34:2553
49. Tobe Y, Saiki S, Utsumi N, Kusumoto T, Ishii H, Kakiuchi K, Kobiro K, Naemura K (1996) J Am Chem Soc 118:9488
50. Cram DJ, Steinberg H (1951) J Am Chem Soc 73:5691
51. Brown CJ, Farthing AC (1949) Nature 164:915
52. Hopf H, Jones PG, Bubenitschek P, Werner C (1995) Angew Chem Int Ed Engl 34:2367–2368
53. Romero MA, Fallis AG (1994) Tetrahedron Lett 35:4711–4714
54. Chang MH, Masek BB, Dougherty DA (1985) J Am Chem Soc 107:1124
55. Cram DJ, Dewhirst KC (1959) J Am Chem Soc 81:5963
56. Campbell ID, Eglington G, Henderson W, Raphael RA (1966) J Chem Soc Chem Commun: 87
57. Stabb HA, Graf F (1966) Tetrahedron Lett:751
58. Ferrara JD, Tessier-Youngs C, Youngs WJ (1987) Organometallics 6:676
59. Collet A (1987) Tetrahedron 43:5725–5759
60. Boese R, Matzger AJ, Vollhardt KPC (1997) J Am Chem Soc 119:2052–2053
61. Haley MM, Bell ML, English JJ, Johnson CA, Weakley TJR (1997) J Am Chem Soc 119:2956–2957
62. Magnus P, Witty D, Stamford A (1993) Tetrahedron Lett 34:23–26
63. Gleiter R, Hövermann K, Ritter J, Nuber B (1995) Angew Chem Int Ed Engl 34:789–791
64. Malthête J, Collet A (1987) J Am Chem Soc 109:7544
65. Miyahara Y, Inazu T, Yoshino T (1983) Tetrahedron Lett 24:5277–5280
66. Gutche CD (1989) Calixarenes. Royal Society of Chemistry, Cambridge
67. Ikeda A, Shinkai S (1997) Chem Rev 97:1713
68. Goren Z, Biali SE (1990) J Chem Soc Perkin Trans 1:1484
69. Bauer M, Nieger M, Vögtle F (1992) Chem Ber 125:2533–2538
70. Kawase T, Ueda N, Darabi HR, Oda M (1996) Angew Chem Int Ed Engl 35:1556–1558
71. Kammermeier S, Jones PG, Herges R (1996) Angew Chem Int Ed Engl 35:2669–2671

72. Kammermeier S, Jones PG, Herges R (1997) Angew Chem Int Ed Engl 36:2200–2202
73. Wittig G, Rumpler KD (1971) Liebigs Ann Chem 751:1
74. Cram DJ, Kaneda T, Helgeson RC, Brown SB, Knobler CB, Maverick E, Trueblood KN (1985) J Am Chem Soc 107:3645
75. Baldwin KP, Bradshaw ID, Tessier CA, Youngs WJ (1993) Synlett:853
76. Baldwin KP, Simons RS, Rose J, Zimmerman P, Hercules DM, Tessier CA, Youngs WJ (1994) J Chem Soc Chem Commun:1257
77. Kawase T, Darabi HR, Oda M (1996) Angew Chem Int Ed Engl 35:2664–2666
78. Gribble GW, Nutaitis CF (1985) Tetrahedron Lett 26:6023–6026
79. Dopper JH, Wynberg H (1975) J Org Chem 40:1957–1966
80. Scholl R, Meyer K (1932) Chem Ber 65:902
81. Yamamoto K, Harada T, Nakazaki M, Naka T, Kai Y, Harada S, Kasai N (1983) J Am Chem Soc 105:7171
82. Stabb HA, Diederich F (1983) Chem Ber 116:3487
83. Vögtle F, Staab HA (1968) Chem Ber 101:2709
84. Thulin B, Wennerström O (1977) Tetrahedron Lett 11:929
85. Thulin B, Wennerström O (1976) Acta Chem Scand B 30:688
86. Anderson S, Neidlein U, Gramlich V, Diederich F (1995) Angew Chem Int Ed Engl 34:1596
87. Trost BM, Kinson PL (1970) J Am Chem Soc 92:2591–2593
88. Trost BM, Kinson PL, Maier CA, Paul IC (1971) J Am Chem Soc 93:7275–7281
89. Mislow K, Glass MAW, Hopps HB, Simon E, Wahl GH, Jr (1964) J Am Chem Soc 86:1710–1733
90. Hubert AJ, Dale J (1965) J Chem Soc:3160–3170
91. Cram DJ, Reeves RA (1958) J Am Chem Soc 80:3094
92. Rubin Y, Parker TC, Khan SI, Holliman CL, McElvany SW (1996) J Am Chem Soc 118:5308–5309
93. Högberg H-E, Thulin B, Wennerström O (1977) Tetrahedron Lett:931–934
94. Olsson T, Tanner D, Thulin B, Wennerström O (1981) Tetrahedron 37:3491–3495
95. Méric R, Vigneron J-P, Lehn J-M (1993) J Chem Soc Chem Commun:129–131
96. Sendhoff N, Kißener W, Vögtle F, Franken S, Puff H (1988) Chem Ber 121:2179–2185
97. Wu Z, Lee S, Moore JS (1992) J Am Chem Soc 114:8730–8732
98. Bedard TC, Moore JS (1995) J Am Chem Soc 117:10662–10671
99. Vinod TK, Hart H (1988) J Am Chem Soc 110:6574–6575
100. Vinod T, Hart H (1990) J Org Chem 55:881–890
101. Vinod TK, Hart H (1991) J Org Chem 56:5630–5640
102. Vinod TK, Hart, H (1994) Top Curr Chem 172:119
103. Goto K, Tokitoh N, Goto M, Okazaki R (1993) Tetrahedron Lett 34:5605–5608
104. Goto K, Tokitoh N, Okazaki R (1995) Angew Chem Int Ed Engl 34:1124–1126
105. Sekine Y, Brown M, Boekelheide V (1979) J Am Chem Soc 101:3126–3127
106. Sekine Y, Boekelheide V (1981) J Am Chem Soc 103:1777–1785
107. Gleiter R, Kratz D (1993) Acc Chem Res 26:311–318
108. Sakamoto Y, Miyoshi N, Shinmyozu T (1996) Angew Chem Int Ed Engl 35:549–550
109. Sakamoto Y, Miyoshi N, Hirakida M, Kusumoto S, Kawase H, Rudzinski JM, Shinmyozu T (1996) J Am Chem Soc 118:12,267–12,275
110. Ashton PR, Isaacs NS, Kohnke FH, D'Alcontres GS, Stoddart JF (1989) Angew Chem Int Ed Engl 28:1261–1263
111. Kohnke FH, Mathias JP, Stoddart JF (1993) Top Curr Chem 165:1
112. Kissener W, Vögtle F (1985) Angew Chem Int Ed Engl 24:794–795
113. Otsubo T, Tozuka Z, Mizogami S, Sakata Y, Misumi S (1972) Tetrahedron Lett:2927–2930
114. Otsubo T, Mizogami S, Otsubo I, Tozuka Z, Sakagami A, Sakata Y, Misumi S (1973) Bull Chem Soc Jpn 46:3519–3530
115. Misumi S (1983) In: Keehn PM, Rosenfeld SM (eds) Cyclophanes II. Academic Press, New York, p 573
116. Breidenbach S, Ohren S, Nieger M, Vögtle F (1995) J Chem Soc, Chem Commun:1237

117. Breidenbach S, Ohren S, Vögtle F (1996) Chem Eur J 2:832–837
118. Rajca A, Safronov A, Rajca S, Shoemaker R (1997) Angew Chem Int Ed Engl 36: 488–491
119. König B, Heinze J, Meerholz K, de Meijere A (1991) Angew Chem Int Ed Engl 30: 1361–1363
120. Schlicke B, Schlüter A-D, Hauser P, Heinze J (1997) Angew Chem Int Ed Engl 36: 1996–1998
121. Diercks R, Vollhardt KPC (1986) J Am Chem Soc 108:3150–3152
122. Psiorz M, Hopf H (1982) Angew Chem Int Ed Engl 21:623–624
123. Kissener W, Vögtle F (1985) Angew Chem Int Ed Engl 24:222–223
124. Ho DM, Pascal RA Jr (1993) Chem Mater 5:1358–1361
125. Kang HC, Hanson AW, Eaton B, Boekelheide V (1985) J Am Chem Soc 107:1979–1985
126. Saitmacher K, Schulz JE, Nieger M, Vögtle F (1992) J Chem Soc Chem Commun:175–176
127. Ashton PR, Isaacs NS, Kohnke FH, Mathias JP, Stoddart JF (1989) Angew Chem Int Ed Engl 28:1258–1261
128. Cory RM, McPhail CL (1996) Tetrahedron Lett 37:1987–1990
129. Moran JR, Karbach S, Cram DJ (1982) J Am Chem Soc 104:5826–5828
130. Cram DJ (1983) Science 219:1177–1183
131. Vögtle F, Schröder A, Karbach D (1991) Angew Chem Int Ed Engl 30:575–577
132. Ashton PR, Isaacs NS, Kohnke FH, Slawin AMZ, Spencer CM, Stoddart JF, Williams DJ (1988) Angew Chem Int Ed Engl 27:966–969
133. Benkhoff J, Boese R, Klärner F-G (1997) Liebigs Ann/Recueil:501–516
134. Tani K, Seo H, Maeda M, Imagawa K, Nishiwaki N, Ariga M, Tohda Y, Higuchi H, Kuma H (1995) Tetrahedron Lett 36:1883–1886
135. Rajca A, Safronov A, Rajca S, Ross I, C R, Stezowski JJ (1996) J Am Chem Soc 118: 7272–7279
136. Petti MA, Shepodd TJ, Barrans J, R E, Dougherty DA (1988) J Am Chem Soc 110: 6825–6840
137. Schwartz EB, Knobler CB, Cram DJ (1992) J Am Chem Soc 114:10,775–10,784
138. van Loon J-D, Kraft D, Ankoné MJK, Verboom W, Harkema S, Vogt W, Böhmer V, Reinhoudt DN (1990) J Org Chem 55:5176–5179
139. Kraft D, van Loon J-D, Owens M, Verboom W, Vogt W, McKervey MA, Böhmer V, Reinhoudt DN (1990) Tetrahedron Lett 31:4941–4944
140. Peterson BR, Diederich F (1994) Angew Chem Int Ed Engl 33:1625–1628
141. Balaban AT, Harary F (1968) Tetrahedron 24:2505–2516
142. Nikolić S, Trinajstić N, Knop JV, Müller WR, Szymanski K (1990) J Math Chem 4: 357–375
143. Müller M, Petersen J, Strohmaier R, Günther C, Karl N, Müllen K (1996) Angew Chem Int Ed Engl 35:886–888
144. Morgenroth F, Reuther E, Müllen K (1997) Angew Chem Int Ed Engl 36:631–634
145. Iyer VS, Wehmeier M, Brand JD, Keegstra MA, Müllen K (1997) Angew Chem Int Ed Engl 36:1604–1610
146. Rabideau PW, Abdourazak AH, Folsom HE, Marcinow Z, Sygula A, Sygula R (1994) J Am Chem Soc 116:7891–7892
147. Abdourazak AH, Marcinow Z, Sygula A, Sygula R, Rabideau PW (1995) J Am Chem Soc 117:6410–6411
148. Hagen S, Bratcher MS, Erickson MS, Zimmermann G, Scott LT (1997) Angew Chem Int Ed Engl 36:406–408
149. Scott LT, Bratcher MS, Hagen S (1996) J Am Chem Soc 118:8743–8744
150. Bodwell GJ, Bridson JN, Houghton TJ, Kennedy JWJ, Mannion MR (1996) Angew Chem Int Ed Engl 35:1320–1321
151. Seiders TJ, Baldridge KK, Siegel JS (1996) J Am Chem Soc 118:2754–2755
152. Siepen A, Zett A, Vögtle F (1996) Liebigs Ann:757–760
153. Chen C-T, Gantzel P, Siegel JS, Baldridge KK, English RB, Ho DM (1995) Angew Chem Int Ed Engl 34:2657–2660

154. Chen K-T (1993) I. Synthesis of the First Highly Symmetrical, Nonplanar $K_{3,3}$ Molecule. II. Physical Studies of m-Terphenyl Acid: Acidity Constants and Nucleic Base Binding. PhD Dissertation University of California, San Diego
155. Watanabe S, Goto K, Kawashima T, Okazaki R (1995) Tetrahedron Lett 36:7677–7680
156. Watanabe S, Goto K, Kawashima T, Okazaki R (1997) J Am Chem Soc 119:3195–3196
157. Osawa E, Yonemitsu O (eds) (1992) Carbocyclic cage compounds: Chemistry and Applications VCH, New York
158. Norinder U, Wennerström O (1985) J Phys Chem 89:3233–3237
159. Vögtle F, Gross J, Seel C, Nieger M (1992) Angew Chem Int Ed Engl 31:1069–1071
160. Wu Z, Moore JS (1996) Angew Chem Int Ed Engl 35:297–299
161. Wambach L, Vögtle F (1985) Tetrahedron Lett 26:1483–1486
162. Hong J-I, Namgoong SK, Bernardi A, Still WC (1991) J Am Chem Soc 113:5111–5112
163. Borchardt A, Still WC (1994) J Am Chem Soc 116:7467–7468
164. West AP Jr, Van Engen D, Pascal RA Jr (1989) J Am Chem Soc 111:6846–6847
165. Méric R, Lehn J-M, Vigneron J-P (1994) Bull Soc Chim Fr 131:579–583
166. Takeshita M, Nishio S, Shinkai S (1994) J Org Chem 59:4032–4034
167. Gross J, Harder G, Siepen A, Harren J, Vögtle F, Stephan H, Gloe K, Ahlers B, Cammann K, Rissanen K (1996) Chem Eur J 2:1585–1595
168. Gross J, G H, Vögtle F, Stephan H, Gloe K (1995) Angew Chem Int Ed Engl 34:481–484
169. Matsuzawa N, Fukunaga T, Dixon DA (1992) J Phys Chem 96:10,747–10,756
170. Battersby TR (1996) Synthesis and Characterization of High Symmetry Polycyclic Aromatic Molecules. PhD Dissertation. University of California, San Diego
171. Collet A, Dutasta J-P, Lozach B, Canceill J (1993) Top Curr Chem 165:103
172. Canceill J, Collet A (1988) J Chem Soc, Chem Commun:582
173. Canceill J, Cesario M, Collet A, Guilhem J, Pascard C (1985) J Chem Soc, Chem Commun:361
174. Gabard J, Collet A (1981) J Chem Soc, Chem Commun:1137
175. Cram DJ, Tanner ME, Keipert SJ, Knobler CB (1991) J Am Chem Soc 113:8909–8916
176. Tanner ME, Knobler CB, Cram DJ (1992) J Org Chem 57:40–46
177. Böhmer V, Goldmann H, Vogt W, Vicens J, Asfari Z (1989) Tetrahedron Lett 30:1391–1394
178. Araki K, Sisido K, Hisaichi K, Shinkai S (1993) Tetrahedron Lett 34:8297–8300
179. Blanda MT, Griswold KE (1994) J Org Chem 59:4313–4315
180. Ulrich G, Ziessel R (1994) Tetrahedron Lett 35:6299–6302
181. Timmerman P, Verboom W, van Veggel FCJM, van Hoorn WP, Reinhoudt DN (1994) Angew Chem Int Ed Engl 33:1292–1295
182. Timmerman P, Verboom W, van Veggel FCJM, van Duynhoven JPM, Reinhoudt DN (1994) Angew Chem Int Ed Engl 33:2345–2348
183. Sherman JC (1995) Tetrahedron 51: 3395–3422
184. Cram DJ, Cram JM (1994) Container molecules and their guests. Royal Society of Chemistry, Cambridge
185. Cram DJ, Karbach S, Kim YH, Baczynskyj L, Kalleymeyn GW (1985) J Am Chem Soc 107:2575
186. Cram DJ, Karbach S, Kim YH, Baczynskyj L, Marti K, Sampson RM, Kalleymeyn GW (1988) J Am Chem Soc 110:2554–2560
187. Warmuth R (1997) Angew Chem Int Ed Engl 36:1347
188. Cram DJ, Tanner ME, Thomas R (1991) Angew Chem Int Ed Engl 30:1024–1027
189. Arimura T, Matsumoto S, Teshima O, Nagasaki T, Shinkai S (1991) Tetrahedron Lett 32:5111–5114
190. Kroto HW, Heath JR, O'Brien SC, Curl RF, Smalley RE (1985) Nature 318:162–163
191. Krätschmer W, Lamb LD, Fostiropoulos K, Huffman DR (1990) Nature 347:354–358
192. Thilgen C, Herrmann A, Diederich F (1997) Helv Chim Acta 80:183–199
193. Li Q, Wudl F, Thilgen C, Whetten RL, Diederich F (1992) J Am Chem Soc 114:3994–3996
194. Ettl R, Chao I, Diederich F, Whetten RL (1991) Nature 353:149–153
195. Schill G (1971) Catenanes, rotaxanes and knots. Academic Press, New York
196. Amabilino DB, Stoddart JF (1995) Chem Rev 95:2725–2828

197. Wasserman E (1960) J Am Chem Soc 82:4433–4434
198. Schill G (1967) Chem Ber 100:2021–2037
199. Chambron J-C, Dietrich-Buchecker CO, Sauvage J-P (1993) Top Curr Chem 165:131–162
200. Sauvage J-P (1990) Acc Chem Res 23:319–327
201. Dietrich-Buchecker CO, Sauvage J-P (1987) Chem Rev 87:795–810
202. Dietrich-Buchecker CO, Sauvage J-P (1983) Tetrahedron Lett 24: 5095–5098
203. Dietrich-Buchecker CO, Sauvage J-P (1984) J Am Chem Soc 106:3043–3045
204. Sauvage J-P, Weiss J (1985) J Am Chem Soc 107:6108–6110
205. Dietrich-Buchecker CO, Frommberger B, Lüer I, Sauvage J-P, Vögtle F (1993) Angew Chem Int Ed Engl 32:1434–1437
206. Ortholand J-Y, Slawin AMZ, Spencer N, Stoddart JF, Williams DJ (1989) Angew Chem Int Ed Engl 28:1394–1395
207. Ashton PR, Goodnow TT, Kaifer AE, Reddington MV, Slawin AMZ, Spencer N, Stoddart JF, Vicent C, Williams DJ (1989) Angew Chem Int Ed Engl 28:1396–1399
208. Amabilino DB, Ashton PR, Reder AS, Spencer N, Stoddart JF (1994) Angew Chem Int Ed Engl 33:1286–1290
209. Amabilino DB, Ashton PR, Reder AS, Spencer N, Stoddart JF (1994) Angew Chem Int Ed Engl 33:433–437
210. Hunter CA (1992) J Am Chem Soc 114:5303–5311
211. Vögtle F, Meier S, Hoss R (1992) Angew Chem Int Ed Engl 31:1619–1622
212. Jäger R, Vögtle F (1997) Angew Chem Int Ed Engl 36:930–944
213. Dietrich-Buchecker CO, Sauvage J-P (1989) Angew Chem Int Ed Engl 28:189–192
214. Dietrich-Buchecker CO, Guilhem J, Pascard C, Sauvage J-P (1990) Angew Chem Int Ed Engl 29:1154–1156
215. Nierengarten J-F, Dietrich-Buchecker CO, Sauvage J-P (1994) J Am Chem Soc 116: 375–376
216. Simmons HE III, Maggio JE (1981) Tetrahedron Lett 22:287–290
217. Benner SA, Maggio JE, Simmons HE III (1981) J Am Chem Soc 103:1581–1582
218. Paquette LA, Vazeux M (1981) Tetrahedron Lett 22:291–294
219. Paquette LA, Williams RV, Vazcux M, Browne AR (1984) J Org Chem 49:2194–2197
220. Kuck D, Schuster A (1988) Angew Chem Int Ed Engl 27:1192–1194
221. Kuck D (1996) Synlett:949–965
222. Walba DM, Richards RM, Haltiwanger RC (1982) J Am Chem Soc 104:3219–3221
223. Pascal RA, Jr, Grossman RB, Van Engen D (1987) J Am Chem Soc 109:6878–6880
224. Lofthagen M, Chadha R, Siegel JS (1991) J Am Chem Soc 113:8785–8790
225. Lofthagen M, Siegel JS (1995) J Org Chem 60:2885–2890
226. O'Krongly D, Denmeade SR, Chiang MY, Breslow R (1985) J Am Chem Soc 107: 5544–5545
227. Berscheid R, M N, Vögtle F (1992) Chem Ber 125:1687–1695
228. Friedrichsen B, Whitlock HW (1989) J Am Chem Soc 111:9132–9134
229. Friedrichsen BP, Powell DR, Whitlock HW (1990) J Am Chem Soc 112:8931–8941
230. Hart H (1993) Pure Appl Chem 65:27–34
231. Beck A, Bleicher MH, Crowe DW (1969) Excursions into Mathematics. Worth, New York

Modern Routes to Extended Aromatic Compounds

Stefan Hagen* · Henning Hopf

* Institut für Organische Chemie der Technischen Universität Braunschweig, Hagenring 30,
  D-38106 Braunschweig, Germany
  E-mail: shagen@tu-bs.de

Aromatic compounds and structures have recently attracted a lot of interest because of their importance in material sciences, the design of molecular devices, metabolism and other biological processes. Although the C-C bond formation is now extensively explored, the accessibility of parent and functionalized aromatics is limited to relatively few examples, especially for high molecular mass polycyclic aromatic hydrocarbons (PAHs). Therefore, this review will concentrate on new developments in the preparation of PAHs in the last decade including pyrolytic approaches to bowl-shaped fullerene fragments and other aromatic compounds, photocyclization and metathesis reactions of stilbenes, [2 + 2 + 2] cyclotrimerizations of arynes and alkynes as well as modern Diels-Alder, electrophilic cyclization, and other reactions towards condensed aromatic π-systems. Approaches to symmetrical and unsymmetric (hetero)biaryls as well as other transition metal-catalyzed cyclization reactions will also be discussed briefly. It will be shown that a great variety of synthetic procedures has been developed to prepare PAHs with very interesting molecular structures and/or high efficiency, selectivity, and functionality. The review presents 59 schemes and 318 references.

Keywords: Arenes, polycyclic, biaryls, preparation, review.

Topics in Current Chemistry, Vol. 196
© Springer Verlag Berlin Heidelberg 1998

Abbreviations

Ac	acetyl
Ar	aryl
Bu	butyl
cod	1,5-cyclooctadiene
Cp	cyclopentadienyl
DA	Diels-Alder reaction
dba	dibenzylideneacetone
DDQ	2,3-dichloro-5,6-dicyanobenzoquinone
DMA	*N,N*-dimethylacetamide
DME	1,2-dimethoxyethane
DMF	*N,N*-dimethylformamide
DMG	directing metallation group
DNA	de(s)oxyribonucleic acid
dppp	1,3-bis(diphenylphosphino)propane
EPR	electron paramagnetic resonance (spectroscopy)
Et	ethyl
FMOC	fluorenylmethoxycarbonyl
FP	flow pyrolysis
FVP	flash vacuum pyrolysis
HMPA	hexamethylphosphoric triamide
ISNA	International Symposium on Novel Aromatics
ISPAC	International Symposium on Polycyclic Aromatic Compounds
L	ligand
Me	methyl
NMR	nuclear magnetic resonance (spectroscopy)

NLO   non-linear optical
PAH   polycyclic aromatic hydrocarbon
PDC   pyridinium dichromate
Ph   phenyl
Pr   propyl
py   pyridine
TBAF   tetra-*n*-butylammonium fluoride
Tf   triflate (trifluoromethanesulfonate)
TFA   trifluoroacetic acid
THF   tetrahydrofuran
TMS   trimethylsilyl
$\Delta G^{\ddagger}_{inv}$ free enthalpy of activation for the barrier of inversion

1
Introduction

Aromatic compounds have not only been of academic interest ever since organic chemistry became a scientific discipline in the first half of the nineteenth century but they are also important products in numerous hydrocarbon technologies, e.g. the catalytic hydrocracking of petroleum to produce gasoline, pyrolytic processes used in the formation of lower olefins and soot or the carbonization of coal in coke production [1]. The structures of benzene and polycyclic aromatic hydrocarbons (PAHs) can be found in many industrial products such as polymers [2], specialized dyes and luminescence materials [3], liquid crystals and other mesogenic materials [4]. Furthermore, the intrinsic (electronic) properties of aromatic compounds promoted their use in the design of organic conductors [5], solar cells [6], photo- and electroluminescent devices [3, 7], optically active polymers [8], non-linear optical (NLO) materials [9], and in many other fields of research.

In general, the majority of aromatic compounds are characterized by their delocalized electrons, a rigid and well-defined geometric structure, a relatively high thermal, photochemical and chemical inertness as well as the possibility of regioselective functionalization [10]. Thus, it is not surprising, that aromatic and heteroaromatic substructures are major parts of interesting compounds in which a defined and functionalized shape plays a key role, like in calixarenes [11], crown ethers [12], catalysis with an asymmetric binaphthyl backbone as ligands for enantioselective catalysis [13], cyclophanes [14], self-assembled supramolecular complexes [15], metal complexes as catalysts in stereoselective polymer synthesis [16] and most types of cage compounds [17]. The perfect cages themselves, the recently discovered fullerenes and nanotubes [18], represent polycyclic aromatic all-carbon compounds with a highly curved, closed surface. These hollow structures have also been found to form endohedral (inclusion) complexes with metals or inert gases [18]. Furthermore, the characteristics mentioned above, make aromatics very suitable for the design of molecular devices [19], such as catenanes and rotaxanes [20], Jøergensen's chiroptical molecular switch [21], Kelly's molecular brake [22] and others. Furthermore, as it has been known for a long time, aromatic structures and substructures are found in numerous natural products.

Domestic and natural combustion of coal, wood and other organic materials supply a steady source of PAHs. The environmental ubiquity of several aromatic representatives together with their procarcenogenic and mutagenic properties also created a great deal of interest in the formation of PAHs in fuel-rich flames as well as in the emission, occurrence and analysis of PAHs combined with aspects of metabolism, deoxyribonucleic acid (DNA)-adduct formation, biomarkers, exposure and risk assessment [23].

Beside the practical importance of aromatic compounds, there has always been an interest in more or less theoretical problems like the scope, limitation and effects of electron delocalization in aromatic compounds (the "aromaticity problem"). These investigations were strongly encouraged by the discovery of fullerene formation in a carbon plasma [18], in fuel-rich flames [24] or by the pyrolytic transformation of PAHs [25] together with a variety of the as yet potential application of these aromatic carbon cage compounds [18]. New selective C-C bond formation reactions as well as mechanisms of the rearrangement in carbon skeletons have been studied.

Although millions of tons of aromatic compounds are produced by the chemical industry as intermediates or byproducts every year, the accessibility of parent and functionalized aromatics is limited to relatively few examples, especially for high molecular PAHs. Therefore, in preparative aromatic chemistry it is a common situation that structurally rather simple starting materials or reaction intermediates are commercially not available and/or have not been synthesized previously. For the reasons mentioned above, there is a constant interest in the synthesis of analytically pure samples of known and still unknown PAHs with or without functionalization.

The topic "Polycyclic Aromatic Hydrocarbons" has been reviewed by different authors in the last decades [26]. The most complete summary of classical synthetic routes to PAHs may still be found in Eric Clar's *"Polycyclic Hydrocarbons"* [26a]. Most recently, Ronald G. Harvey reviewed the syntheses and characteristics of PAHs up to seven rings in a very comprehensive fashion [26f] in his book *"Polycyclic Aromatic Compounds"*. Recent advances in the synthesis of PAHs are published in all major journals of organic chemistry, in *Polycyclic Aromatic Hydrocarbons* [27], and presented during the regular *ISNA* [28] and *ISPAC* [29] conferences.

This contribution will focus on *new developments* in the key step for the synthesis of PAHs: the C-C bond formation as one of the most fundamental and challenging processes in organic and organometallic chemistry. In addition, these new synthetic approaches often give access to substituted aromatic structures. The authors attempt to cover the literature until the end of 1996. Important contributions in the synthesis of PAHs during the work on this review in the beginning of 1997 will also be considered whenever possible. The functionalization of aromatic compounds [10] and the chemistry of fullerenes [18] are beyond the limits of this contribution. As mentioned above, this review will concentrate on *modern* routes to preferably large condensed PAHs. Classical approaches towards aromatic hydrocarbons like the Pschorr, Elbs, Hayworth,

McMurry, Diels-Alder, Scholl, and Wittig reactions as well as *intra*molecular cyclodehydration procedures of carboxylic acids, oxidation of hydrogenated aromatics and reduction of quinones and phenols have already been extensively reviewed [26, 30] and will be mentioned only in few exemplified syntheses of new parent aromatic systems or in those cases when significant modifications result in much improved yields and/or milder reaction conditions.

2
Synthesis of PAHs by Thermal Conversions

The conversion of appropriate precursors to condensed PAHs at elevated temperatures is the classical synthesis of numerous pure aromatic compounds. Depending on the reactivity of the starting materials temperatures between 300 °C and 1300 °C have been applied [31]. The reactions are normally performed in an inert quartz tube placed in a furnace with resistance heating (see Scheme 1). The precursor is transferred into the gas phase in a temperature controlled evaporation zone and swept through the pyrolysis tube by a carrier gas and/or a pressure gradient produced by a vacuum pump. The process has been referred to as flash vacuum pyrolysis (FVP) with typical reaction times of 2–50 ms or as flow pyrolysis (FP, typical reaction times 0.2–2 s) if the conversion is carried out under normal pressure. The pyrolysis products condense immediately behind the oven in a cold trap.

The design of the precursor as well as the pyrolysis technique have to be optimized to obtain satisfactory selectivities and yields in the thermal conversions. The starting material has to be stable enough to be transferred into the gas phase without decomposition at temperatures up to 400 °C (depending on the vapor pressure and therefore on the molecular weight and structure) but also to possess an inherent reactivity which can be exploited to yield the desired products with high selectivities and yields under conditions as mild as possible. Thus, enthalpies of activation of 200–400 kJ*mol^{-1} [31, 32] are representative for pyrolyses as synthetic procedures with moderate or high yields. Numerous special techniques have been developed to transfer thermally labile products into the gas phase (spray pyrolysis [33], solvent assisted sublimation techniques

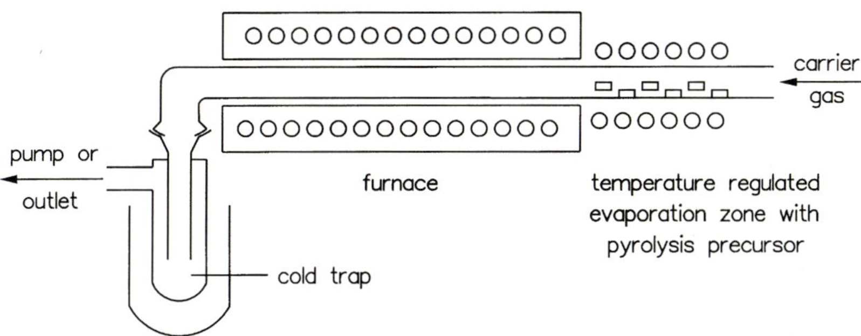

Scheme 1. General apparatus for flash vacuum pyrolysis and/or flow pyrolysis [10]

[34]) or to trap highly reactive pyrolysis products (matrix isolation [35]). Monomolecular reaction conditions resulting from high dilution techniques and short reaction times are often required to suppress the formation of byproducts resulting from *inter*molecular reactions. The preparation of PAHs by thermal condensation reactions between two or more molecules has been largely abandoned in the past decades due to the low selectivity in the product formation.

2.1
Cyclodehydrogenation and Related Reactions

The thermal cyclodehydrogenation involves the formation of an additional ring in a not fully condensed PAH and the abstraction of two hydrogen atoms formerly connected to those carbon atoms which comprise the new C-C bond. Ring formation and hydrogen abstraction can take place in a consecutive or a parallel fashion.

The thermal disrotatory [$_\pi$6] electrocyclization of *cis*-1,3,5-hexatriene systems has been extensively employed for the synthesis of cyclic hydrocarbons. The average enthalpy of activation is in the range of about 120 kJ*mol^{-1} [36]. The incorporation of two of the hexatriene double bonds in phenyl rings (stilbene, 1) stabilizes the precursor significantly and necessitates temperatures of 1050 °C to obtain a 30% yield of phenanthrene (2, see Scheme 2, [37]). An enthalpy of activation of (250 ± 20) kJ*mol^{-1} was estimated for the conversion of 9,9′-bifluorenylidene (3) to benz[e]indeno[1,2,3-hi]acephenanthrylene (4), a reaction that is accompanied by the radical initiated isomerization of 3 to dibenzo[g,p]chrysene (5, Scheme 2, [38]). It is assumed that both reactions 1 → 2 and 3 → 4 are initiated by an electrocyclic ring closure forming a 4a,4b-dihydrophenanthrene (1a) intermediate.

If the [$_\pi$6] electrocyclization is impeded for steric or electronic reasons, the yields of thermal cyclodehydrogenation reactions normally decrease to a few or even less than one percent [38–41]. Nevertheless, the interesting bowl-shaped fullerene fragments diindeno[1,2,3,4-*defg*;1′,2′,3′,4′-*mnop*]chrysene (6, [38 d, e]) and triacenaphtho[3,2,1,8-*cdefg*;3′,2′,1′,8′-*ijklm*;3″,2″,1″,8″-*opqra*]triphenylene (7, [39]) could be obtained from 4 and decacyclene (8), respectively, for the first time (see Scheme 3).

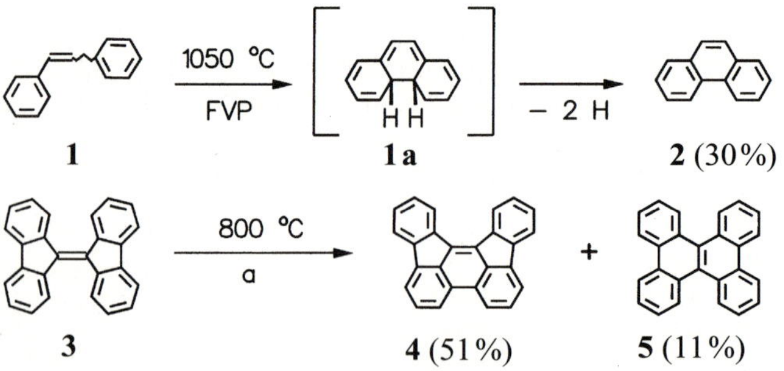

Scheme 2. Thermal cyclodehydrogenation reactions of stilbene type aromatics [37, 38]; a) toluene (*p* = 7 mbar) as carrier gas and radical scavenger

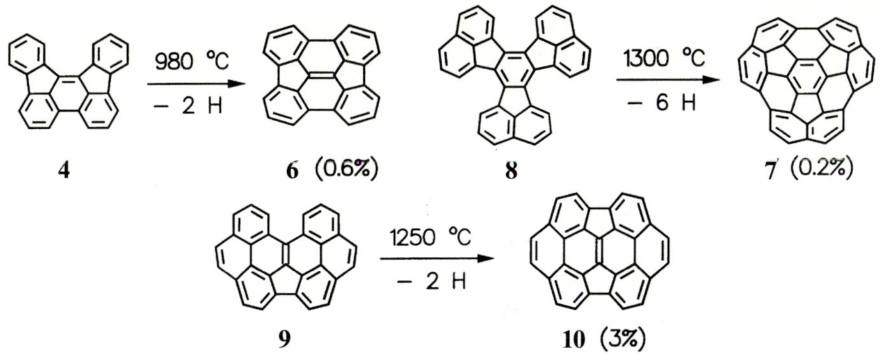

Scheme 3. Syntheses of bowl-shaped fullerene fragments by thermal cyclodehydrogenation using the flash vacuum pyrolysis methodology [38, 39]

The thermal cyclodehydrogenation is not limited to the formation of six-membered carbon rings, new five-membered rings were obtained under the same conditions with similar yields; e.g. the ring closure of dibenzo[*de,hi*]-fluoreno[2,1,9,8,7-*mnopqr*]naphthacene (**9**) to the fullerene-C_{78} subunit difluoreno[2,1,9,8,7-*defghi*;2',1',9',8',7'-*mnopqr*]naphthacene (**10**, Scheme 3, [40]). It is assumed that the cyclization is initiated by random H-abstraction from the polycyclic hydrocarbons and that the aryl radicals generated in the fjord regions undergo a radical cycloaddition reaction to form the new C-C bond in the more condensed PAH [38 – 41].

Consequently, significantly higher product selectivities and yields were obtained at somewhat lower pyrolysis temperatures, when the aryl radicals were generated intentionally by the homolysis of CX bonds (X = halogens, alkyl, carbonyl) having lower bond dissociation enthalpies than $C_{aryl}H$ bonds [32b]. Pioneering work was accomplished for example by Schaden's synthesis of pyracylene (**11**, see Scheme 4, [42]) from pyrenediones and mechanistic studies by Brown and Eastwood [31e] in the pyrolysis of anhydrides and carbonic acid derivatives. Appropriate precursors with halogen substituents in the fjord regions or at *ortho*-positions have been applied by several other research groups to synthesize PAH in significantly higher yields (compare yields for **2** and **6** in Scheme 4 with those given in Scheme 2 and 3, [31f, g, 37, 43 – 45]). The bowl-shaped fullerene fragments dibenzo[*a,g*]corannulene (**12**, [31 g, 46]) and benz-[5,6]-*as*-indaceno[3,2,1,8,7-*mnopqr*]indeno[4,3,2,1-*cdef*]chrysene (**13**, [40, 45]) could be obtained by this approach from the dibrominated 7,10-diphenylfluoranthene **14a** and the tribrominated benzo[*c*]naphtho[2,1-p]chrysene **15a**, respectively, where the pyrolysis of the unsubstituted parent PAH **14b** and **15b** failed completely [41, 47].

Bromine and chlorine substituents have been employed in most syntheses as a compromise of thermal stability and reactivity of the reactants necessary in the pyrolysis procedure as outlined above. Needless to say that these halogen substituents must normally be introduced at those carbon atoms which eventually participate in the new C-C bond. If this is not possible for synthetic reasons, the substitution in other positions can provide an interesting alternative. It has

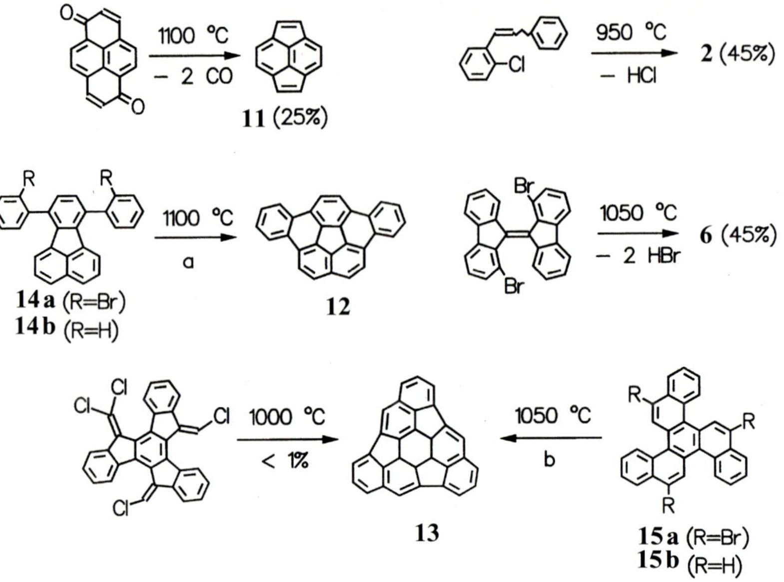

Scheme 4. Syntheses of bowl-shaped and other PAHs by FVP by elimination of suitable leaving groups; a) R=Br, –2 HBr, 12:10%; R=H, 12:0%; b) R=Br, –3 HBr, 13: 7.5–9%; R=H, 13: 0% [31 g, 37, 40, 42 a, b, 45 – 47]

been shown very recently by labelling studies, that a 1,2-hydrogen shift in aryl radicals can indeed take place at temperatures above 1000 °C, forcing the hydrogen atom out of the sterically crowded fjord region and creating a radical at the desired position (e.g. 15a → 13, Scheme 4, [40, 48]). The thermal homolysis of a $C(sp^2)C(sp^3)$ bond to create the desired aryl radicals in the synthetic approach to highly condensed PAHs has also been reported very recently, but it seems to be inferior compared to the C-halogen scission [40].

2.2
Sulfone Pyrolysis

The thermal elimination of sulfur dioxide from mainly dialkyl sulfones is a well established and documented preparative method in the formation of C-C bonds [31 b]). Partly saturated aromatic hydrocarbons, obtained in the primary step, can undergo further thermal or catalytic dehydrogenation reactions resulting in fully unsaturated PAHs as illustrated for the syntheses of corannulene (16, [49]) and [7]circulene (17, [50]) in Scheme 5.

The most important application of sulfone pyrolysis lies in the selective formation of cyclophanes and cage compounds (e.g. 18 and 19 in Scheme 6, [51, 52]) from relatively conveniently available bisbenzyl sulfones. The preparative results of a sulfone pyrolysis strongly depend on the structure and the number of the sulfone groups present in the molecule and can vary between a few percent and almost quantitative yields [31 b, 50 – 52]).

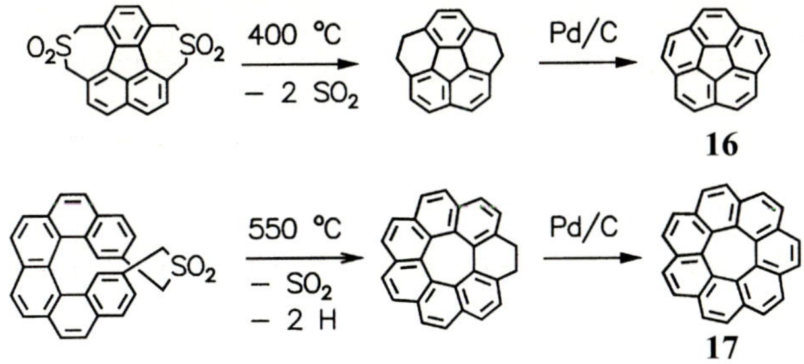

Scheme 5. Syntheses of the circulenes **16** and **17** from suitable sulfones by flash vacuum pyrolysis and aromatization [49, 50]

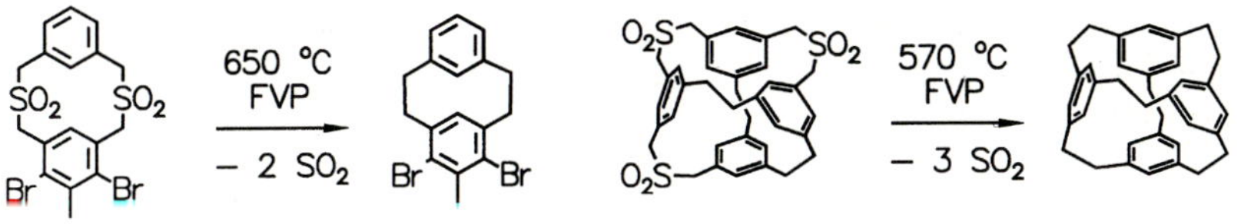

Scheme 6. Syntheses of cyclophanes by sulfone pyrolysis [51, 52]

2.3
Carbene Insertion Reactions

The reversible rearrangement of terminal acetylenes to vinylidenes under FVP conditions was discovered by Brown's group in 1972 [31e, 53]. Shortly afterwards, the synthetic feasibility of the reactive carbene insertion was demonstrated by the formation of phenanthrene (**2**) in the pyrolysis of 2-ethynylbiphenyl (**20**) by an *intra*molecular CH insertion via the transient carbene (see Scheme 7).

Rather than being a scientific curiosity, this approach has widely been used during the last decade to prepare condensed PAHs [54–56], that were hitherto either not available at all or only by extended synthetic routes; the most inspiring example probably being the convenient and short synthesis of corannulene (**16**) from 7,10-diethynylfluoranthene (**21**, Scheme 7, [54a]). The reversibility of this vinylidene carbene insertion has been established recently [54f].

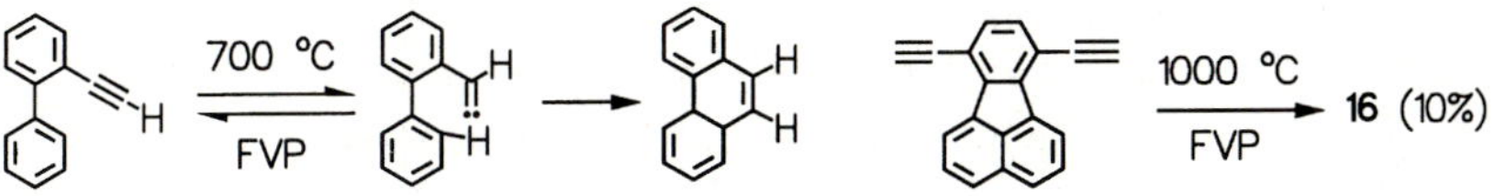

Scheme 7. General method for the preparation of PAHs by insertion reactions of carbenes generated by the thermal acetylene to vinylidene-carbene rearrangement [53b, 54a]

The relatively low thermal stability of the acetylene precursors inspired the search for a more stable, masked ethynyl group that can be quantitatively converted into acetylenes in the gas phase of the pyrolysis apparatus. Presently, the state of the art consists in the substitution of ethynyl groups by chloroethenyl substituents [54b–f, 55, 56]). The latter show a higher thermal stability and are conveniently available from acetyl derivatives by reaction with PCl₅ or from trimethylsilyl (TMS)-substituted acetylenes by treatment with hydrochloric acid in glacial acetic acid (see Scheme 8).

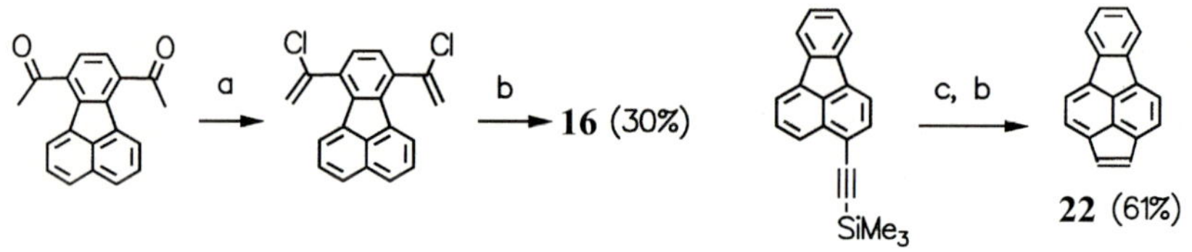

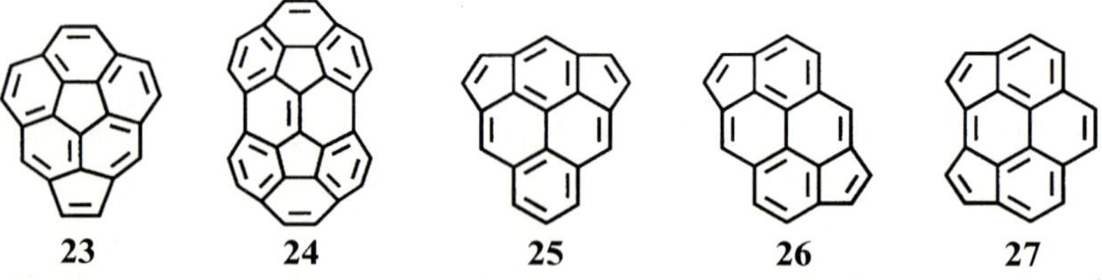

Scheme 8. Flash vacuum pyrolysis of chloroethenyl compounds [31 g, 54 c, d]: a) PCl₅; b) FVP, 1100 °C; c) HCl, CH₃COOH

Numerous aromatic compounds including several bowl-shaped fullerene fragments have been prepared by this method, e.g. cyclopenta[*ij*]fluoranthene (**22**, Scheme 8, [54 d, 56 a]), cyclopenta[*bc*]corannulene (**23**, [55 a]) and diacenaphtho[3,2,1,8-*cdefg*;3′,2′,1′,8′-*lmnop*]chrysene (**24**, see Scheme 9, [55 b, c]). Of special importance is this approach for the synthesis of cyclopenta-annelated PAHs, e.g. and the three isomeric dicyclopentapyrenes (**25–27**, Scheme 9, [54 e, 56 b]). Using these reference samples, several cyclopenta-annelated PAHs could be identified as byproducts formed in the incomplete oxidation of hydrocarbons in fuel rich flames [57].

Other masked ethynyl groups including dihaloethenyls [54 b], ethenyl ethers [58] and terminally substituted acetylenes [59] have also been applied in the pyrolytic preparation of PAHs, presumably also via carbene intermediates [60]. Furthermore, the combination of the different synthetic strategies is also possible, as demonstrated by the first successful synthesis of benzocorannulene (**28**, Scheme 10, [54 g]) and other PAHs.

The thermal formation of an alkylcarbene followed by CH insertion has been used as a key step in the pyrolysis of cyclobuta[*de*]naphthalene (**29**) and its analogs (see Scheme 10, [61]). The starting methoxy(1-naphthyl)methyltrimethylsilane (**30**) is easily prepared and allows the synthesis of numerous derivatives of **29**.

Scheme 9. Selected PAHs prepared by flash vacuum pyrolysis [40, 54 e, 55, 56 b]

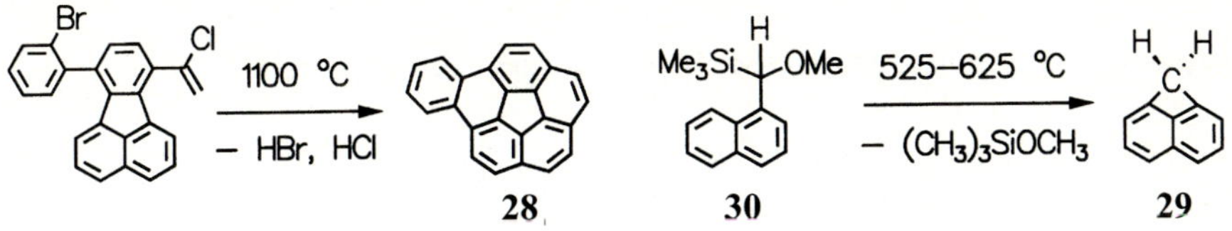

Scheme 10. Flash vacuum pyrolysis: combinatorial approach in the synthesis of benzocorannulene (**28**, [54 g]) and preparation of aromatics with four membered rings (e.g. **29**, [61])

2.4
Thermal Rearrangements of Carbon Skeletons

The thermal automerization and rearrangement reactions of PAHs have been widely investigated during the past two decades (for examples see refs. [31 e, g, 62–64]). The main objective was to understand the processes of formation of aromatic hydrocarbons in fuel rich flames and the mechanisms of transformation of the PAHs that have been observed at these elevated temperatures. In most cases, thermally initiated rearrangement reactions in the carbon skeletons of PAHs require high enthalpies of activation resulting in low product selectivities and poor overall yields. Because the expected products are often more effectively prepared by conventional routes, this approach has been used as a synthetic tool only in a few cases, e.g. the synthesis of azulenes [65] and the rearrangement of bifluorenylidenes to benzenoid hydrocarbons [38].

A remarkable transformation is the synthesis of anthracene (**31 a**) and its heteroanalogous compound acridine (**31 b**) from dibenzo[*a,e*]cyclooctadiene (**32 a**), and 5,6,11,12-tetrahydrodibenzo[*b,f*]azoicene (**32 b**), respectively. The proposed general mechanism for this reaction involves splitting of one bridge of the starting material, an *intra*molecular cycloaddition sequence and the formal loss of the other bridge (see Scheme 11 [66]).

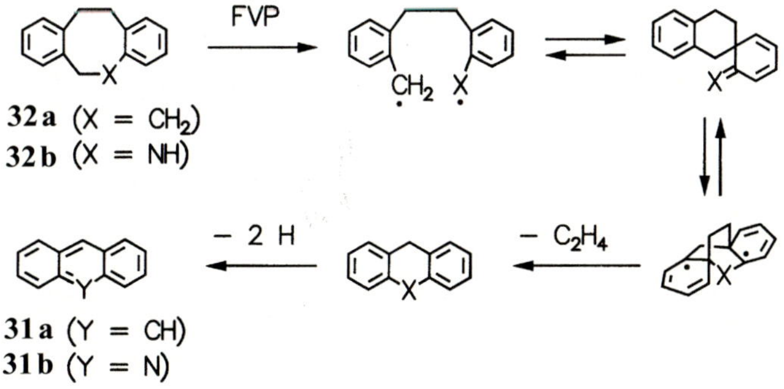

32a (X = CH$_2$)
32b (X = NH)

31a (Y = CH)
31b (Y = N)

Scheme 11. Novel syntheses of anthracene (**31 a**, R=CH: 900 °C, 25 %) and acridine (**31 b**, R=N: 750 °C, 75 %) by flash vacuum pyrolysis [66]

3
Photocyclization Reactions of Stilbene Type Compounds

The photolytic ring closure of stilbene type compounds has been widely employed as a versatile method in the preparation of condensed aromatic and heteroaromatic compounds [67]. According to orbital symmetry conservation rules [68], the reaction is a reversible conrotatory $[_\pi 6]$ electrocyclization of the *cis*-isomer (e. g. *cis*-stilbene (**1**), Scheme 12) to the *trans*-4a,4b-dihydrophenanthrene (**1b**). Because the *cis/trans*-isomerization of stilbenes is sufficiently fast under the reaction conditions applied, the *trans*-isomer of the stilbenes can also be used. Depending on concentration, structure, and solvent, the stilbene type compounds can also undergo a [2 + 2] cycloaddition reaction leading to aryl-substituted cyclobutanes [69].

The *trans*-4a,4b-dihydrophenanthrene (**1b**) intermediates have to be removed from the equilibrium by aromatization to the corresponding phenanthrenes (e. g. **2**) either by cleavage of a suitable leaving group at the 4a-position in **1b** or, most commonly, by an oxidation agent; the generally accepted method employs the use of air plus a catalytic amount of iodine. This procedure has been applied as a key step in numerous syntheses of interesting PAHs, such as helicenes [70] as well as the first syntheses of [7]phenacene (**34**, [71]) and [7]circulene (**17**, [72]) respectively (see Scheme 5, 13).

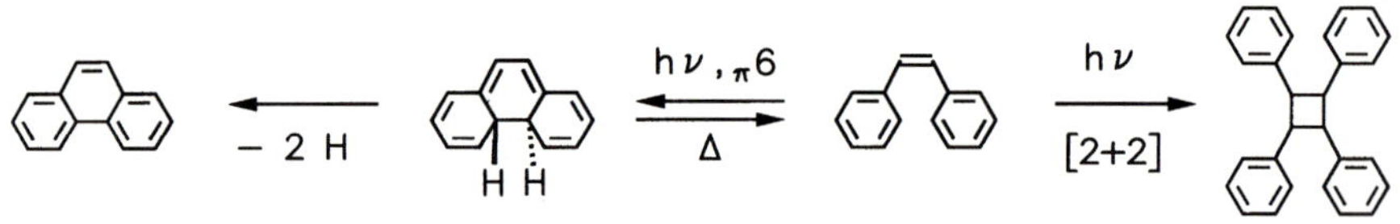

Scheme 12. Different photochemical reaction pathways of *cis*-stilbene (**1**): (i) to phenanthrene (**2**) by conrotatory $[_\pi 6]$ electrocyclization, and (ii) to tetraphenylcyclobutane (**33**) by a [2 + 2] photocycloaddition [67 a – d, 68]

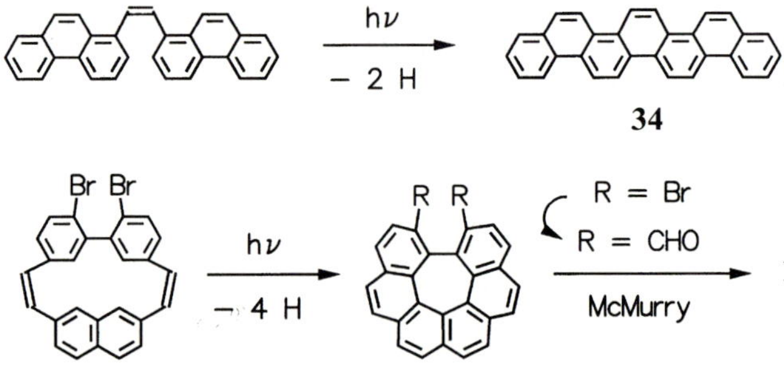

Scheme 13. Photocyclization reactions as key steps in the syntheses of [7] phenancene (**34**, [71]) and [7] circulene (**17**, [72])

Depending on the chemical reactivity of the stilbene derivatives other oxidizing agents such as diphenyl selenide [73] and, with superior results, the recently discovered system iodide/propylene oxide in an inert gas atmosphere [74] have been applied. The absence of air diminishes photooxidative side reactions and the trapping of photogenerated hydrogen iodine with propylene oxide prevents the reduction of the double bonds in the stilbenes. An alternative to the complete aromatization is the partial aromatization to other dihydrophenanthrenes by isomerization (hydrogen shifts, [67a–c]).

Symmetrical and unsymmetrical stilbenes can be conveniently synthesized [67d]; approaches most often applied are the Wittig [75,76], the Siegrist [77] and the Heck reaction ([78, 79] see also Sect. 4.3.5) as well as the McMurry coupling [80]. Another major advantage of photolytic ring closures in stilbenoid compounds is, that in cases when several reactions between different carbon atoms are possible, the preferred reaction path can be safely predicted with only a few exceptions. In general, high selectivities for cyclization reactions will be obtained only for those C centers which are to be linked (r, s) and that have the highest value for the sum of the free valencies in the first excited state (ΣF_{rs}^*), whereas ΣF_{rs}^* should be preferably greater than 1 [67d, 81]. For example, the photocyclodehydrogenation of the stilbene **35** could, in principle, have led to ten different products, but the PAH **36** was formed in an excellent yield of 93% as a key intermediate in the synthesis of circumanthracene (**37**, [82], see Scheme 14). In those cases where the less favored reaction path is desired, the introduction of bromine can direct the way along which the photocyclization should proceed. It has been found, that a bromine substituent blocks effectively the position it occupies as well as the adjacent one [83].

A topological reaction enforcement enabled the oxidative photocyclization of 9,9′-bifluorenylidenes for the first time ([73], see Scheme 15). The two fluorene

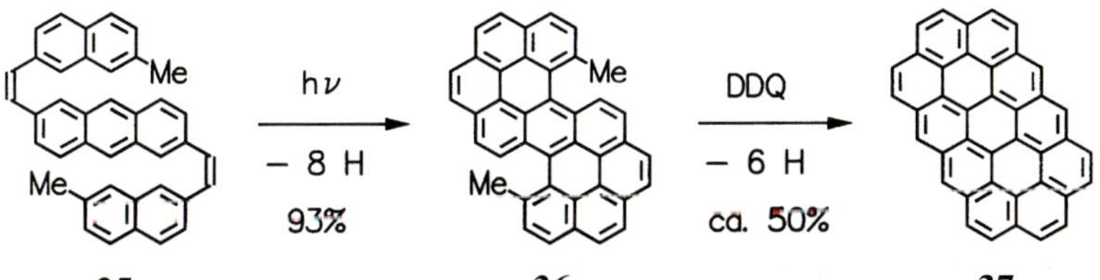

Scheme 14. Ring closure reactions in the synthesis of circumanthracene (**37**, [82])

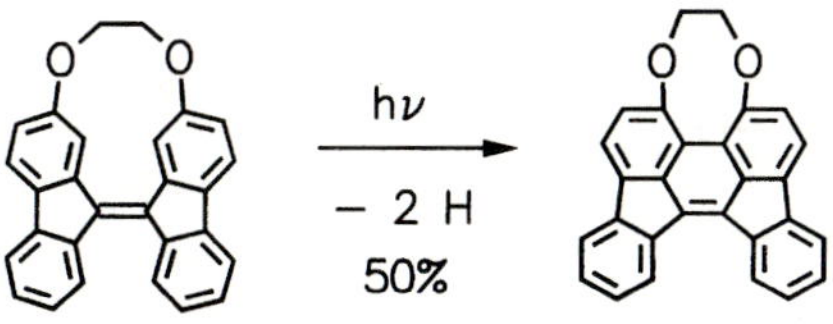

Scheme 15. Photocyclization of tethered bifluorenylidenes [73]

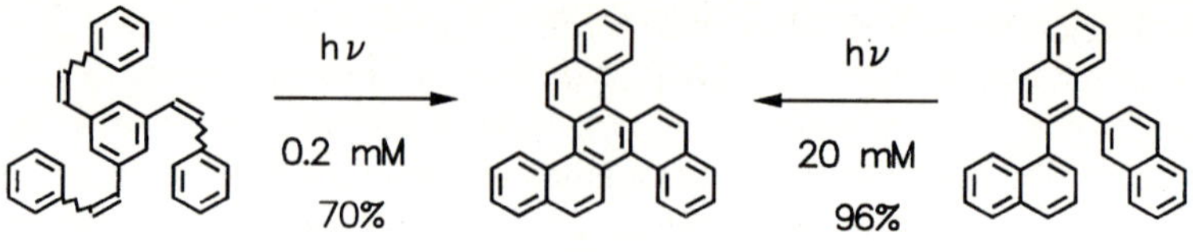

Scheme 16. Different ranges of concentration that have been used in selected photocyclization reactions to benzo[*c*]naphtho[2,1-*p*]chrysene (15b, [67d, 84c, 85])

units were linked together with an alkoxy tether and presumably forced into a favorable geometry for ring closure, providing an interesting route to polycyclic systems, where the parent aromatic hydrocarbon fails completely to react in the course of the photochemical conversion.

Irradiation of *ortho*-diarylarenes (e.g. *o*-terphenyls) results exclusively in condensed PAH as products from photocyclization reactions (e.g. triphenylenes, see [84]) and, of course, no [2+2]cycloaddition is observed. This fact enables the use of higher concentrations of *ortho*-diarylarenes compared to stilbenes in a photocyclization reaction as outlined in Scheme 16 for the synthesis of benzo[*c*]naphtho[2,1-*p*]chrysene (15b) from 1,3,5-tristyrylbenzene (38, c ≤0.2 mmol l⁻¹ [85]) or 1,2′:1′,2″-ternaphthalene (39, c=20 mmol l⁻¹ [84c]) respectively. If higher concentrations of 38 were applied, [2+2] cycloaddition reactions yielding multibridged cyclophanes were observed almost exclusively [85].

Although the [2+2] photocycloaddition is preferably prevented in the synthesis of PAHs, it has been proven to be a very feasible approach in the selective synthesis of *syn*-[2.2]cyclophanes [86, 87]. Topological reaction control in solution has been achieved for a multistep *intra*molecular [2+2] cycloaddition reaction in the photochemical formation of [*n*]-ladderanes from *pseudo-gem*-bis(polyene) substituted [2.2]paracyclophanes [88]. The probably most well-known example of an *intra*molecular photocycloaddition process is one of the

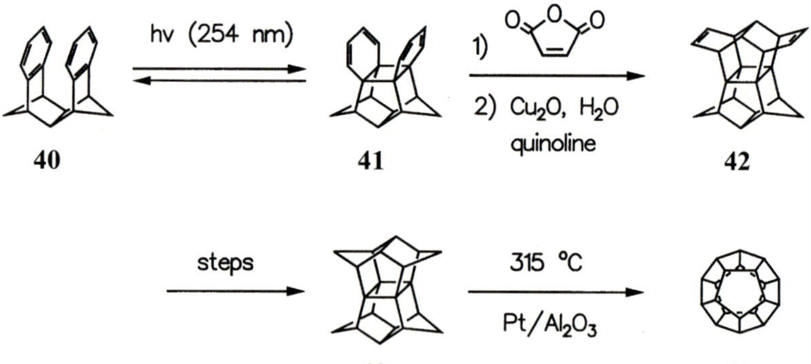

Scheme 17. Syntheses of pagodane (43, [89]) and dodecahedrane (44, [90])

key steps in the syntheses of pagodanes [89]. The rigid face-to-face orientation of the two benzene units in the dibenzo-annelated isodrin **40** allows a benzo/benzo [6 + 6] photocycloaddition to the metastable hydrocarbon **41**, that itself can be converted to the [2.2.1.1]pagodadiene (**42**) by a simple two step domino Diels-Alder/decarboxylation procedure (see Scheme 17). The [1.1.1.1]pagodane (**43**) resulting from **42** in a multi-step sequence can be converted to the dodecahedrane (**44**) by vapor-phase isomerization on 0.1% Pt/Al$_2$O$_3$ in 8% yield [90, 91].

Topological reaction control in the solid state is also most likely the key step in the synthesis of the fully conjugated tube-shaped hydrocarbon (**45**) from tetradehydrodianthracene (**46**, see Scheme 18, [92a]). While the reaction fails in solution under various conditions, the double bonds are favorably arranged for an *inter*molecular [2 + 2] cycloaddition reaction in the crystalline state. The photochemically induced reactions of **46** with a number of cycloalkenes and benzene also resulted in a variety of aromatic metathesis products [92b, c].

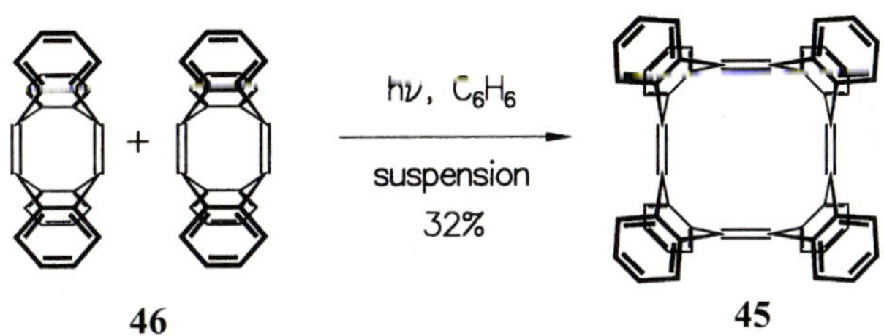

Scheme 18. *Inter*molecular [2 + 2] cycloaddition of **46** in the solid state [92a]

4
Aryl-Aryl Coupling and Condensation Reactions

Biaryls are important components in many natural products, pharmaceuticals, polymers, liquid crystals, crown ethers, chiral catalysts, stiff linear spacers and other functional molecules [93]. The normally low regioselectivity in aromatic substitution reactions of bi- and polyaryls often necessitates the coupling of the already more or less functionalized aryl moieties in the design of the target biaryl product. By this approach, the problem of the regioselectivity is transferred to the synthesis of suitably substituted aromatics in combination with an appropriate aryl coupling methodology, allowing regioselective, chemoselective, and if necessary, enantioselective formation of the desired aryl-aryl bond.

The very large variety of aryl-aryl coupling reactions developed in the last decades exceeds the scope of this article. An excellent review on the formation of biaryls with special attention to the synthesis of biaryl natural products was published by Bringmann et al. [93a] in 1990, further reviews on the Stille [94, 95a], Suzuki [95], and Heck [78, 79] coupling reactions appeared more recently. This survey will present some selected examples of aryl-aryl coupling and condensation reactions described in recent years with special attention on the formation of large polycyclic aromatic systems. We are trying to cover the most popular synthetic principles towards biaryl systems including modern approaches using transition metal catalysis as well as reactions, that have been developed many decades ago, but are still potentially useful for the synthesis of high molecular PAHs.

4.1
Reductive Coupling Reactions

4.1.1
Aryl-Aryl Coupling Reactions

The reductive dimerization of aryl halides with zero-valent copper (Ullmann coupling [96]) is the method of choice for the synthesis of symmetrical biaryls from electron-deficient aryl iodides (see Scheme 19). Although the coupling positions are unequivocally predetermined by the substitution pattern of the halogen substituents, the application for the synthesis of highly functional-ized biaryls has been limited because of the high reaction temperatures often required and the poor yields in the synthesis of unsymmetrical biaryls. There-fore, the very recent discovery, that copper(I) thiophene-2-carboxylate can induce the Ullmann-like *homo*coupling of selected aryl iodides and bromides at ambient temperatures could be of significant synthetic utility [97].

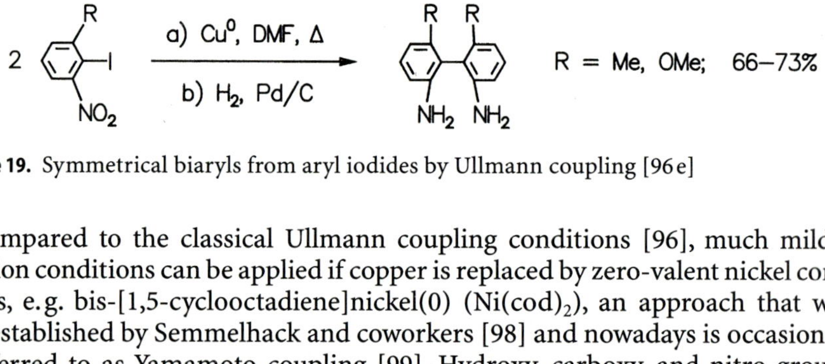

$$2 \quad \text{Ar–I} \quad \xrightarrow[\text{b) } H_2, \text{ Pd/C}]{\text{a) Cu}^0, \text{ DMF, } \Delta} \quad \text{Ar–Ar} \qquad R = Me, OMe; \quad 66{-}73\%$$

Scheme 19. Symmetrical biaryls from aryl iodides by Ullmann coupling [96e]

Compared to the classical Ullmann coupling conditions [96], much milder reaction conditions can be applied if copper is replaced by zero-valent nickel com-plexes, e.g. bis-[1,5-cyclooctadiene]nickel(0) (Ni(cod)$_2$), an approach that was first established by Semmelhack and coworkers [98] and nowadays is occasional-ly referred to as Yamamoto coupling [99]. Hydroxy, carboxy, and nitro groups generally interfere with the reaction, while good results have been obtained in the presence of amino and cyano substituents [93a]. The sensitive and expensive nickel complexes have been substituted in several recent applications by catalytic amounts of Ni(0) species which are generated *in situ* from Ni(II) halides or Ni(II) complexes by reduction with Zn, NaH or electrochemically [3b, 93a, 100, 101]. Excellent yields for *homo*coupling reactions have been obtained by this approach at temperatures of 20–70 °C (see Scheme 20). Moderate yields (45–62%) were even reported for unsymmetrical coupling reactions [3b].

Although the classical Wurtz–Fittig synthesis has been widely replaced by procedures using milder reaction conditions (see following chapters), alkali metals like potassium in dimethoxyethane (DME) have been used in the dime-rization of 1,1′binaphthyls (e.g. **47**, see Scheme 21) to perylene derivatives (e.g.

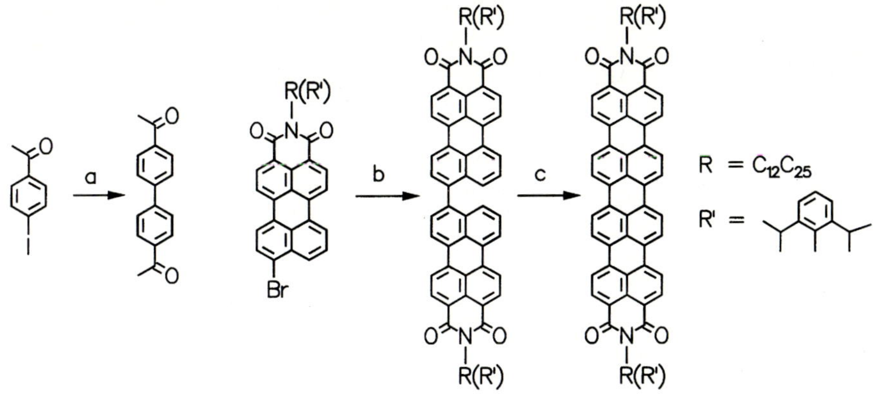

Scheme 20. *Homo*coupling reactions with zerovalent nickel: a) $NiBr_2$, Zn, KI, DMF, HMPA, 50 °C, 3 h, 96 % [100]; b) $Ni(cod)_2$, DMF, 60 °C, 2 d, 83–89 %; and oxidative cyclodehydrogenation to extended π-systems: c) KOH, EtOH, ox., R:37 %, R':83 % [101]

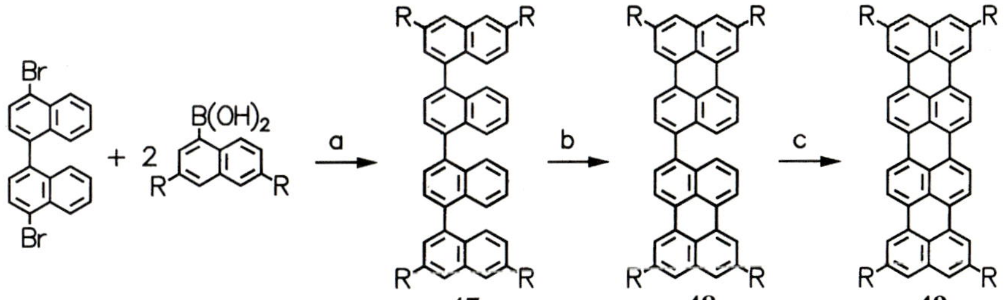

Scheme 21. Reductive and oxidative aryl-aryl coupling reactions in the synthesis of quarterrylene **49**: R = *t*Bu; a) 3 mol % $Pd(PPh_3)_4$, K_2CO_3, toluene, 3 d, reflux, 74 %; b) K, DME, 7 d, rt, $CdCl_2$, 48 %; c) $AlCl_3$, $CuCl_2$, CS_2, 8 h, rt, 48 % [102]

48 [102]). The quarterrylene (**49**) and other *oligo(peri*-naphthylene)s, which are large condensed PAHs with interesting fluorescence behavior, have been synthesized in an overall three-step aryl-aryl coupling sequence from suitably substituted naphthalenes. Electron paramagnetic resonance (EPR) and nuclear magnetic resonance (NMR) investigations indicate, that this reaction proceeds via the dianions of **47** and **48**, which were oxidized with cadmium chloride to the neutral hydrocarbons.

A one-pot synthesis of phenanthrenes from *ortho*-alkoxyarene aldehydes and ketones has been accomplished using $TiCl_3$-Li-THF [103]. The 2,2′dialkoxystilbenes, resulting from a classical McMurry coupling [80], are subjected to a reductive dealkoxylation and under the reaction conditions applied [103] form the *intra*molecular CC coupling products in moderate overall yields (25–38 %).

A highly valued methodology which leads to controlled syntheses of biaryls in symmetrical and unsymmetrical cases is the aryl-aryl bond formation via

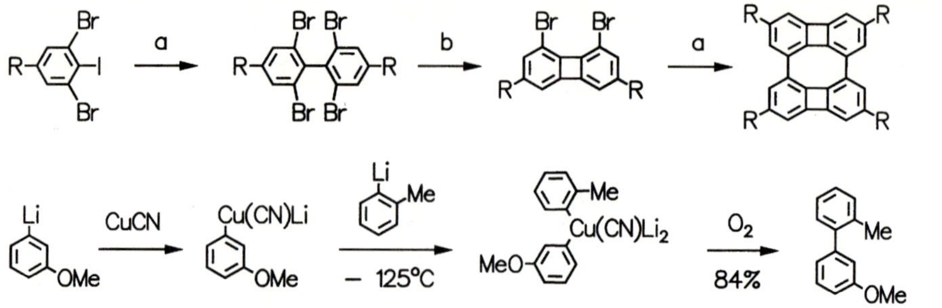

Scheme 22. Aryl-aryl bond formation via arylcuprates: a) (1) *n*-BuLi, ether; (2) CuBr$_2$ (30%), [104]; b) (1) *n*-BuLi, THF; (2) CuCN; (3) O$_2$, –78 °C (30%) [105]

aryl cuprates. In general, the aryllithiums, generated by lithium/bromine or lithium/iodine exchange, were oxidized with Cu(II) salts or were converted to symmetrical diarylcyanocuprates with CuCN which in turn were oxidized with oxygen or air (see Scheme 22 [104, 105]). Although this approach has been preferentially applied to the synthesis of symmetrical biaryls, cross-coupling can be achieved in high yields if an initially formed "low order" arylcyanocuprate reacts with a second equivalent of another aryllithium followed by oxidation with oxygen at –125 °C in 2-methyltetrahydrofuran [105].

4.1.2
Additional Reductive Coupling Reactions for the Synthesis of PAHs

An unprecedented lithium-induced cyclization reaction of tribenzocyclotriyne (**50**) to the hydrocarbon **51** has recently been reported [106]. The proposed aromatic dianion **52** cyclizes and than abstracts protons from the solvent tetrahydrofuran (THF) and, after further reduction with lithium metal, the resulting intermediate is quenched with methanol to give **51** in 60% yield (see Scheme 23).

Low valent titanium reagents have been used in the first "low temperature" synthesis of the fullerene fragment dimethyl corannulene (**53**) from the 1,6-bis(bromomethyl)-7,10-bis(1-bromoethyl)fluoranthene (**54**, see Scheme 24 [107]). Obviously, the high thermodynamic driving forces for the reductive elimination of bromine as well as for the aromatization with 2,3-dichloro-5,6-dicyanobenzoquinone (DDQ) in the course of the reaction via the relatively unstrained *cis/trans*-dimethyltetrahydrocorannulenes enabled the formation of

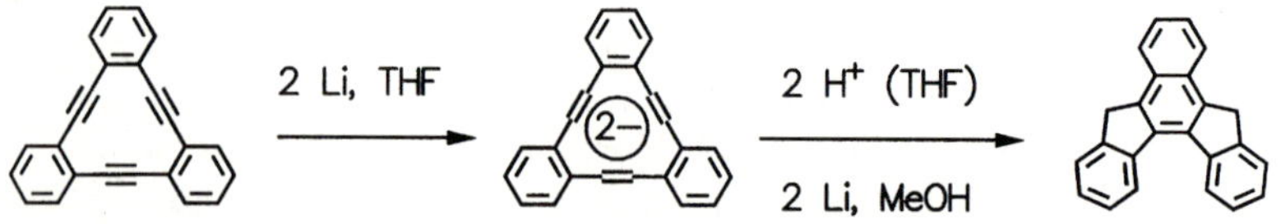

Scheme 23. *Intra*molecular lithium-induced cyclization of cyclic arylalkynes [106]

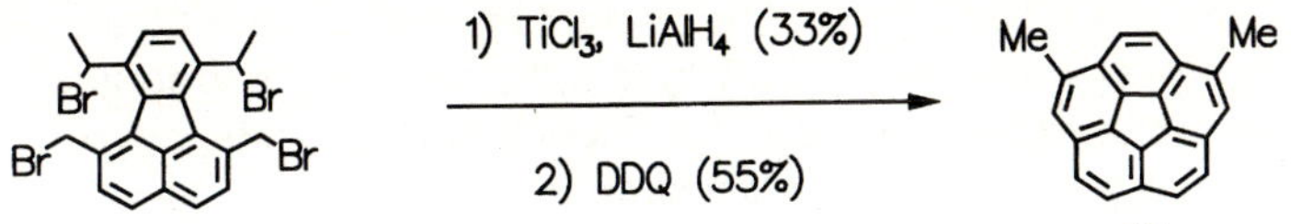

Scheme 24. "Low temperature" synthesis of a functionalized corannulene (**53**, [107])

the highly strained, bowl-shaped PAH **53** under relatively mild conditions, compared to the pyrolytic approaches (see Sect. 2).

4.2
Oxidative Aryl-Aryl Coupling and Condensation Reactions

4.2.1
Intermolecular Reactions

Although the *inter*molecular oxidative coupling of aromatic compounds by thermal dehydrogenation under pyrolysis conditions supplied a major source of PAHs in the beginning and the middle of the century, it has largely been abandoned because of its low selectivity and yield [108]. In general, the same statement could be made for *inter*molecular coupling reactions catalyzed by $AlCl_3$ (Scholl-type reaction, see Scheme 25) but a few noteworthy exceptions have still been found recently, e.g. the formation of condensed dimers **55–57** from the highly symmetrical corannulene (**16**, C_5-symmetry [109]) and coronene (**58**, C_6-symmetry [110]), a reaction that presumably proceeds via the corresponding biaryls. Hydrogen transfer reactions are often observed under these conditions as indicated by the formation of dihydrocorannulene as a byproduct in the conversion of **16**.

The ^{1}H NMR spectrum of **55** indicates, that the hydrocarbon exists in two nonplanar conformations; the coalescence temperature was found to be $283 \pm 5\,K$ in deuterated bromobenzene corresponding to an inversion barrier of

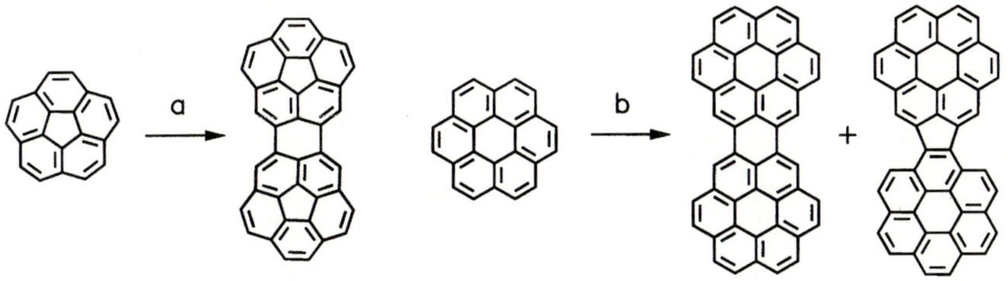

Scheme 25. Scholl-type reactions: a) $AlCl_3$, 1,2-dichloroethane, 3 h, room temperature, 20 % [109]; b) $AlCl_3/NaCl$ melt, 10 min, 160 °C, ratio **56 : 57** ≅ 80 : 20 [110 a, b]

$\Delta G^{\ddagger}_{inv} = 13.2 \pm 0.5$ kcal/mol [109]. The planar benzo[1,2,3–bc:4,5,6-$b'c'$]dicoronene (**56**) represents the largest PAH which so far has been unambiguously identified in coal tar as well as the largest *peri*condensed aromatic hydrocarbon for which a crystal structure has been determined [110d].

Other oxidizing agents such as CoF_3/CF_3COOH, $Tl(OCOCF_3)_3$, $AlCl_3/NaCl$ and bulk electrolysis have been used in the stepwise oxidative dimerization of 7,12-diphenylbenzo[k]fluoranthene (**59**) to the tetraphenyl substituted PAH **60** (see Scheme 26). The regioselectivity in this reaction is not unexpected because of the high electron density at the bond forming sites [7].

Oxidative biomimetic aryl coupling reactions of electron-rich (phenolic) aromatic hydrocarbons have occasionally been reported; for selected examples see the reviews by Bringmann et al. [93a] and the review on the ellagitannin chemistry by Quideau and Feldman [93b]. Oxidative *homo*coupling has also been achieved with palladium(II) acetate via the 5-arylpalladium complexes [78d], but stoichiometric amounts of the palladium reagents are often required and acetylation can occur as a side reaction [93a].

4.2.2
Intramolecular Reactions

The classical oxidative, *intra*molecular formation of aryl-aryl bonds by thermal cyclodehydrogenation and photocyclization reactions have already been discussed in Sects. 1 and 2 respectively. Compared to the thermal approach, higher yields have often been obtained using classical hydrogen transfer catalysts such as highly dispersed platinum or palladium on suitable supports or Friedel–Crafts type catalysts like the classical $AlCl_3/NaCl$ melt (Scholl reaction, see Scheme 26 and 27 [7, 38 d, e, 108]).

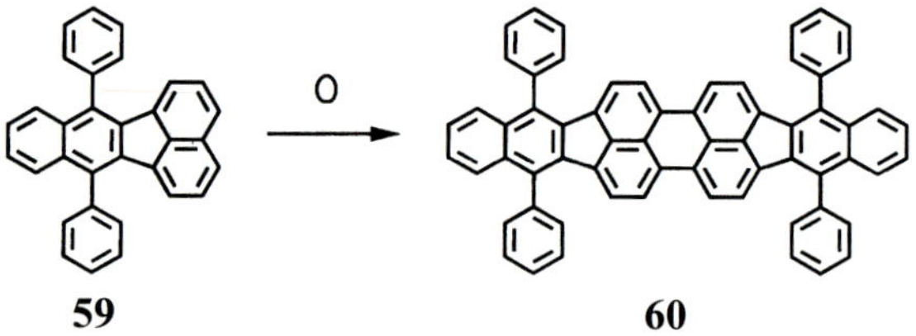

Scheme 26. Oxidative *homo*coupling of fluoranthene derivative **59** [7]: "O" = CoF_3/TFA, $AlCl_3/NaCl$, $Tl(OCOCF_3)_3$ or bulk electrolysis

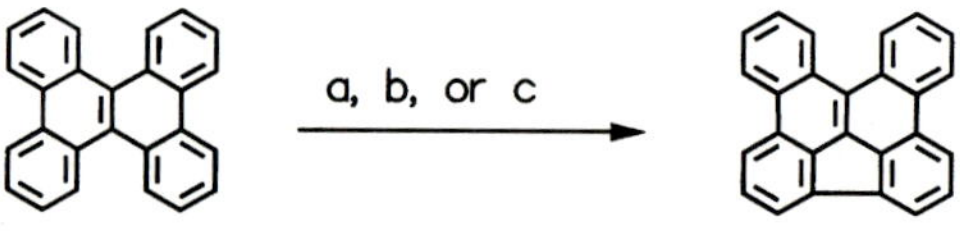

Scheme 27. Formation of five-membered rings by different cyclodehydrogenation procedures: a) FP, 900 °C (< 10%); b) 0.7% Pt/SiO_2, 700 °C, **61** (30%), **5** (49% recovered); c) $AlCl_3/NaCl$, 145 °C, 40 s (58%) [38d, e]

The drastic conditions of the Scholl reaction can be circumvented in some cases by using $AlCl_3$ in CS_2 together with cupric halides or triflates as oxidizing agents. This relatively mild approach was presumably first applied by Kovacic and Koch [111] and has been further improved by Müllen et al. [102b, 112] for the syntheses of planar benzenoid PAHs with remarkably high molecular weights (e.g. **49** and **62**, see Scheme 21 and 28).

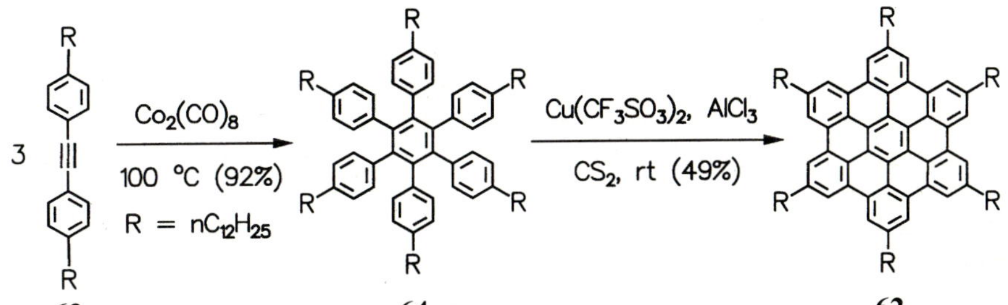

Scheme 28. Cobalt catalyzed [2 + 2 + 2] cyclotrimerization of the diaryl alkyne (**63**) tothehexaphenylbenzene (**64**) and modern Scholl-type reaction in the synthesis of hexabenzocoronene (**62**, [112])

Other oxidants like thallium(III) oxide, vanadium(V) oxyfluoride, palladium(II) acetate, and ruthenium(IV) tetrakis(trifluoracetate) have been developed as powerful tools for the *intra*molecular biaryl coupling reaction [7, 93, 113]. Nevertheless, DDQ is still one of the most versatile reagents in oxidative coupling reactions (see Scheme 14 and 29 [82, 114]). The highly strained dioxa[8](2,7)pyrenophane (**65**), portraying an overall curvature of nearly 90° for the pyrene subunit, was finally obtained from the *meta*-cyclophanediene (**66**) by dehydrogenation with DDQ in refluxing benzene in 67% yield [114].

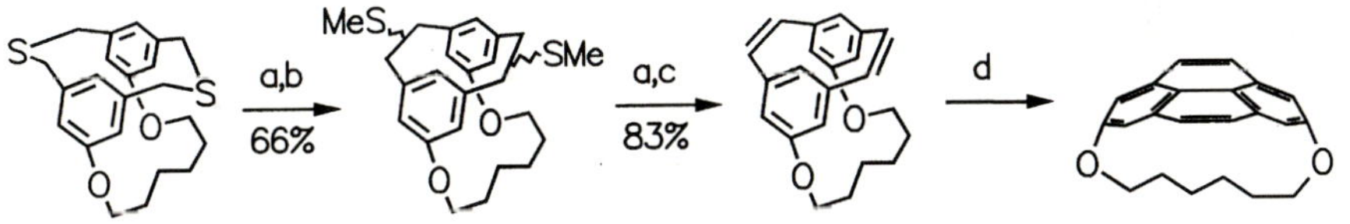

Scheme 29. Reaction sequence to the curved dioxa[8](2,7)pyrenophane (**65**, [114]): a) $(MeO)_2CHBF_4$, CH_2Cl_2, room temperature, 2 h; b) *t*BuOK, THF, 3 h; c) *t*BuOK, THF/ *t*BuOH = 1/1, room temperature, 3 h; d) DDQ, benzene, Δ, 12 h, **65** (67%), **66** (26%)

4.3
Intermolecular Cross Coupling Reactions between Aromatic Electrophiles and Nucleophiles

While there exists a relatively wide range of *homo*coupling reactions available for the *inter*molecular formation of aryl-aryl bonds (e.g. Scheme 20 and 25), the regioselectivity of *inter*molecular cross-coupling reactions can be substantially limited because of statistical competition with the *homo*coupling reactions, e.g. in the cross coupling of two (different) aryl halides. Therefore it is significantly more effective to couple an electrophilic with an nucleophilic aryl component.

4.3.1
Oxazoline Coupling (Meyer's Coupling)

Arylic methoxy groups can be replaced in an uncatalyzed nucleophilic substitution by Grignard reagents. A carboxylic functional group in *ortho*-position to the methoxy group seems to be obligatory to (i) increase the electrophilicity at the carbon atom connected to the methoxy group and to (ii) function as a complex ligand in an *ortho*-directing fashion. Most common is an oxazoline group synthesized from a carboxylic acid, thionyl chloride and 2-amino-2-methylpropanol which can be hydrolyzed and reduced after the coupling reaction (Meyer's procedure, see Scheme 30 [93 a, 115]). An interesting alternative consists in the synthesis of 1,1′-binaphthyl-2-carboxylates from 1-methoxy-2-naphthoic esters and the desired 1-naphthyl Grignard reagent. This method also allows the synthesis of chiral 1,1′-binaphthyls with moderate enantiomeric excesses (Fuson's procedure [116]).

4.3.2
Transition Metal Catalyzed Grignard Coupling Reactions

Since the discovery of phosphanenickel complexes being suitable catalysts to promote the cross-coupling of aryl Grignard reagents with organic halides [117–119] this reaction has been widely employed in the formation of unsymmetrical carbocyclic and heterocyclic biaryl compounds (Kumada coupling, see Scheme 31 [84b, 118]). The more expensive analogous palladium catalysts have also been used in some cases. Derivatives of aldehydes, ketones, and carboxylic acids are tolerated if the aggressive Grignard compounds are replaced by less reactive organometallic zinc, aluminum or zirconium reagents [93 a, 118 a, 120].

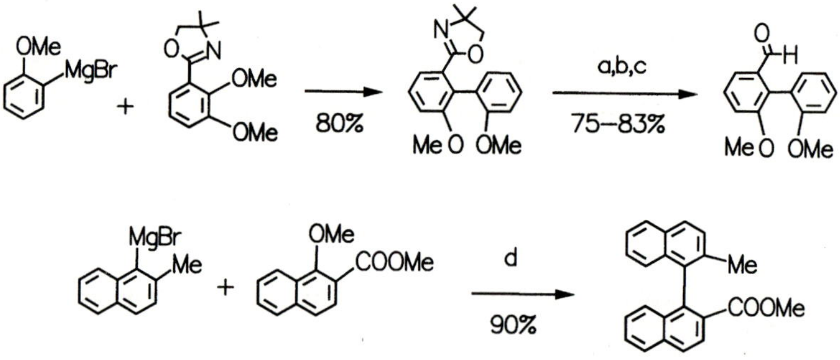

Scheme 30. Coupling of aryl electrophiles with aryl Grignard reagents: Meyer's coupling [115]: a) MeOTf, b) $NaBH_4$, c) $(COOH)_2$; Fuson's procedure [116b]: d) Et_2O, C_6H_6

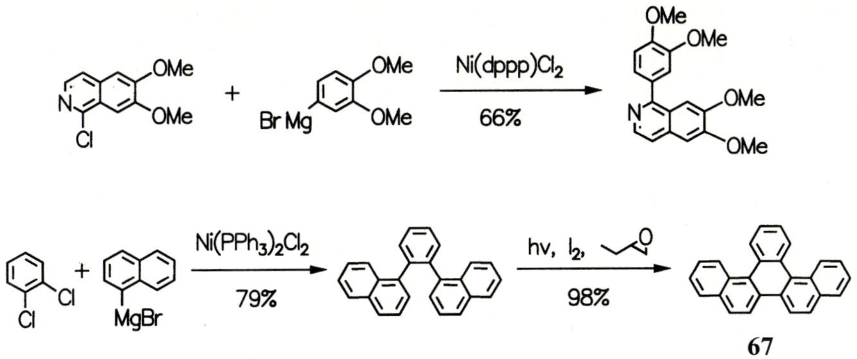

Scheme 31. Kumada coupling reactions of (hetero)aryl halides with aryl Grignard reagents [84b, 118] and oxidative photocyclization towards benzo[s]picene (**67**, [84b])

4.3.3
Suzuki and Stille Coupling

Naturally, the strongly nucleophilic lithium, Grignard and copper reagents will not tolerate sensitive functionalities which may be imperative in a specific total synthesis sequence. In addition, these reagents are sometimes difficult to handle because of their air and moisture sensitivity. On the other hand, arylboronic acids (Suzuki coupling [95, 121–123]) and arylstannanes (Stille coupling [94, 96, 124]) have a significantly smaller difference between the electronegativities of carbon and boron or carbon and tin, respectively, and therefore, these reagents may indeed well tolerate many different functional groups in carbocyclic and heterocyclic arenes. These compounds are readily available as well as environmentally less toxic and easier to handle than most organometallics. Therefore, the palladium-catalyzed cross-coupling of an aryl halide or triflate with an arylboronic acid or an arylstannane has become the method of choice in the preparation of unsymmetrical biaryls and is also often applied in the approach towards other functionalized arenes.

For the Suzuki coupling reaction, the arylboronic acids [125] are generally prepared from lithium or Grignard reagents with boron halides or borates. Aryl bromides have been used most often as the electrophile although aryl iodides are even more reactive [95b]. While aryl triflates and arenediazonium salts [126] have also been employed occasionally, aryl chlorides are usually not reactive enough to participate in the coupling process except for nitrogen-containing electron-deficient heteroaryl chlorides.

Tetrakis(triphenylphosphane)palladium(0) is the catalyst of choice, but palladium(II) acetate and other Pd(II) and Pd(0) sources have also been used. The intensively studied catalytic cycle requires two equivalents of base and the reaction is often carried out under phase transfer conditions. However, newly developed palladacycles from palladium(II) acetate and selected triarylphosphanes show much higher turnover numbers, do not decompose significantly at the

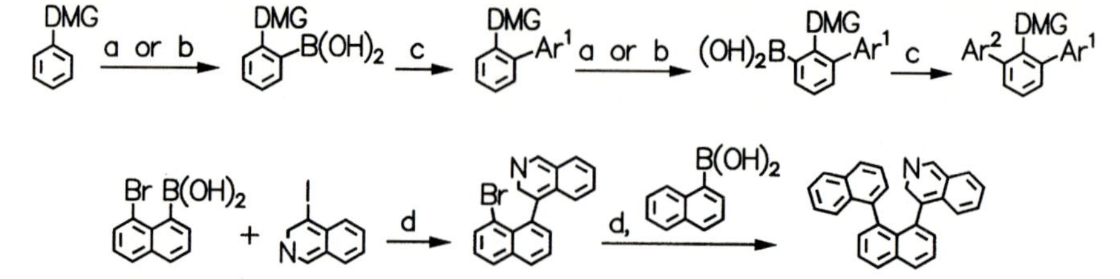

Scheme 32. Repetitive Suzuki coupling reactions in the synthesis of poly(hetero)aryls [122a, e]:
a) DMG (directing metallation group) = $CONiPr_2$ or $OCONEt_2$, (1) RLi/TMSCl, (2) BBr_3, (3)
H^+; b) DMG = OMOM or $NHCO_2tBu$, (1) $RLi/B(OMe)_3$, (2) H^+; c) Ar^1Br or Ar^2Br, $Pd(PPh_3)_4$,
Na_2CO_3, toluene; d) $Pd(PPh_3)_4$, KOH, nBu_4Br, 1,2-dimethoxyethane

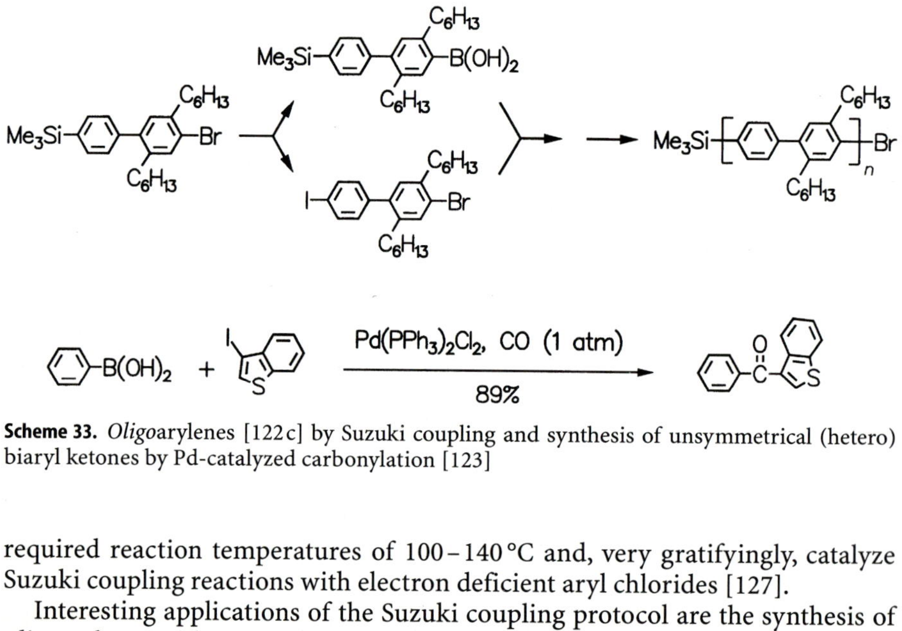

Scheme 33. *Oligo*arylenes [122c] by Suzuki coupling and synthesis of unsymmetrical (hetero)
biaryl ketones by Pd-catalyzed carbonylation [123]

required reaction temperatures of 100–140 °C and, very gratifyingly, catalyze
Suzuki coupling reactions with electron deficient aryl chlorides [127].

Interesting applications of the Suzuki coupling protocol are the synthesis of
*oligo*arylenes with precise length and defined functional groups at both termini
of the chain as well as the preparation of unsymmetrical biaryl ketones by the
Pd-catalyzed cross-coupling reactions of arylboronic acids with iodoarenes (see
Scheme 33).

A very comprehensive and critical review on the Stille reaction covering the
literature up to 1995 appeared recently [94a]. Although the Stille reaction can
take place in the presence of many useful functional groups (e.g. alcohol, ester,
nitro, acetal, ketone, and aldehyde [94, 124b]), it does not seem to be favored over
the Suzuki or Kumada coupling for formation of biaryls, presumably because of
the lower yields sometimes obtained in the Stille procedure [124]. In general, the
Stille procedure is far more often applied for the introduction of other unsaturated groups (e.g. ethenyl, allyl, alkynyl) into aromatic systems than for aryl and

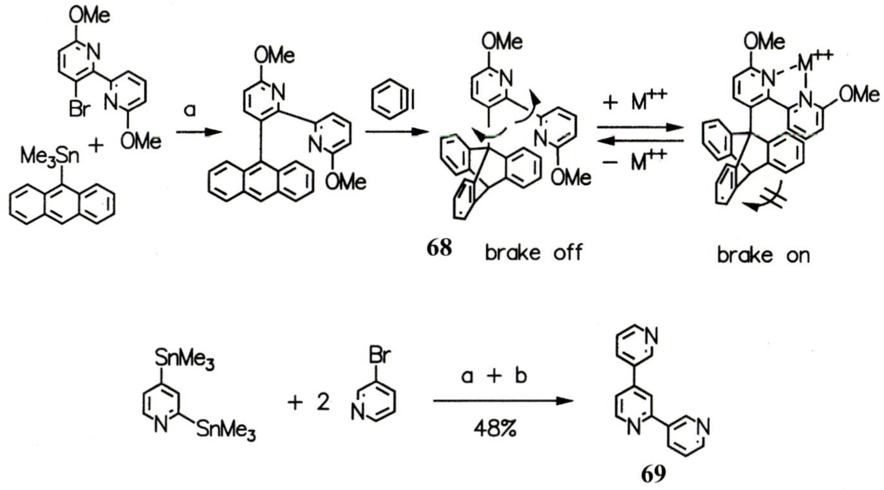

Scheme 34. Stille coupling of (hetero)aryl halides with (hetero)aryltin reagents: Kelly's molecular brake (**68**, [22]) and nicotelline (**69**, [124a]): a) Pd(PPh$_3$)$_4$, b) xylene, Δ

heteroaryl groups. The syntheses of Kelly's molecular brake (**68**, [22]) and nicotelline (**69**, [124a]) are depicted as typical examples for (hetero)aryl-heteroaryl Stille coupling reactions in Scheme 34.

4.3.4
Silicon Reagents

Disilanes have been used in stoichiometric amounts as reductants in the palladium-catalyzed coupling of electron deficient aryl chlorides. By this procedure, trimellitic anhydride acid chloride (**70**) was converted to the high temperature polymer precursor biphenyl dianhydride (**71**) in 85% yield with concomitant decarbonylation (see Scheme 35 [128]).

Aryl chlorosilanes were found to undergo palladium-catalyzed cross coupling reaction with aryl chlorides in the presence of a fluoride salt giving unsymmetrical biaryls (see Scheme 36 [129]). A variety of functional groups (cyano, acetyl, fluoro, trifluoromethyl) are tolerated in this reaction. While the

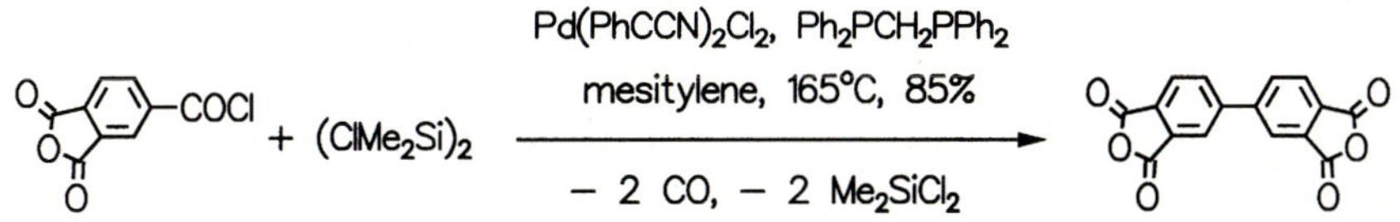

Scheme 35. Pd-catalyzed biaryl coupling of aroyl chlorides with decarbonylation [128]

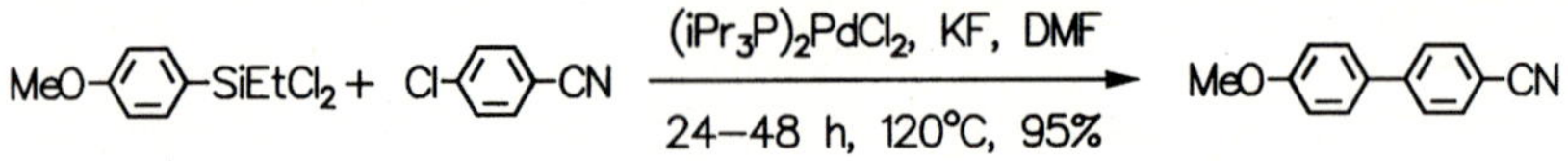

Scheme 36. Pd-catalyzed cross-coupling of arylsilanes with aryl chlorides [129]

reaction proceeds smoothly in good yields for aryl chlorides containing electron-withdrawing groups, the results are unsatisfactory for aryl chlorides with electron-donating substituents.

4.3.5
Heck-Type Reactions

The palladium-catalyzed arylation of alkenes and arenes offers one of the conceptually most intelligent solutions for the synthesis of PAHs from the appropriate aryl halides or triflates. The reaction is usually carried out in a polar solvent (*N,N*-dimethylformamide (DMF) or *N,N*-dimethylacetamide (DMA)) at relatively high temperatures (100–170 °C) with a base to trap the acid formed, and a phase transfer catalyst. Most often Pd(II) acetate or a Pd(II) complex are used, that are converted to a catalytically active Pd(0) species in the course of the reaction. The mechanisms of these Heck reactions have been discussed widely throughout the pertinent literature [78, 79].

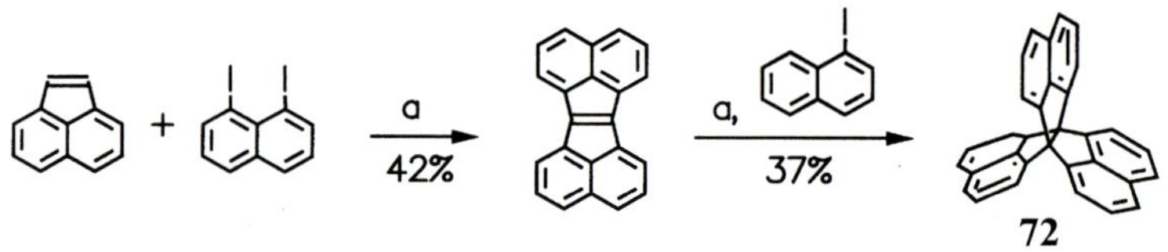

Scheme 37. Dyker's synthesis of hexaarylethane **72** by Heck-type reactions [130c]: a) 2 mol% Pd(OAc)$_2$, K$_2$CO$_3$, *n*-BuN$_4$Br, DMF, N$_2$, 100 °C

The *inter*molecular palladium-catalyzed arylation of alkenes (classical Heck reaction) has been used e.g. to synthesize the highly strained hexaarylethane derivative **72** in a very straight-forward fashion (see Scheme 37 [130]). The X-ray structure investigations of **72** and similar propellanes revealed an elongation of the central CC single bond to ca. 162 pm [130c].

Domino coupling reactions of aryl halides with norbornene and its derivatives provide a simple route to PAHs. In a four component sequence, norbornene (**73**) is arylated with an excess of iodobenzene to the terphenyl **74**, that can be converted to the benz[*e*]pyrenes **75** and **76** by classical aromatic conversion reactions [131]. The domino sequence is a consequence of the fact that the five-membered intermediate palladacycle **77a** is able to add a second molecule of iodobenzene (**77a** → **77b**), and the intermediate arylpalladium halide resulting from reductive elimination in **77b** can even add a third molecule of iodobenzene before the final elimination of the Pd(0) complex PdL$_2$ occurs (see Scheme 38).

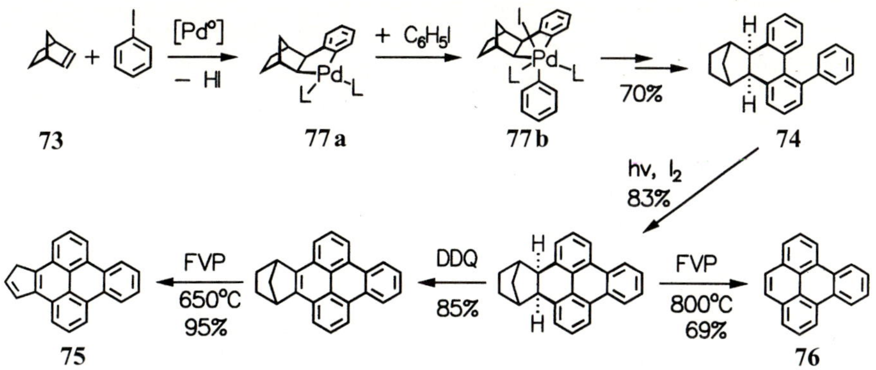

Scheme 38. Preparation of the benz[*e*]pyrenes **75** and **76** by domino Heck reactions [131]

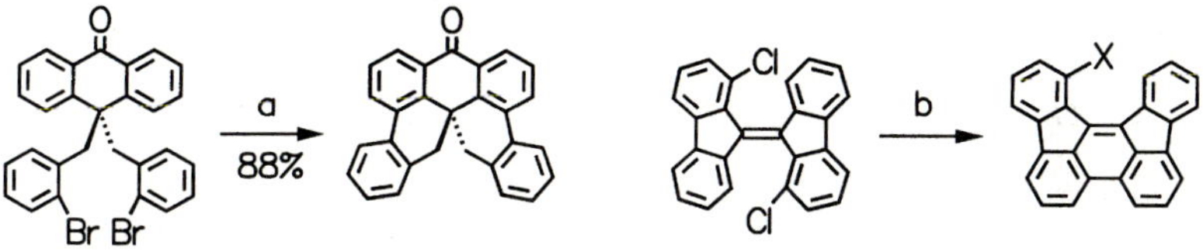

Scheme 39. *Intra*molecular Heck-type cyclization reactions of aryl halides with arenes: a) Pd(OAc)$_2$, DMF, K$_2$CO$_3$, Bu$_4$NBr, 140 °C, 36 h [132b]; b) Pd(OAc)$_2$, DMF, K$_2$CO$_3$, nBu$_4$NHSO$_4$, 120 °C, 51 h, X=Cl (3.2 %), X=H (5.6 %) [133]

The formation of six-membered ring carbocycles is illustrated in Scheme 39 [132, 133]. The reaction proceeds well with aryl bromides, iodides, and triflates especially when containing additional electron-withdrawing groups, while aryl chlorides are converted in low yields only. It has been demonstrated, that for less reactive *p*-methoxyaryl derivatives the addition of lithium iodide improves the results significantly [132].

5
[2 + 2 + 2] Cyclotrimerization Reactions of Alkynes and Arynes

The [2 + 2 + 2] cyclotrimerization of alkynes and its analogous arynes [134] leads directly to aromatic hydrocarbons and can be considered as special cases of [2 + 2 + 2] cycloaddition reactions of unsaturated compounds [135, 136]. Especially the catalyzed cyclotrimerization reactions are often highly regio- and stereoselective and have been established as valuable methods for the syntheses of highly substituted benzenes, biphenylenes, triphenylenes, and in the approach towards natural products.

Although the [2 + 2 + 2] cyclotrimerization is symmetry allowed and highly exothermic in most cases, the uncatalyzed, purely thermal reaction has been reported only in a few cases, presumably because of entropic effects and/or high

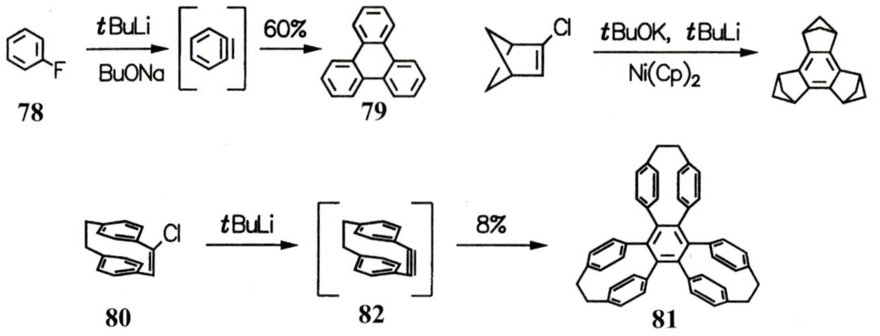

Scheme 40. *Inter*molecular [2 + 2 + 2] cyclotrimerization of reactive aryne and alkyne intermediates [137c, 138, 139c]

enthalpies of activation. The exemplary formation of triphenylenes from *o*-dihalogenated benzenes via the aryne intermediates ([137], see e.g. **78** → **79**) and the trimerization of 1-chloro-[2.2]paracyclophane-1-ene (**80**) to the so-called trifoliaphane (**81**) via the highly strained [2.2]paracyclophane-1-yne (**82**, [138]) are illustrated in Scheme 40.

In the majority of the catalyzed [2 + 2 + 2] cyclotrimerization reactions CpCo(CO)$_2$ has been used as the catalyst of choice [135, 140, 141]), but Ni(0) and Ni(II) catalysts [139] have also been employed in some occasions. The synthesis can be carried out in a purely *inter*molecular fashion (see Scheme 28 and 40) as well as in a partly or completely *intra*molecular way (see Scheme 41 and 42).

Numerous theoretically interesting hydrocarbons, like iptycenes [137b], starphenylenes (see Scheme 41 [140]) and highly strained threefold bicycloannelated benzenes (see Scheme 40 [139a, c]) were synthesized by this very versatile approach. In addition, when alkynes are brought into a cocyclization reaction with alkyl and aryl nitriles or ω-alkynyl isocyanates validated routes to heterocycles become available as it was demonstrated e.g. for the total synthesis

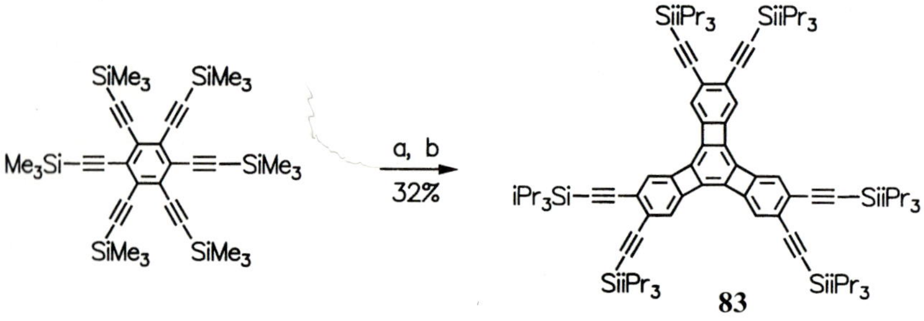

Scheme 41. *Intra*molecular [2 + 2 + 2] cyclotrimerization in the synthesis of the extended biphenylenes **83** and **84** [140b]: a) KF*2H$_2$O, 18-crown-6, DME; b) CpCo(CO)$_2$, toluene, iPrSi-C≡C-C≡C-C≡C-SiiPr (7 eq.), Δ, $h\nu$, 16 h; c) Bu$_4$NF, THF, Toluene, 23 °C, 30 min; d) Me$_3$Si-C≡C-SiMe$_3$, CpCo(CO)$_2$, THF, Δ, $h\nu$, 16 h

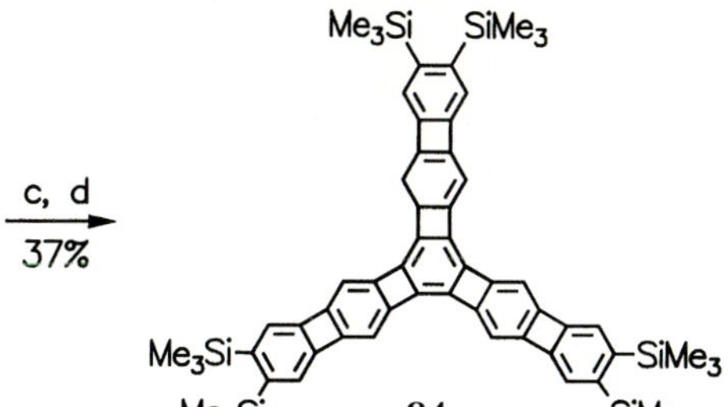

Scheme 41 (continued)

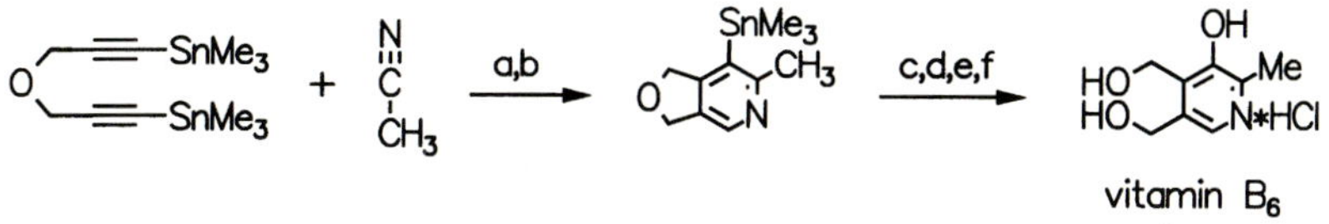

Scheme 42. Preparation of vitamin B$_6$ by [2+2+2] cocyclotrimerization with acetonitrile [135]: a) CpCo(CO)$_2$; b) SiO$_2$, 45%; c) I$_2$; d) NaOMe, 37%; e) HBr; f) AgCl, 68%

of alkaloids [141a], camptothecin [135], and vitamin B$_6$ (see Scheme 42 [135]). A variety of functional groups like alkyl, aryl, ethenyl, CO$_2$R, CH$_2$OH, CH$_2$OR, COR, C=NOR, NR$_2$, SR, and SiMe$_3$ are tolerated in the catalyzed cyclotrimerization [135].

6
Modern Diels-Alder Reactions

The Diels-Alder reaction is a very valuable approach in classical annelation sequences of PAHs and is intensely documented in the reviews on the synthesis of aromatic hydrocarbons [26]. Therefore, we will describe herein only some few, selected, modern applications of the Diels-Alder reaction that caught our interest. To avoid any misunderstanding, the authors will absolutely not pass any judgement on a specific synthetic method by using the terms "classical" and "modern".

A number of new diaryne equivalents have been developed by Hart et al. [137b, 142] in recent years. These very versatile reagents enable the synthesis of large polycyclic hydrocarbons in very few steps. The pentiptycene **85**, for exam-

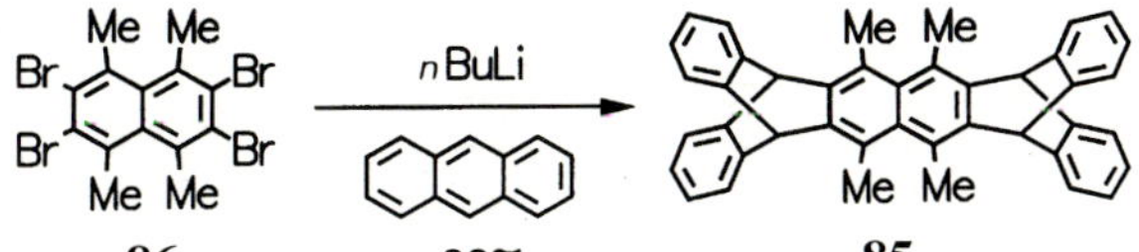

Scheme 43. Synthesis of the pentiptycene **85** by a typical Diels-Alder reaction of aryne intermediates with anthracene [137b]

ple, can be formed in a one-pot reaction in good yields by addition of n-BuLi to a solution of 2,3,6,7-tetrabromo-1,4,5,8-tetramethylnaphthalene (86) trapping the aryne intermediate with anthracene (see Scheme 43, [137b]).

The bisadduct 87, synthesized from anthracene bis-(1,4-epoxide) (88) and tetraphenylcyclone (89), is a synthetic equivalent of the [1,2-c:4,5-c']benzodifuran 90 (R=H). Adduct 87 reacts thermally with dienophiles including some exceedingly weak ones to give linear acene derivatives in moderate to very good yields (see Scheme 44 [142b]).

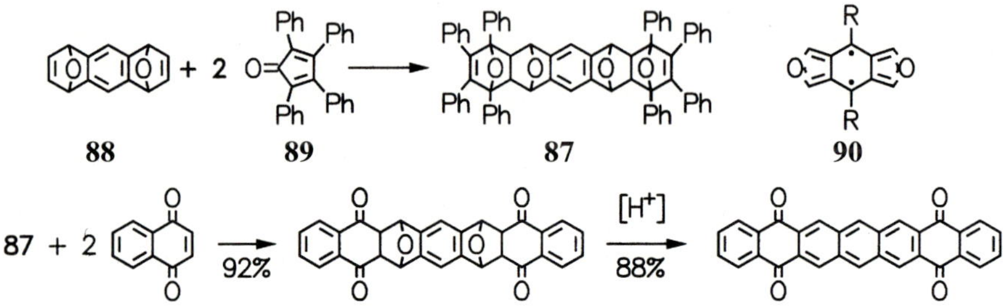

Scheme 44. Diels-Alder reaction with bis-dienes and bis-dienophiles [142b]

In general, two strategies are possible to synthesize macrocycles or linear ladder polymers by a Diels-Alder reaction [143–145]. One possibility entertains the reaction of bis-dienophiles with bis-dienes (route **A**, see Scheme 45, [143]). Again, the in-situ synthesized bis-diene 90 (R=hexyl) serves often as a versatile precursor in the preparation of high molecular aromatic hydrocarbons. If curved bis-dienophiles are brought into a repetitive reaction with curved bis-dienes, sterically rigid macrocycles with well defined cavities and belt-like structures become available [144]. On the other hand, the reaction of hydrocarbons ensuing a diene as well as a dienophile moiety in a planar arrangement also allows the synthesis of linear, aromatic, polymer structures with promising electro-optical characteristics (route **B**, see Scheme 45 [143c]).

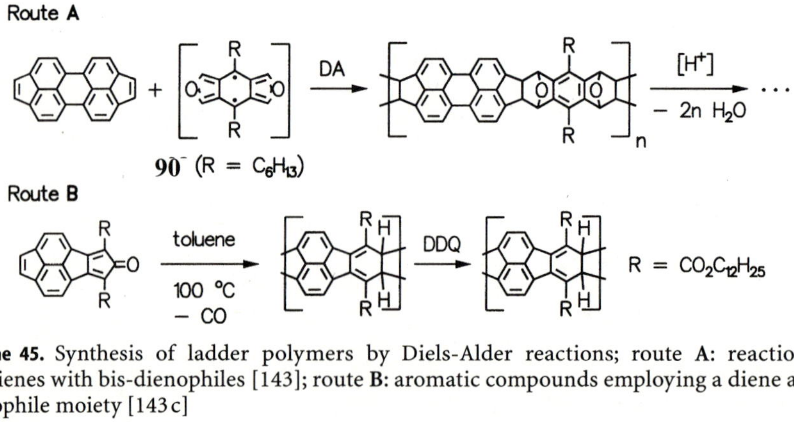

Scheme 45. Synthesis of ladder polymers by Diels-Alder reactions; route **A**: reaction of bis-dienes with bis-dienophiles [143]; route **B**: aromatic compounds employing a diene and a dienophile moiety [143c]

Photocyclization reactions were hitherto the only convenient approach towards helicenes (see Sect. 4). A new method, that promises the cost-effective synthesis of larger quantities of these interesting hydrocarbons, is based on the reaction of benzoquinone with *p*-dialkenylbenzenes producing [5]helicenes, or with 2,7-dialkenylnaphthalenes resulting in functionalized [6]helicenes (**91**) respectively (see Scheme 46 [146]). The yield for this Diels-Alder reaction increases significantly from 6 to 47 % when R = H is replaced by a methoxy group (R = OMe).

A series of longitudinally twisted, phenyl substituted and/or benzo-amelated acenes were synthesized by Pascal's group in the last decade [147]. In general, tetracyclone (**89**), acecyclone and hexaphenylisobenzofuran have been used as dienes, whereas 9-bromophenanthrene or derivatives of anthranilic acid may serve as aryne equivalents (dienophile). The acenes are highly distorted from planarity with an end-to-end twist of the anthracene moiety of 62.8° in deca-phenylanthracene (**92**, see Scheme 47 [147c]). Although only a very meager overall yield of 0.1 % could be obtained for **92**, the reaction protocol is very versatile and gives in general moderate results in very short synthetic approaches.

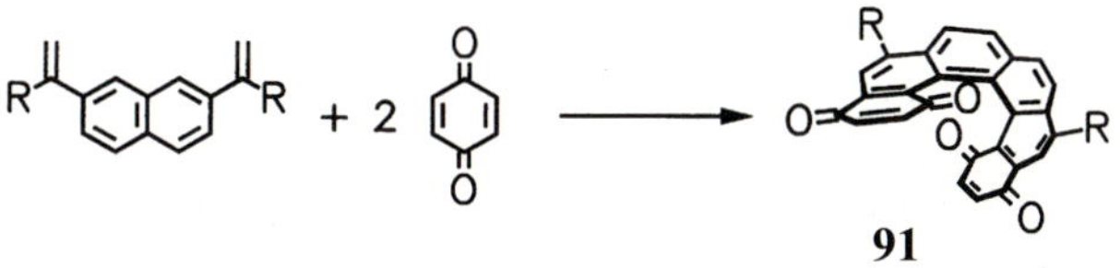

91

Scheme 46. Efficient synthesis of [6]helicenes (**91**) by double Diels-Alder reaction [146]

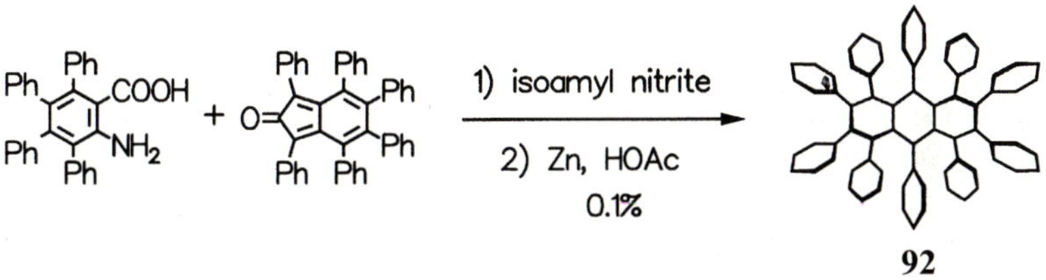

92

Scheme 47. Pascal's approach to a highly, longitudinally twisted acene (**92**, double bonds omitted for clarity [147c])

An interesting alternative to the synthesis of hexaarylbenzenes by [2+2+2] cyclotrimerization (see Sect. 5 and Scheme 28) is the Diels-Alder reaction of diarylalkynes with tetracyclone (**89**), very similar to Pascal's approach towards fully phenylated PAHs (compare Scheme 47 and 48). The conveniently accessible aryldialkyne **93** allows the stepwise construction of very high molecular polyaryl **94**. The phenyl rings in **94** are now arranged in such a favorable fashion, that by application of proven cyclodehydrogenation procedures (see Sect. 4.2.2) **94** can be completely converted to the graphite sheet type PAH **95** in excellent yields [148]. Similar reaction protocols have been applied by Müllen et al. also for the synthesis of other graphite substructures, the so-called series of superacenes [148].

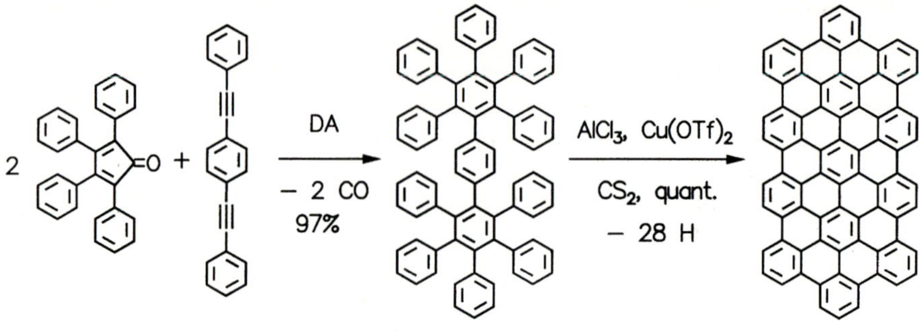

89   **93**   **94**   **95** $C_{78}H_{26}$

Scheme 48. Short approach to large graphite substructures (e.g. **95**) by Diels-Alder (DA) reactions with arylalkynes and a modern variation of the Scholl reaction [148]

7
Directed Electrophilic Cyclization Reactions

The substitution of PAHs with reagents bearing the electrophilic center at a carbon atom, often suffers from the drawback of low regioselectivity or possible multiple substitution steps. The Friedel-Crafts acetylation of phenanthrene, for example, always results in a varying mixture of all five possible acetylphenanthrene isomers [10] making it impossible to use this reaction type for an effective annelation procedure towards PAHs with higher molecular weights. On the other hand, bromine derivatives of aromatic hydrocarbons are readily available in most cases in high yields and regioselectivities through direct electrophilic bromination. Therefore, the following procedure towards benz[*def*]indeno-[1,2,3-*hi*]chrysene (**96**, Scheme 49) was developed by Cho and Harvey [149]. The sequence starts with the bromo-lithium exchange of 6-bromo benzo-[def]chrysene (**97**) followed by the reaction with cyclohexene oxide to the alcohol **98**. The *intra*molecular acid-catalyzed cyclodehydration of the corresponding 2-arylcyclohexanone **99** is now usually directed to the desired position because of different delocalization energies of the intermediates or in most cases simply because of the lack of alternatives. Finally, dehydrogenation with DDQ gives the target indeno annelated PAH **96** in good overall yields for this multistep sequence.

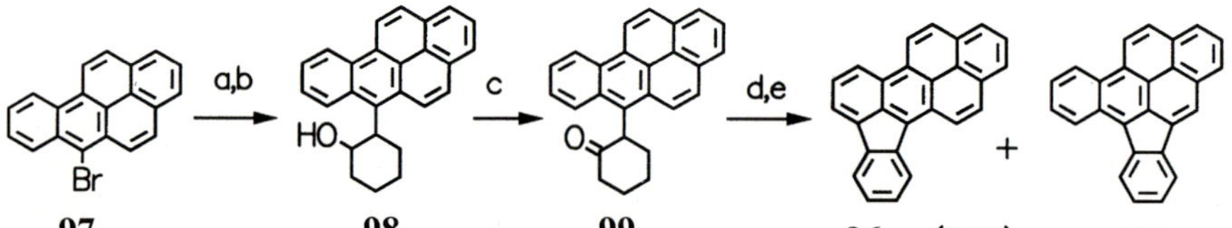

97   **98**   **99**   **96**   (92%)   2%

Scheme 49. General new synthesis of indeno-annelated PAHs [149a]: a) PhLi; b) cyclohexene oxide, 61%; c) PDC, 75%; d) (HPO$_3$)$_n$; e) DDQ

Dibenzofulvene (**100**), which is a byproduct of fluorenylmethoxycarbonyl (FMOC) deprotection in peptide synthesis, undergoes Michael addition reactions with carbanions. Conversion of **100** with the silyl enol ether of cyclohexanone (**101**) and a desilylation reagent furnishes directly the ketone **102**, that now can be subjected to the already described acid mediated *intra*molecular cyclodehydration procedure followed by aromatization. The product is again a polycyclic fluoranthene (**103**), that can be considered as a naphtho annelated fluorene in this sequence (see Scheme 50 [150]).

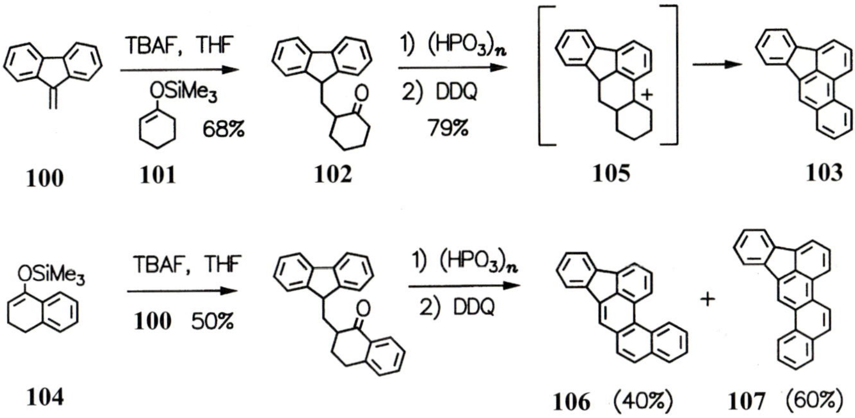

Scheme 50. Synthesis of fluoranthenes by *intra*molecular cyclodehydration [150]

The formation of isomers has been reported if the silyl enol ether of 1-tctralone (**104**) is used instead of the corresponding cyclohexanone derivative. It is assumed, that the intermediate cation **105** can undergo a double Wagner-Meerwein rearrangement, which does not influence the overall result in the reaction towards **103**, but finally furnishes the two isomeric PAHs **106** and **107** starting from **104**.

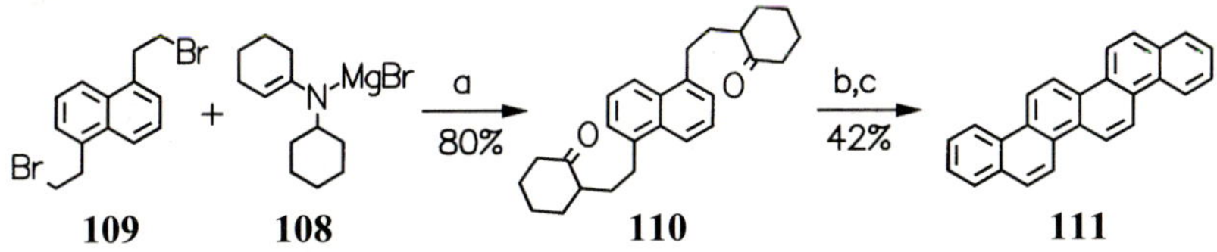

Scheme 51. General annelation procedure using enamine derivatives of cyclohexanone [151]: a) dioxane, 18 h, reflux; b) CH$_3$SO$_3$H; c) 10% Pd/C, triglyme, reflux

Besides the ene oxide and the silyl enol ether discussed above the enamine derivatives of cyclohexanone can be used as a cyclohexanone equivalent as well in these novel annelation procedures [151]. Alkylation of the enamine **108** with 1,5-bis(2′-bromoethyl)naphthalene (**109**) afforded the expected diketone **110** in 80% yield (see Scheme 51). Cyclodehydration of **110** with methanesulfonic acid

followed by dehydrogenation with palladium on charcoal in refluxing triglyme
finally yielded the desired benzo[c]picene (**111**, 42% yield from **110**). The cyclization step has been proven to be very regioselective. This largely general
approach provides convenient synthetic access to a number of PAHs previously
available only via strenuous multistep sequences [26e, 151].

It has been known for several decades, that alkyl aryl ketones with a methyl
or a methylene group in β-position to the keto functionality can undergo a
threefold condensation reaction resulting in 1,3,5-triarylsubstituted benzenes.
This transformation is catalyzed in most cases by Brönsted or Lewis acids, but
bases have also been applied occasionally. If the steric demand is not too high,
1-indanones and acenaphthenones can be "trimerized" to the corresponding
truxene and decacyclene derivatives (e.g. **112** → **113**) respectively (see Scheme 52

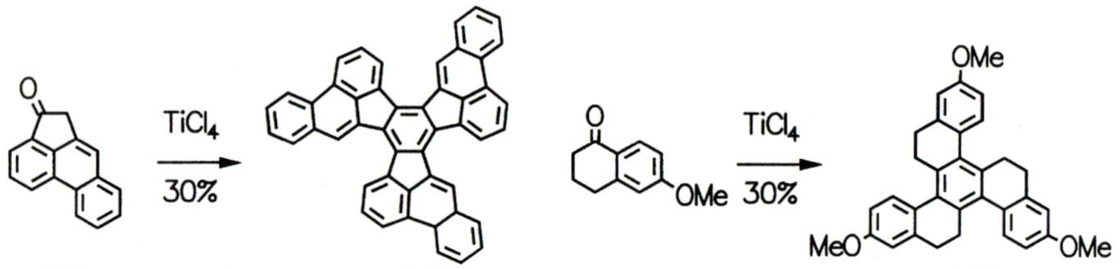

Scheme 52. Syntheses of high molecular PAHs with C_3-symmetry by Lewis acid catalyzed
"trimerization" of 1-indanones, 1-tetralones, and 1-acenaphthenones [109,152]

[109, 152]). Surprisingly, the TiCl$_4$-catalyzed conversion of 1-tetralone results only
in the formation of a dimer [84c], while 6-methoxy-1-tetralone can be converted to the 3,9,15-trimethoxy-5,6,11,12,17,18-hexahydrobenzo[c]naphtho[2,1-p]-
chrysene (**114**) in moderate yields [152a].

A versatile method for the synthesis of highly functionalized *cata*-annela-
ted PAHs and their thiophene analogs has been reported by Swager et al.
[153]. The non-fused skeletal ring system was synthesized by Pd-catalyzed
Suzuki- or Negishi-type cross coupling protocols (see Sect. 4.3). The ring-
forming step in the *o*-(4-alkoxyphenylethynyl)biphenyl group is initiated by
strong electrophiles such as trifluoroacetic acid (TFA) or iodonium tetrafluoroborate. The *para*-alkoxy functionality stabilizes the positive charge in
the ethynyl group and therefore does not only enhance the reactivity but also
directs the primary attack of the electrophile (*E*) in such a fashion that
only phenanthrene ring systems (**115**) and no fulvenes are being formed (see
Scheme 53).

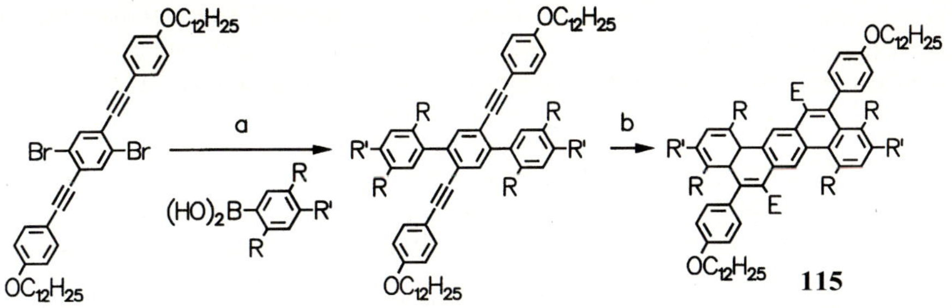

Scheme 53. *Cata*-annelated PAH (115) by Suzuki coupling and *intra*molecular directed electrophilic substitution reactions of alkynes [153]: R = H, Me, OMe; R′ = H, Me; a) Pd(dba)$_2$, Ph$_3$P, KOH, PhNO$_2$, H$_2$O, 82–91%; b) TFA (E = H) or I(py)$_2$BF$_4$ (E = I), 96–99%

8
Miscellaneous

8.1
Centrohexaindane

Benzoannelated triquinacenes [154], fenestranes [155], and centrohexaquinacenes [156] are nonplanar aromatic hydrocarbons with an interesting molecular architecture. Extensive preparative work has been accomplished especially by Kuck and his coworkers [154–156] to explore different routes towards these compounds. Although the syntheses usually do not involve formation of aryl-aryl bonds or even condensed aromatic systems, we have chosen to depict three successful routes that all result in the centrohexaindane 116 in Scheme 54, whereas the probably most convenient route A is illustrated in more detail [156]. The final reaction steps along routes A and B are typical aromatic electrophilic alkylation reactions, and along route C a threefold cyclodehydrogenation cata-

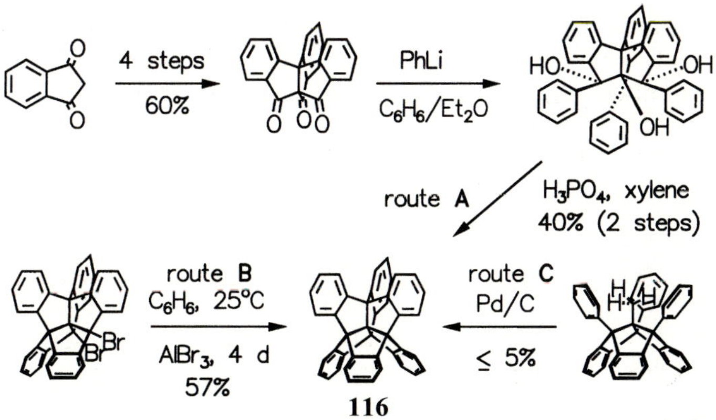

Scheme 54. Synthetic routes to centrohexaindane (116) developed by Kuck et al. [156]

lyzed by palladium on charcoal (see also Sect. 2.1 and 4.2) was utilized to prepare the target hydrocarbon **116**. Interestingly, **116** is one of the very few *topologically* nonplanar molecules that corresponds to the complete graph K_5 [156].

8.2
Anions of PAHs as Synthetic Precursors

A variety of PAHs can be transformed into their corresponding anions with high selectivities, either by deprotonation or by reduction with alkali metals [157]. From these findings Cornelisse et al. [157d] developed a convenient gram scale synthesis of benzo[*ghi*]perylene (**117**) and coronene (**58**). The reduction of the perylene precursor **118** to the doubly charged anion is accomplished with sodium metal under ultrasound irradiation in THF. The subsequent addition of bromoacetaldehyde diethyl acetal furnishes the bay-region substituted 3-peryleneacetaldehyde diethyl acetal (**119**) in 83% yield (Scheme 55). The target PAH **117** was then obtained by sonification of the intermediate **119** in methanol/sulfuric acid. The same reaction protocol has also been used for the transformation of **117** → **58**.

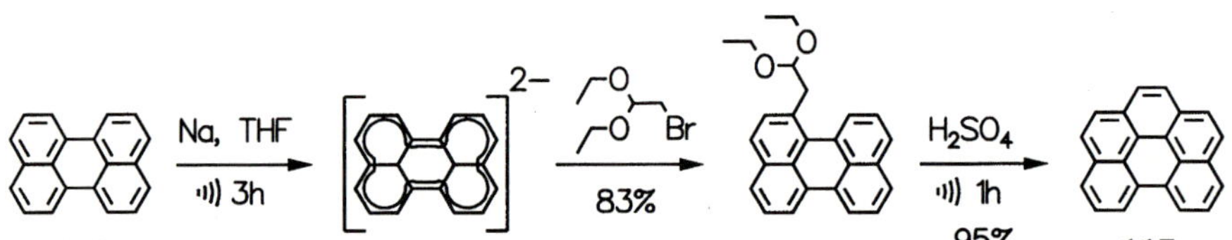

Scheme 55. Novel annelation procedure for perylene (**118**) via its anion [157 d]

8.3
Ring Transformations of Heterocyclic Compounds

An advantageous method for the conversion of 2- and 4-methyl substituted pyridinium and quinolinium salts (**120**) into their corresponding 2,4,6-triarylphenyl derivatives (**121**) in good yields has been developed by Zimmermann [158]. The anhydrobases of **121** were discussed as the key intermediates of this pyrylium ring transformation; they attack the pyrylium cation **122** in the initial step as carbon nucleophiles of the enamine type (see Scheme 56). The reaction

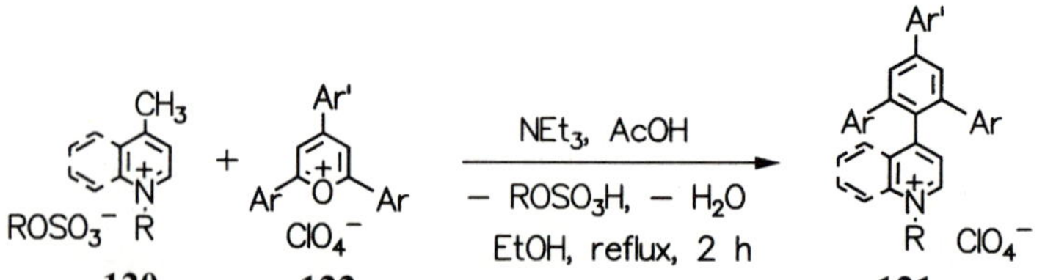

Scheme 56. Ring transformations of (thio)pyrylium salts (**122**) with methyl substituted pyridinium and quinolinium salts (**120**, [158]): Ar, Ar′ = 4-(H, Me, MeO, Cl, Br)C_6H_4, 90–96%

also proceeds well with the corresponding thiopyrylium cations or appropriate bispyridinium salts.

8.4
The Dötz Reaction

Acetylenes can undergo a number of thermal and transition metal promoted cycloaddition reactions. Besides the $[2+2+2]$ cycloaddition (see Sect. 5) the reaction of acetylenes with late transition metal (so-called "Fischer") carbenes is noteworthy for the synthesis of highly and regioselectively functionalized naphthalene derivatives (Dötz reaction), while the co-cycloaddition of acetylenes with alkenes and carbon monoxide gives cyclopentenones (Pauson-Khand reaction) [159, 160].

In a general illustration of the Dötz reaction a terminal or internal alkyne reacts with a carbene **123** and one carbonyl ligand at a $[Cr(CO)_3]$ template in a formal $[3+2+1]$ cycloaddition reaction producing a chromium-complexed naphthol (**124**) under mild reaction conditions via the vinylketene intermediate **125** (see Scheme 57). Terminal alkynes ($R_1C \equiv CR_2$; $R_1 = H, R_2 \neq H$) react with total regioselectivity, while the regiocontrol in the reaction course of internal alkynes

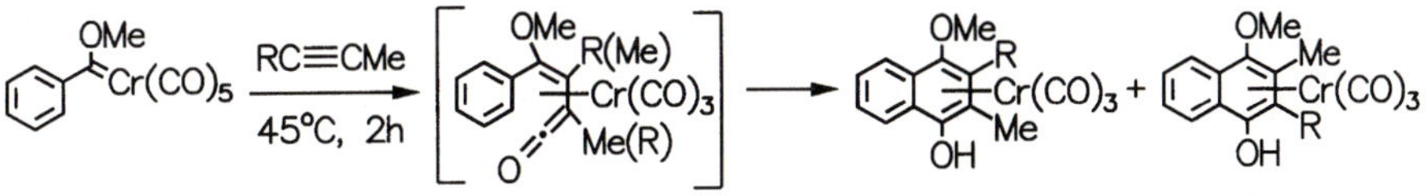

Scheme 57. General scheme for the synthesis of 1-naphthols (**124**) by the Dötz reaction [159 c, 160]: R = Et: 56%; **124a/124b** = 2.5 : 1; R = t-Bu: 59%; **124a/124b** = 14 : 1

(R_1, $R_2 \neq H$) depends almost entirely on the difference in size of the substituents R_1 and R_2 and rarely on electronic effects [159 c]. The high regioselectivity and functional group tolerance predestines the Dötz reaction for the synthesis of various natural products [159, 161 b, c].

When the arylcarbene(pentacarbonyl)chromium complex bears a (phenyl) alkyne at the *ortho*-position of the aryl group, an *intra*molecular alkyne-carbene chelate (**126**) can be formed by low temperature photodecarbonylation. The tetracarbonylchromium chelate **126** obviously dimerizes and furnishes finally a densely functionalized centrosymmetric chrysene (**127**, see Scheme 58 [160 a, d]).

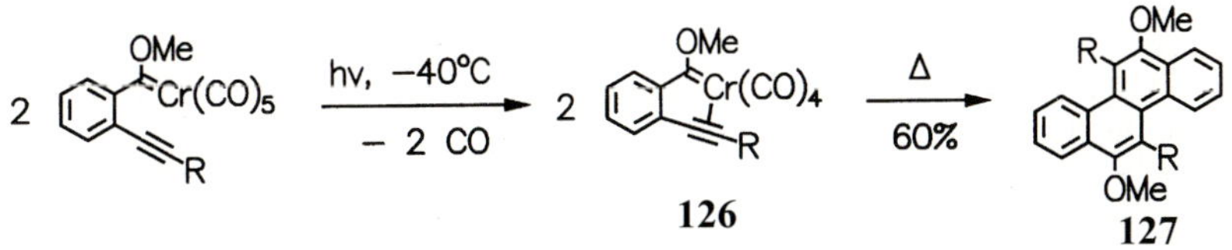

Scheme 58. Novel synthesis of centrosymmetric chrysenes (**127**) by dimerization of *intra*molecularly stabilized alkyne-carbene chromium chelate complexes [160 a, d]

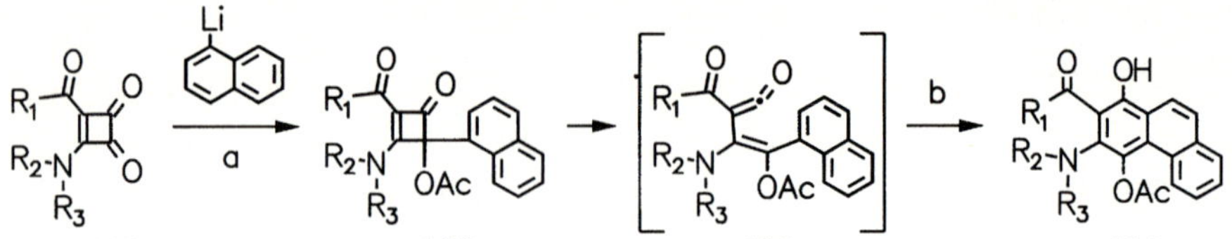

Scheme 59. Novel annelation procedure of cyclobutenones (**129**) via vinylketenes: a) Ac$_2$O, 79%; b) mesitylene, reflux, 10 h, R$_1$=Ph, R$_2$,R$_3$=-piperidinyl, 91% [161a]

An analogous vinylketene intermediate (**127**, see Schemes 57 and 59) as proposed for the Dötz reaction has been assumed in the so-called cyclobutenedione methodology [161]. The key intermediate is a 4-aryl or 4-alkenyl substituted 2-cyclobutenone (**128**) that can be obtained e.g. by the reaction of the 3-cyclobutene-1,2-dione (**129**) with the appropriate lithium reagent or Stille coupling with 4-chloro-3-cyclobutenone. Thermal cyclobutenone ring opening to the vinylketene **130** followed by electrocyclization furnishes the highly substituted aromatic compound **131** (see Scheme 59).

Acknowledgments. The Fonds der Chemischen Industrie is gratefully acknowledged for a Liebig-Stipendium for S. Hagen. We would also like to thank A. Hagen and B. König for their comments and proofreading.

9
References

1. For the resources, formation and application of aromatic compounds in important industrial processes see: Franck HG, Stadelhofer JW (1987) Industrielle Aromatenchemie. Springer, Berlin Heidelberg New York
2. Tour JM (1996) Chem Rev 96:537; b) Dagani R (1996) C&EN March 4:22
3. a) Schlichting P, Rohr U, Müllen K (1997) Liebigs Ann/Recueil 395; b) Holtrup FO, Müller GRJ, Quante H, Feyter S de, Schryver FC de, Müllen K (1997) Chem Eur J 3:219
4. Chandrasekhar S (1993) Liq Cryst 14:3; b) Stegemeyer H (guest ed.) (1994) Liquid crystals. In Baumgärtel H, Franck EU, Grünbein W (eds) Topics in Phys Chem Vol. 3, Steinkopff, Darmstadt, Springer, New York
5. Gama V, Henriques RT, Bonfait G, Almeida M, Meetsma A, Smaalen S van, Boer JL de (1992) J Am Chem Soc 114:1986; b) Dias CB, Santos IC, Gama V, Heriques RT, Almeida M, Pouget JP (1993) Synth Methods 56:1688
6. Hiramoto M, Kishigami Y, Yokoyama M (1990) Chem Lett 119
7. Debad JD, Morris JC, Magnus P, Bard AJ (1997) J Org Chem 62:530
8. Fiesel R, Huber J, Scherf U (1996) Angew Chem 108:2233; Angew Chem Int Ed Engl 35:2111
9. Burland DM, Miller RD, Walsh CA (1994) Chem Rev 94:31; b) Marks TJ, Ratner MA (1995) Angew Chem 107:167; Angew Chem Int Ed Engl 34:155; c) Bahl A, Grahn W, Stadler S, Feiner F, Bourhill G, Bräuchle C, Reisner A, Jones PG (1995) Angew Chem 107:1587; Angew Chem Int Ed Engl 34:1485
10. Taylor R (1990) Electrophilic aromatic substitution. John Wiley, Chichester
11. Vögtle V (1992) Supramolekulare Chemie, Eine Einführung. Teubner, Stuttgart, p 219
12. Lehn JM (1988) Angew Chem 100:91; Angew Chem Int Ed Engl 27:89

13. Gladiali S, Fabbri D (1997) Chem Ber/Recueil 130:543 and references cited therein
14. a) Keehn PM, Rosenfeld SM (eds) (1983) Cyclophanes (Organic chemistry, a series of monographs 45 I/II, Academic Press, New York; b) Vögtle F (1993) Cyclophane chemistry. John Wiley, Chichester, UK; c) König B (1998) Top Curr Chem this issue
15. Lawrence DS, Jiang T, Levett M (1995) Chem Rev 95:2229
16. For a review on the stereoselective olefin polymerization with chiral metallocene catalysts see: Brintzinger HH, Fischer D, Mülhaupt R, Rieger B, Waymouth R (1995) Angew Chem 107:1255; Angew Chem Int Ed Engl 34:1143
17. a) Cram, DJ (1992) Nature 356:29; b) Vögtle F (1992) Supramolekulare Chemie, Eine Einführung. Teubner, Stuttgart, p 45
18. a) Kroto HW, Heath JR, O'Brien SC, Curl RF, Smalley RE (1985) Nature 318:162; b) Krätschmer W, Lamb LD, Fostiropoulos K, Huffman DR (1990) Nature 347:354; c) Edelmann FT (1995) Angew Chem 107:1071; Angew Chem Int Ed Engl 34:981; d) Saunders M, Cross RJ, Jimenez-Vazquez HA, Shimshi R, Khong A (1996) Science 271:1693; e) Hirsch A (1994) The Chemistry of the fullerenes. Thieme, Stuttgart
19. Fabrizzi L, Poggi A (1995) Chem Soc Rev 24:197
20. Anelli PL, Ashton PR, Ballardini R, Balzani V, Delgado M, Gandolfi MT, Goodnow TT, Kaifer AE, Philp D, Pietraszkiewicz M, Prodi L, Reddington MV, Slawin AMZ, Spencer N, Stoddart JF, Vicent C, Williams DJ (1992) J Am Chem Soc 114:193
21. Jørgensen M, Lerstrup K, Frederiksen P, Bjørnholm T, Sommer-Larsen P, Schaumburg K, Brunfeldt K, Bechgaard K (1993) J Org Chem 58:2785
22. Kelly TR, Bowyer MC, Bhaskar KV, Bebbington D, Garcia A, Lang F, Kim MH, Jette MP (1994) J Am Chem Soc 116:3657
23. Proceedings of the 15th International Symposium on Polycyclic Aromatic Compounds, Belgirate, Italy, 19.–22. 09. 1995 in Polycyclic Arom Comp (1996) 9–11
24. a) Gerhardt P, Löffler S, Homann KH (1987) Chem Phys Lett 137:306; b) Pope CJ, Marr JA, Howard JB (1993) J Chem Phys 97:11001
25. Taylor R, Langley GJ, Kroto HW, Walton DRM (1993) Nature 366:728; b) Crowley C, Taylor R, Kroto HW, Walton DRM, Cheng PC, Scott LT (1996) Synth Methods 77:17
26. Clar E (1964) Polycyclic hydrocarbons. Academic Press, New York, Vol. I/II; b) Dias JR (1988) Handbook of polycyclic hydrocarbons (Physical sciences data 30) Elsevier, Amsterdam; c) Fetzer JC (1989) Org Prep Proc Int 21:47; d) Harvey RG (1991) Polycyclic aromatic hydrocarbons, chemistry and carcinogenicity, Cambridge University Press, Cambridge; e) Zander M (1995) Polycyclische Aromaten, Kohlenwasserstoffe und Fullerene. Teubner, Stuttgart; f) Harvey RG (1997) Polycyclic aromatic hydrocarbons. Wiley-VCH, New York
27. Polycyclic aromatic compounds, J Int Soc Polycyclic Aromatic Compounds, Gordon and Breach, ISSN 1040–6638
28. International Symposium on Novel Aromatics (ISNA), ISNA-8th, Braunschweig, Germany, 30. 07.–04. 08. 1995; lectures of this ISNA conference was published in: (1996) Pure Appl Chem 68:209
29. International Symposium on Polycyclic Aromatic Compounds (ISPAC), ISPAC-15th, Belgirate, Italy, 19.–22. 09. 1995
30. Blome H, Clar E, Grundmann C (1981) Aufbau polycyclischer aromatischer Kohlenwasserstoffe. In: Houben-Weyl (Arene und Arine), Georg Thieme, Stuttgart, V/2b:359
31. For reviews on pyrolysis see: a) Seybold G (1977) Angew Chem 89:377; Angew Chem Int Ed Engl 16:365; b) Vögtle F, Rossa L (1979) Angew Chem 91:534; Angew Chem Int Ed Engl 18:514; c) Brown RFC (1980) Pyrolytic methods in organic chemistry (Organic Chemistry Series, Vol.41), Academic Press, New York p 115; d) Karpf M (1986) Angew Chem 98:413; Angew Chem Int Ed Engl 25:414; e) Brown RFC, Eastwood FW (1993) Synlett; f) Rabideau PW, Sygula A (1996) Acc Chem Res 29:235; g) Scott LT (1996) Pure Appl Chem 68:291
32. Golden DM, Spokes GN, Benson SW (1973) Angew Chem 85:602; Angew Chem Int Ed Engl 12:534; b) Berkowitz J, Ellison GB, Gutman D (1994) J Phys Chem 98:2744
33. Clancy M, Hawkins DG, Hesabi MM, Meth-Cohn O, Rhouati S (1982) J Chem Res (S) 78; b) Rubin Y, Lin SS, Knobler CB, Anthony J, Boldi AM, Diederich F (1991) J Am Chem Soc 113:6943

34. Magrath J, Fowler FW (1988) Tetrahedron Lett 29:2171
35. For matrix isolation studies see e. g. a) Pimentel GC (1975) Angew Chem 87:220; Angew Chem Int Ed Engl 14:199; b) Morawietz J, Sander W (1996) Liebigs Ann 2029
36. Houk KN, Li Y, Evanseck JD (1992) Angew. Chem. 104:711; Angew Chem, Int Ed Engl 31:682 and references cited therein
37. Plater MJ (1994) Tetrahedron Lett 35:801
38. a) Alder RW, Whittaker G (1975) J Chem Soc Perkin Trans II:712; b) Hagen S, Nuechter U, Nuechter M, Zimmermann G (1994) Tetrahedron Lett 35:7013; c) Hagen S, Christoph H, Zimmermann G (1995) Tetrahedron 51:6961; d) Hagen S, Nuechter U, Nuechter M, Zimmermann G (1995) Polycyclic Arom Comp 4:209; e) Hagen S (1995) Dissertation, Universität Leipzig
39. Scott LT, Bratcher MS, Hagen S (1996) J Am Chem Soc 118:8743
40. Hagen S, Bratcher MS, Erickson MS, Zimmermann G, Scott LT (1997) Angew Chem 109:407; Angew Chem Int Ed Engl 36:406
41. Mehta G, Raghava Sharma GV, Krishna Kumar MA, Vedavyasa TV, Jemmis ED (1995) J Chem Soc Perkin Trans I:2529
42. a) Schaden G (1983) J Org Chem 48:5385; b) Schaden G (1977) Angew Chem 89:50; Angew Chem Int Ed Engl 16:50; c) for the synthesis of a fullerene-C_{119} from a dimeric fullerene-C_{60} oxide $C_{120}O$ by thermolysis including a formal loss of CO see: Gromov A, Ballenweg S, Giesa S, Lebedkin S, Hull WE, Krätschmer W (1997) Chem Phys Lett 267:460
43. Plater MJ (1994) Tetrahedron Lett 35:6147
44. a) Tanga MJ, Bupp JE (1993) J Org Chem 58:4173; b) Mehta G, Shah SR, Ravikumar K (1993) J Chem Soc Chem Commun 1006; c) Clayton MD, Rabideau PW (1997) Tetrahedron Lett 38:741
45. Abdourazak AH, Marcinov Z, Sygula A, Sygula R, Rabideau P W (1995) J Am Chem Soc 117:6410
46. Bratcher MS, Scott LT (1994) 207th ACS National Meeting, Organic Division (San Diego) Poster No. 420
47. Faust R, Vollhardt KPC (1992) Seventh International Symposium on Novel Aromatic Compounds (Victoria, Canada), Poster No. 61
48. Brooks MA, Scott LT (1997) 213th ACS National Meeting, Organic Division (San Francisco) Poster No. 205
49. Borchardt A, Fuchicello A, Kilway KV, Baldridge KK, Siegel JS (1992) J Am Chem Soc 114:1921
50. Yamamoto K, Sonobe H, Matsubara H, Sato M, Okamoto S, Kitaaura K (1996) Angew Chem 108:69; Angew Chem Int Ed Engl 35:69
51. Mitchell RH, Iyer VS, Khalifa N, Mahadevan R, Venugopalan S, Weerawarna SA, Zhou P (1995) J Am Chem Soc 117:1514 and literature cited therein
52. a) Vögtle F, Groß J, Seel C, Nieger M (1992) Angew Chem 104:1112; Angew Chem Int Ed Engl 31:1069; b) Vögtle F, Seel C, Berscheid R, Groß J, Windscheif PM (1994) Architecture of new concave host molecules. In: Wipff G (ed) Computational approaches in supramolecular chemistry, NATO ASI Ser. Ser. C 426:311; c) Groß J, Harder G, Vögtle F, Stephan H, Gloe K (1995) Angew Chem 107:523; Angew Chem Int Ed Engl 34:481
53. a) Brown RFC, Harrington KJ (1972) J Chem Soc Chem Commun 1175; b) Brown RFC, Harrington KJ, McMullen GL (1974) J Chem Soc Chem Commun 123
54. a) Scott LT, Hashemi MM, Meyer DT, Warren HB (1991) J Am Chem Soc 113:7082; b) Scott LT, Hashemi MM, Bratcher MS (1992) J Am Chem Soc 114:1920; c) Scott LT, Cheng PC, Bratcher MS (1992) Seventh International Symposium on Novel Aromatic Compounds (Victoria, Canada), Poster No. 64 ; d) Necula A (1996) PhD thesis, Boston College; e) Scott LT, Necula A (1996) J Org Chem 61:386; f) Scott LT, Necula A (1997) Tetrahedron Lett 38:1877; g) McComas C, Bratcher MS, Scott LT (1996) 211th Acs National Meeting, Division of Chemical Education (New Orleans)
55. a) Abdourazak AH, Sygula A, Rabideau PW (1993) J Am Chem Soc 115:3010; b) Rabideau PW, Abdourazak AH, Folsom HE, Marcinow Z, Sygula A, Sygula R (1994) J Am Chem Soc 116:7891; c) Clayton MD, Marcinov Z, Rabideau PW (1996) J Org Chem 61:6052

56. a) Sarobe M, Snoeijer JD, Jenneskens LW, Slagt MQ, Zwikker JW (1995) Tetrahedron Lett 36:8489; b) Sarobe M, Flink S, Jenneskens LW, Poecke BLA van, Zwikker JW (1995) J Chem Soc Chem Commun 2415; c) Sarobe M, Snoeijer JD, Jenneskens LW, Zwikker JW, Wesseling J (1995) Tetrahedron Lett 36:9565; d) Sarobe M, Zwikker JW, Snoeijer JD, Wiersum UE, Jenneskens LW (1994) Chem Soc Chem Commun 89; e) Sarobe M, Jenneskens LW, Wesseling J, Snoeijer JD, Zwikker JW, Wiersum UE (1997) Liebigs Ann/Recueil 1207

57. a) Pope CJ, Marr JA, Howard JB (1993) J Phys Chem 97:11001; b) Marr JA, Giovane LM, Longwell JP, Howard JB, Lafleur AL (1994) Combust Sci Technol 101:301; c) Lafleur AL, Howard JB, Tahizadeh K, Plummer EF, Scott LT, Necula A, Swallow KC (1996) J Phys Chem 100:17421

58. Liu CZ, Rabideau PW (1996) Tetrahedron Lett 37:3437

59. Zimmermann G, Nuechter U, Hagen S, Nuechter M (1994) Tetrahedron Lett 35:4747

60. In fact, the formation of benzene from hexa-1,3-diene-5-yne is possible by at least three different mechanistic pathways: (i) at temperatures below 500–600°C preferably by an electrocyclization via the isobenzene followed by a 1,3-hydrogen shift, (ii) at temperatures above 500–600 °C preferably by the discussed carbene mechanism, and (iii) in the presence of hydrogen radicals by an intramolecular radical cyclization sequence. For further discussion including labelling studies see Nüchter U, Zimmermann G, Francke V, Hopf H (1997) Liebigs Ann/Recueil 1505 and the references cited herein.

61. Engler TA, Shechter H (1982) Tetrahedron Lett 23:2715

62. a) Sarobe M, Jenneskens LW, Wiersum UE (1996) Tetrahedron Lett 37:1121; b) Neilen RHG, Wiersum UE (1996) J Chem Soc Chem Commun 149

63. Brown RFC, Coulston KJ, Eastwood FW (1996) Tetrahedron Lett 37:6819

64. Tomioka H, Kobayashi N (1991) Bull Chem Soc Jpn 64:327

65. a) Dehmlow EV, Balschukat D (1985) Chem Ber 118:3805; b) for a review concerning the thermal rearrangement of azulene, naphthalene and other PAHs see: Scott LT (1992) Acc Chem Res 15:52

66. a) Trahanovsky WS, Lee SK, Fennell JW (1995) J Org Chem 60:8410; b) Trahanovsky WS, Lee SK (1996) Synthesis 1085

67. a) Muszkat KA (1980) Top Curr Chem 88:89; b) Mallory FB, Mallory CW (1984) Org React 30:1; c) Laarhoven WH (1989) Org Photochem 10:163; d) Meier H (1992) Angew Chem 104:1425; Angew Chem Int Ed Engl 31:1399; e) Tominaga Y, Castle RN (1996) J Heterocycl Chem 33:523

68. Fleming I (1976) Frontier orbitals and organic chemical reactions. John Wiley, London

69. Crimmins MT, Reinhold TL (1993) Org React 44:297

70. For a review on helicenes see: Oremek G, Seiffert U, Janecka A (1987) Chem Ztg 111:69

71. Mallory FB, Butler KE, Evans AC, Mallory CW (1996) Tetrahedron Lett 37:7173

72. a) Yamamoto K, Harada T, Okamoto Y, Chikamatsu H, Nakazaki M, Kai Y, Nakao T, Tanaka M, Harada S, Kasai N (1988) J Am Chem Soc 110:3578; b) [7.7]Circulene was also obtained by the same group using two photolytic ring closures in one of the key steps: Yamamoto K, Saitho Y, Iwaki D, Ooka T (1991) Angew Chem 103:1202; Angew Chem Int Ed Engl 30:1173

73. Kuo CH, Tsau MH, Weng DTC, Lee GH, Peng SM, Luh TY, Biedermann PU, Agranat I (1995) J Org Chem 60:7380

74. a) Liu L, Yang B, Katz TJ, Poindexter MK (1991) J Org Chem 56:3769; b) Yang CX, Harvey RG (1993) J Org Chem 58:4155

75. For recent reviews on the Wittig reaction see: a) Bestmann HJ, Vostrowsky O (1983) Topics Curr Chem 109:85; b) Pommer H, Thieme PC (1983) Topics Curr Chem 109:165, c) Maryanoff BE, Reitz AB (1989) Chem Rev 89:863; d) Vedejs E, Peterson MJ (1994) Topics Stereochem 21:1; e) Walker BJ (1995) Organophosphorus Chem 26:265; f) Nicolaou KC, Härter MW, Gunzner JL, Nadin A (1997) Liebigs Ann/Recueil 1283

76. For applications of the bis-Wittig reaction in the synthesis of PAHs see e.g. a) Vollhardt KPC (1975) Synthesis 765; b) Minsky A, Rabinovitz M (1983) Synthesis 497; c) Yang CX, Yang DTC, Harvey RG (1992) Synlett 799

77. For references on the Siegrist reaction see ref. [67d] and literature cited therein as well as: a) Siegrist AE (1967) Helv Chim Acta 50:906; b) Nair V, Jahnke TS (1984) Synthesis 424

78. For reviews on Heck-type reactions see: a) Heck RF (1982) Org React 27:345; b) Ryabov AD (1985) Synthesis 233; c) de Meijere A, Meyer FE (1994) Angew Chem 106:2473; Angew Chem Int Ed Engl 33:2479; d) Bräse S, de Meijere A (1997) Palladium-catalyzed coupling of organyl halides to alkenes – the Heck reaction. In: Stang PJ, Diederich F (eds) Metal-catalyzed cross-coupling reactions. Wiley-VCH, Weinheim, p 99

79. For recent reviews on Palladium reagents see: a) Tsuji J (1995) Palladium reagents and catalysts: innovations in organic synthesis. Wiley, Chichester; b) Heck RF (1985) Palladium reagents in organic syntheses. Academic Press, New York

80. For reviews on the McMurry reaction see: a) Welzel P (1983) Nachr Chem Techn Lab 31:814; b) McMurry JE (1989) Chem Rev 89:1513

81. a) Scholz M, Dietz F, Mühlstädt M (1967) Z Chem 7:329; b) Scholz M, Mühlstädt M, Dietz F (1967) Tetrahedron Lett 7:665; c) Laarhoven WH, Cuppen TJHM, Nivard RJF (1968) Rec Trav Chim 87:687

82. Broene RD, Diederich F (1991) Tetrahedron Lett 32:5227

83. Sudhakar A, Katz TJ, Yang BW (1986) J Am Chem Soc 108:2790

84. a) Gundermann KD, Romahn E, Zander M (1992) Z Naturforsch 47b:1764; b) Tang XQ, Harvey RG (1995) J Org Chem 60:3568; c) Hagen S, Scott LT (1996) J Org Chem 61:7198

85. Winter W, Langjahr U, Meier H, Merkuschew J, Juriew J (1984) Chem Ber 117:2452

86. Nakamura Y, Tsuihiji T, Mita T, Minowa T, Tobita S, Shizuka H, Nishimura J (1996) J Am Chem Soc 118:1006

87. For the photodimerization of acenaphthylenes and related structures see: a) Haga N, Takayanagi H, Tokumaru K (1997) J Org Chem 62:3734; b) Magnus P, Morris JC, Lynch V (1997) Synthesis 506

88. a) Greiving H, Hopf H, Jones PG, Bubenitschek P, Desvergne JP, Bouas-Laurent H (1994) J Chem Soc Chem Commun 1075; b) Hopf H, Greiving H, Jones PG, Bubenitschek P (1995) Angew Chem 107:742; Angew Chem Int Ed Engl 34:685

89. a) Fessner WD, Sedelmeier G, Spurr PR, Rihs G, Prinzbach H (1987) J Am Chem Soc 109:4626; b) Murty BARC, Spurr PR, Pinkos R, Grund C, Fessner WD, Hunkler D, Fritz H, Roth WR, Prinzbach H (1987) Chimia 41:32

90. Fessner WD, Murty BARC, Wörth J, Hunkler D, Fritz H, Prinzbach H, Roth WR, Ragué Schleyer P von, McEwen AB, Maier WF (1987) Angew Chem 99:484; Angew Chem Int Ed Engl 26:451

91. The first synthesis of dodecahedrane was accomplished by: Paquette LA, Balogh DW (1982) J Am Chem Soc 104:774

92. a) Kammermeier S, Jones PG, Herges R (1996) Angew Chem 108:2834; Angew Chem Int Ed Engl 35:2669; b) Kammermeier S, Herges R (1996) Angew Chem 108:470; Angew Chem Int Ed Engl 35:417; c) Sauer J, Breu J, Holland U, Herges R, Neumann H, Kammermeier S (1997) Liebigs Ann/Recueil 1473 and references cited therein

93. a) Bringmann G, Walter R, Weirich R (1990) Angew Chem 102:1006; Angew Chem Int Ed Engl 29:977; b) Quideau S, Feldman KS (1996) Chem Rev 96:475

94. For recent reviews on the Stille reaction see: a) Farina V, Krishnamurthy V, Scott WJ (1997) Org React 50:1; b) Stille JK (1986) Angew Chem 98:504; Angew Chem Int Ed Engl 25:508

95. For recent reviews on the Suzuki coupling see: a) Martin AR, Yang Y (1993) Acta Chem Scand 47:221; b) Miyaura N, Suzuki A (1995) Chem Rev 95:2457

96. For references on the classical Ullmann coupling see ref. [93a] and: a) Fanta PE (1964) Chem Rev 64:613; b) Fanta PE (1974) Synthesis 9; c) Sainsbury M (1980) Tetrahedron Lett 36:3327; d) Nelson TD, Meyers AI (1994) Tetrahedron Lett 35:3259; e) Cereghetti M, Schmid R, Schönholzer P, Ragoet A (1996) Tetrahedron Lett 37:5343

97. Zhang S, Zhang D, Liebeskind LS (1997) J Org Chem 62:2312

98. Semmelhack MF, Helquist PM, Jones LD (1971) J Am Chem Soc 93:5908

99. Yamamoto T, Morita A, Miyazaki Y, Maruyama T, Wakayama H, Zhou Z, Nakamura Y, Kanbara T (1992) Macromolecules 25:1214

100. Takagi K, Hayama N, Inokawa S (1980) Bull Chem Soc Jpn 53:3691

101. Quante H, Müllen K (1995) Angew Chem 107:1487; Angew Chem Int Ed Engl 34:1323

102. a) Koch KH, Müllen K (1991) Chem Ber 124:2091; b) Müller M, Mauermann-Düll H, Wagner M, Enkelmann V, Müllen K (1995) Angew Chem 107:1751; Angew Chem Int Ed Engl 34:1583; c) For the use of the potassium-graphite intercalation compound C_8K in ring closure reactions of benzil see: Tamarkin D, Rabinovitz M (1987) J Org Chem 52:3472

103. Banerji A, Nayak SK (1991) J Chem Soc Chem Commun 1432

104. a) Rajca A, Safronov A, Rajca S, Ross CR, Stezowski JJ (1996) J Am Chem Soc 118:7272; b) Rajca A, Safronov A, Rajca S, Shoemaker R (1997) Angew Chem 109:504; Angew Chem Int Ed Engl 36:488

105. Lipshutz BH, Siegmann K, Garcia E, Kayser F (1993) J Am Chem Soc 115:9276

106. a) Malaba D, Djebli A, Chen L, Zarate EA, Tessier CA, Youngs WJ (1993) Organometallics 12:1266; b) Bradshaw JD, Solooki D, Tessier CA, Youngs WJ (1994) J Am Chem Soc 116:3177

107. Seiders TJ, Baldridge KK, Siegel JS (1996) J Am Chem Soc 118:2754

108. Blome H, Clar E, Grundmann C (1981) Aufbau polycyclischer aromatischer Kohlenwasserstoffe. In: Houben-Weyl (Arene und Arine), Georg Thieme, Stuttgart, V/2b:359

109. Hagen S, McMahon B, Scott LT (1996) unpublished results

110. a) Zander M, Franke W (1958) Chem Ber 91:2794; b) Zander M, Friedrichsen W (1992) Z Naturforsch 74b:1314; c) Zander M (1995) Polycyclic Arom Comp 7:209; d) Goddard R, Haenel MW, Herndon WC, Krüger C, Zander M (1995) J Am Chem Soc 117:30

111. Kovacic P, Koch FW (1965) J Org Chem 30:3176

112. Stabel A, Herwig P, Müllen K, Rabe JP (1995) Angew Chem 107:1768; Angew Chem Int Ed Engl 34:1609

113. Mohr B, Enkelmann V, Wegner G (1994) J Org Chem 59:635

114. Bodwell GJ, Bridson JN, Houghton TJ, Kennedy JWJ, Mannion MR (1996) Angew Chem 108:1418; Angew Chem Int Ed Engl 35:1320

115. Meyers AI, Flanagan ME (1992) Org Synth 71:107

116. a) Fuson RC, Wassmundt FW (1956) J Am Chem Soc 78:5409; b) Hotta H, Suzuki T, Miyano S (1990) Chem Lett 143

117. The crosscoupling of aryl Grignard reagents with organic halides is generally referred to as Kharasch reaction. See: Kharasch MS, Fields EK (1941) J Am Chem Soc 63:2316

118. a) Kumada M (1980) Pure Appl Chem 52:669; b) Tamao K, Komada S, Nakajima I, Kumada M, Minato A, Suzuki K (1982) Tetrahedron 38:3347; c) Ibuki E, Ozasa S, Fujioka Y, Mizutani H (1982) Bull Chem Soc Jpn 55:845

119. Negishi E (1982) Acc Chem Res 15:340

120. Erdik E (1992) Tetrahedron 48:9577

121. For more general references concerning the Suzuki coupling see e.g.: a) Miyaura N, Yamada K, Suginome H, Suzuki A (1985) J Am Chem Soc 107:972; b) Miyaura N, Ishiyama T, Sasaki H, Ishikawa M, Satoh M, Suzuki A (1989) J Am Chem Soc 111:315; c) Smith GB, Dezeny GC, Hughes DL, King AO, Verhoeven TR (1994) J Org Chem 59:8151

122. For the application of the Suzuki coupling in the synthesis of biaryls see e.g. ref. [3b, 9c, 93, 95, 102a] and a) Sharp MJ, Cheng W, Sniekus V (1987) Tetrahedron Lett 28:5093; b) Wang D, Haseltine J (1994) J Heterocycl Chem 31:1637; c) Liess P, Hensel V, Schlüter AD (1996) Liebigs Ann 1037; d) Harre K, Enkelmann V, Schulze M, Bunz UHF (1996) Chem Ber 129:1323; e) Zoltewicz JA, Maier NM, Lavieri S, Ghiviriga I, Abboud KA, Fabian WMF (1997) Tetrahedron 53:5379

123. Ishiyama T, Kizaki H, Miyaura N, Suzuki A (1993) Tetrahedron Lett 34:7595

124. For the application of the Stille coupling in the synthesis of biaryls see e.g. ref. [22, 93–95, 123] and e.g.: a) Yamamoto Y, Azuma Y, Mitoh H (1986) Synthesis 564; b) Stille JK, Echavarren AM, Williams RM, Hendrix JA (1992) Org Synth 71:97; c) Boese R, Bräunlich G, Gotteland JP, Hwang JT, Troll C, Vollhardt KPC (1996) Angew Chem 108:1100; Angew Chem Int Ed Engl 35:995

125. Arylboronic acids are most often applied. An interesting one pot, two-step procedure for the synthesis of unsymmetrical biaryls directly from aryl halides or triflates via the arylboronic esters has been reported most recently: Ishiyama T, Itoh Y, Kitano T, Miyaura N (1997) Tetrahedron Lett 38:3447

126. Sengupta S, Bhattacharyya S (1997) J Org Chem 62:3405
127. a) Herrmann WA, Broßmer C, Öfele K, Reisinger CP, Priermeier T, Beller M, Fischer H (1995) Angew Chem 107:1989; Angew Chem Int Ed Engl 34:1844; b) Beller M, Fischer H, Herrmann WA, Öfele K, Broßmer C (1995) Angew Chem 107:1992; Angew Chem Int Ed Engl 34:1848
128. Krafft TE, Rich JD, McDermott PJ (1990) J Org Chem 55:5430
129. Gouda KI, Hagiwara E, Hatanaka Y, Hiyama T (1996) J Org Chem 61:7232
130. a) Dyker G (1991) Tetrahedron Lett 32:7241; b) Dyker G (1993) J Org Chem 58:234; c) Dyker G, Körning J, Bubenitschek P, Jones PG (1997) Liebigs Ann/Recueil 203; d) Dyker G, Kerl T, Jones PG, Dix I, Bubenitschek P (1997) J Chem Soc Perkin Trans 1:1163
131. Albrecht K, Reiser O, Weber M, Knieriem B, Meijere A de (1994) Tetrahedron 50:383
132. a) Rice JE, Cai ZW (1993) J Org Chem 58:11415; b) Gonzalez JJ, Garcia N, Gomez-Lor B, Echavarren AM (1997) J Org Chem 62:1286 and references cited therein; c) For selected examples of the synthesis of PAH by Heck-type reactions see also ref. [78, 79, 130–133]
133. Pogodin S, Biedermann PU, Agranat I (1997) J Org Chem 62:2285
134. A recent collection of references for the generation of arynes is given in: Sato H, Isono N, Miyoshi I, Mori M (1996) Tetrahedron 52:8143
135. For a review on cobalt-assisted [2+2+2] cycloaddition reactions see: Vollhardt KPC (1984) Angew Chem 96:525; Angew Chem Int Ed Engl 23:539
136. For the rarely reported [2+2+1] cycloaddition reactions of alkynes to fulvenes see: O'Connor JM, Hiibner K, Merwin R, Gantzel PK, Fong BS (1997) J Am Chem Soc 119:3631 and references cited herein
137. a) Carey JG, Millar IT (1959) J Chem Soc 3144; b) Hart H, Shamouilian S, Takehira Y (1981) J Org Chem 46:4427; c) Fossatelli M, Brandsma L (1992) Synthesis 756
138. a) Psiorz M, Hopf H (1982) Angew Chem 94:639; Angew Chem Int Ed Engl 21:623; b) an attempted synthesis of a so-called tribenzotrifoliaphane from benzo annelated [2.2]paracyclophane-1,9-dienes by a domino-type Heck reaction of benzo[2.2]paracyclophanene with two molecules of 9-bromo-1,2-benzo[2.2]paracyclophanene actually yielded an isomer of dihydrotribenzotrifoliaphane with a bicyclo[3.3.0]octadiene ring system fusing three [2.2]paracyclophanene skeletons: de Meijere A, Rauch K, Albrecht K unpublished results. Cf. König B, de Meijere A (1998) Synlett: in press; König, B Topics Curr Chem: this volume
139. a) Komatsu K, Akamatsu H, Jinbu Y, Okamoto K (1988) J Am Chem Soc 110:633; b) Sato Y, Nishimata T, Mori M (1994) J Org Chem 59:6133; c) Frank NL, Baldridge KK, Siegel JS (1995) J Am Chem Soc 117:2102
140. a) Nambu M, Mohler DL, Hardcastle K, Baldridge KK, Siegel JS (1993) J Am Chem Soc 115:6138; b) Boese R, Matzger AJ, Mohler DL, Vollhardt KPC (1995) Angew Chem 107:1630; Angew Chem Int Ed Engl 34:1478
141. a) Saa C, Crotts DD, Hsu G, Vollhardt KPC (1994) Synlett 487; b) Cruciani P, Aubert C, Malacria M (1996) Synlett 105
142. a) Hart H, Lai CY, Nwokogu G, Shamouilian S, Teuerstein A, Zlotogorski C (1980) J Am Chem Soc 102:6649; b) Luo J, Hart H (1988) J Org Chem 53:1343
143. a) Kohnke FH, Slawin AMZ, Stoddart JF, Williams DJ (1987) Angew Chem 99:941; Angew Chem Int Ed Engl 26:892; b) Kohnke FH, Stoddart JF (1989) Pure Appl Chem 61:1581; c) Benkhoff J, Boese R, Klärner FG, Wigger AE (1994) Tetrahedron Lett 35:73
144. a) Schlüter AD, Löffler M, Enkelmann V (1994) Nature 368:831; b) Löffler M, Schlüter AD, Geßler K, Saenger W, Toussaint JM, Bredas JL (1994) Angew Chem 106:2281; Angew Chem Int Ed Engl 33:2209; c) Schlicke B, Schirmer H, Schlüter AD (1995) Adv Mater 7:544; d) For other examples of aromatic ladder polymers see: Grüner J, Hamer PJ, Friend RH, Huber HJ, Scherf U, Holmes AB (1994) Adv Mater 6:748; e) For syntheses towards beltenes see: Graham RJ, Paquette LA (1995) J Org Chem 60:5770
145. For the Diels Alder synthesis of other linear acenes see e.g. ref. [143] and a) Thomas AD, Miller LL (1986) J Org Chem 51:4160; b) Wegener S, Müllen K (1991) Chem Ber 124:2101; c) Meier H, Rose B, Schollmeyer D (1997) Liebigs Ann/Recueil 1173

146. a) Willmore ND, Liu L, Katz TJ (1992) Angew Chem 104:1081; Angew Chem Int Ed Engl 31:1093; b) Nuckolls C, Katz TJ, Castellanos L (1996) J Am Chem Soc 118:3767

147. a) Pascal RA Jr., McMillan WD, Engen D vaan, Eason RG (1987) J Am Chem Soc 109:4660; b) Smyth N, Engen D vaan, Pascal RA Jr. (1990) J Org Chem 55:1937; c) Qiao X, Padula MA, Ho DM, Vogelaar NJ, Schutt CE, Pascal RA Jr. (1996) J Am Chem Soc 118:741

148. a) Morgenroth F, Reuther E, Müllen K (1997) Angew Chem 109:647; Angew Chem Int Ed Engl 36:631; b) Vivekanantan SI, Wehmeier M, Brand JD, Keegstra MA, Müllen K (1997) Angew Chem 109:1676; Angew Chem Int Ed Engl 36:1604; c) Müller M, Vivekanantan SI, Kübel C, Enkelmann, Müllen K (1997) Angew Chem 109:1679; Angew Chem Int Ed Engl 36:1607

149. a) Cho BP, Harvey RG (1987) J Org Chem 52:5668; b) Cho BP, Harvey RG (1987) J Org Chem 52:5679

150. a) Cho BP (1995) Tetrahedron Lett 36:2403; b) Cho BP, Zhou L (1996) Tetrahedron Lett 37:1535

151. Harvey RG, Pataki J, Cortez C, Di Raddo P, Yang CX (1991) J Org Chem 56:1210

152. For recent examples on the "trimerization" of alkyl aryl ketones see: a) Pyrko AN (1992) J Org Chem (Russ.) 28:215; J Org Chem (Engl.) 28:179; b) Plater MJ (1993) Synlett 405; c) Wang L, Shevlin PB (1996) 212th ACS National Meeting, Organic Division (Orlando) Poster No. 345; d) Dehmlow EV, Kelle T (1997) Synth Commun 27:2021; e) Hagen S, Altmann A, Zimmermann G, Dehmlow EV, Kelle T (1998) manuscript in preparation

153. Goldfinger MB, Crawford KB, Swager TM (1997) J Am Chem Soc 119:4578

154. a) Kuck D, Neumann E, Schuster A (1994) Chem Ber 127:151; b) Haag R, Ohlhorst B, Noltemeyer M, Fleischer R, Stalke D, Schuster A, Kuck D, de Meijere A (1995) J Am Chem Soc 117:10474 and references cited therein

155. a) Kuck D, Bögge H (1986) J Am Chem Soc 108:8107; b) Gupta AK, Fu X, Snyder JP, Cook JM (1991) Tetrahedron 47:3665; c) Kuck D (1994) Chem Ber 127:409; d) for a review on benzannelated fenestranes see: Kuck D manuscript in preparation #

156. Kuck D (1997) Liebigs Ann/Recueil 1043 and references cited in this review

157. a) Rabinovitz M (1988) Topics Curr Chem 146:99; b) Ayalon A, Sygula A, Cheng PC, Rabinovitz M, Rabideau PW, Scott LT (1994) Science 265:1065; c) Dijk JTM van, Lugtenburg J, Cornelisse J (1995) J Chem Soc Perkin Trans II:1489; d) Dijk JTM van, Hartwijk A, Bleeker AC, Lugtenburg J, Cornelisse J (1996) J Org Chem 61:1136

158. a) Zimmermann T (1995) J Heterocycl Chem 32:563; b) Zimmermann T (1995) J Heterocycl Chem 32:991; c) Zimmermann T, Schmidt K (1996) J Heterocycl Chem 33:783

159. For reviews of carbene complexes and transition metal mediated cycloaddition reactions of alkynes in organic synthesis see: a) Dötz KH (1984) Angew Chem 96:573; Angew Chem Int Ed Engl 23:587; b) Schore NE (1988) Chem Rev 88:1081; c) Casalnuova JA, Schore NE (1995) Organomettalic cycloaddition reactions of acetylenes. In: Stang PJ, Diederich F (eds) Modern acetylene chemistry, VCH, Weinheim, p 139; d) de Meijere A (1996) Pure Appl Chem 68:61

160. For selected recent applications of alkyne carbene and related complexes in the syntheses of PAH see: a) Dötz KH, Schäfer, Kroll F, Harms K (1992) Angew Chem 104:1257; Angew Chem Int Ed Engl 31:1236; b) Merlic CA, Burns EE, Xu D, Chen SY (1992) J Am Chem Soc 114:8722; c) Merlic CA, Pauly ME (1996) J Am Chem Soc 118:11319; d) Hohmann F, Siemoneit S, Nieger M, Kotila S, Dötz KH (1997) Chem Eur J 3:853

161. For recent references discussing the cyclobutene dione and related methodologies see: a) Sun L, Liebeskind LS (1995) J Org Chem 60:8194; b) Danheiser RL, Casebier DS, Firooznia F (1995) J Org Chem 60:8341; c) Koo S, Liebeskind LS (1995) J Am Chem Soc 117:3389

Carbon Rich Cyclophanes with Unusual Properties – an Update

Burkhard König

Institut für Organische Chemie der Technischen Universität Braunschweig, Hagenring 30, D-38106 Braunschweig, Germany. *E-mail: B.Koenig@tu-bs.de*

Cyclophane chemistry and the research targets in this area have changed over the past two decades. While the unique electronic and structural properties of cyclophanes are still fascinating, more and more attention is now being paid to the function of the molecules, e.g., to the formation of inclusion complexes of large cyclophanes or the use of chiral cyclophanes as auxiliars in asymmetric synthesis. The rigid structures of short bridge cyclophanes has stimulated the synthesis of compounds with extended *p*-systems. Both their interesting physical properties, e.g., in intramolecular electron transfer processes, and their often beautiful structures of high symmetry are alluring. While synthetic approaches to cyclophanes, their reactivity and inclusion ability have been reviewed extensively, applications of the cyclophane scaffold and its properties are scattered. Therefore the author will focus on the recent developments in the chemistry of cyclophanes that have been used as building blocks for functional devices, as ligands for metal coordination and chiral auxiliaries or chiral ligands in synthesis. It will show that a class of compounds considered esoteric 20 years ago is now even reaching industrial applications. A review with 60 schemes and 122 references.

Keywords: Cyclophanes, arene complexes, alkyne complexes, chirality, photochemistry.

Topics in Current Chemistry, Vol. 196
© Springer Verlag Berlin Heidelberg 1998

1
Introduction

More than 90 years after the first reported synthesis of [2.2]metacyclophane [1] and nearly 50 years after the synthesis of [2.2]paracyclophane, [2] cyclophane chemistry is still a field of ongoing research. In the beginning, work had been focused on the development of new synthetic methods yielding cyclophanes and ansa-compounds and the investigation of their physical properties. Later, the scope was extended to the incorporation of heterocycles into phanes and the more sophisticated techniques allowed the synthesis of multibridged and multilayered phanes. All this has been extensively reviewed [3].

The first practical application for cyclophanes was found in host-guest chemistry [4]. Molecular recognition of ions and neutral molecules in functionalized cavities of cyclophanes has been demonstrated for numerous examples. Here, special interest has been attracted by complexes in water since it can directly model molecular recognition events in biological systems. The development of suitable water soluble hosts with high binding selectivity is an active field of research [5] that has recently been reviewed [6].

A different use of inclusion was realized with carceplexes: [7] starting from concave cyclophanes, such as cavitands, carcerands with an inner sphere were synthesized or non-covalently assembled [8]. Reactive molecules, trapped in the interior and so shielded from reaction partners could be investigated by spectroscopy. Cyclobutadiene [9] and benzyne [10] are the most striking examples of highly reactive molecules isolated by this technique.

Calixarenes [11] turned out to be especially useful for practical applications. Starting from readily available *para-tert*-butyl-calix[4]-, -[6]- and -[8]arenes many derivatives have been prepared, and their properties have been determined [12]. Methods for nearly all desired functionalizations of the calixarene core are available by now, so that future work in this area will be directed mainly towards applications. A current area of interest is the use of substituted [1_n]phanes as artificial receptors and sensor molecules [12a].

However, on top of the inclusion phenomena, cyclophanes have more interesting properties to offer. In his first publication on cyclophanes, Cram [13] had already expressed his opinion about some peculiarities to be expected in cyclophanes, which he outlined as follows: a) electronic interaction between aromatic rings placed "face to face", b) the resulting influence on substitution reactions in the aromatic rings by transannular electronic effects, c) intramolecular charge transfer complexes and d) ring strain, steric strain and transannular strain. These effects have been studied on the parent compounds in detail [3].

Over the last 10 years more complex cyclophane systems have become available by modern synthetic methods. Such extended systems make use of the special electronic and steric effects of cyclophanes in physical organic and coordination chemistry in an advanced way. Recent developments in this area will be reviewed in paragraph 2 and 3.

Although the chirality of cyclophanes was recognized more than 50 years ago by Lüttringhaus and Gralheer in ansa compounds [14], chiral cyclophanes have been applied as auxiliaries and ligands in asymmetric synthesis only recently. Some examples of this exciting use of cyclophanes will be given.

Last but not least, rigid cyclophanes can be used to assemble functional groups in defined orientation in space [15]. Such orientation determines in many cases the stereochemistry of their reactions, so that topochemical reaction control [16], previously restricted to the solid state, was expanded into solution. The use of functionalized cyclophanes as stereochemical reaction control elements will be summarized in chapter 5 of this article.

2
Unusual Molecules with Cyclophane Substructures

2.1
Extended Orthogonal π-Systems

[2.2]Paracyclophane 1 is the smallest stable member of the [m,n]paracyclophane series (Scheme 1). The close proximity of the two benzene rings leads to a strong interaction of the π-systems. In the dibenzoannelated analogue 2 two pairs of orthogonal π-systems can be found. Both molecules have been used to synthesize extended π-systems, in which 2 allows a mutually orthogonal arrangement of their subunits.

Traditionally it has been believed that conjugated π-systems are essential for the efficient movement of electrons through molecules. However, it has been suggested that electron transfer through a σ-framework might also be effective

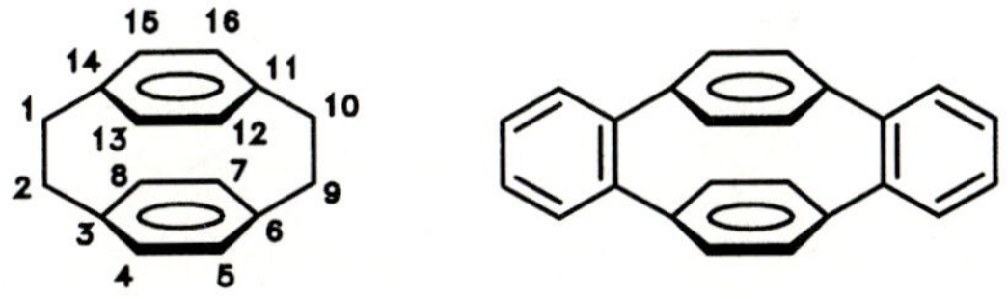

Scheme 1. [2.2]Paracyclophane (1) and dibenzo[2.2]paracyclophane-1,9-diene (2) [13]

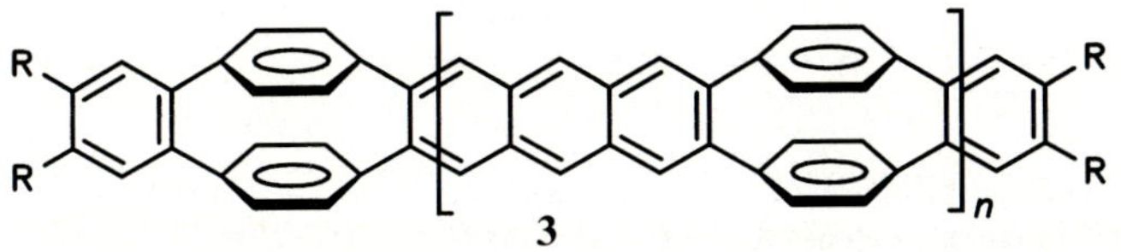

Scheme 2. An extended orthogonal π-system

[17]. The ideal organic model structure of aromatic rings that are not conjugated, but capable of taking up electrons, is represented by an array of bridge-annelated [2.2]paracyclophanes connected through common arene moieties.

Dibenzo[2.2]paracyclophane-1,9-diene (2), a unique molecule with two parallel sets of mutually orthogonal aromatic rings, was first synthesized in 1985 by Wong et al. [18]. The reported five-step synthesis gave the strained hydrocarbon in only 0.5% yield. Later a more versatile and economic approach to dibenzoannelated [2.2]paracyclophanedienes 11 including the parent compound 2 was developed by de Meijere et al. [19]. The dibromo[2.2]paracyclophane-1-ene (6) and the tetrabromo[2.2]paracyclophane-1,9-diene (7) were obtained from commercially available [2.2]paracyclophane (1) in a bromination-dehydrobromination reaction sequence. With compounds 6 and 7 available in large quantities, the palladium catalyzed coupling with alkenes (Heck reaction)

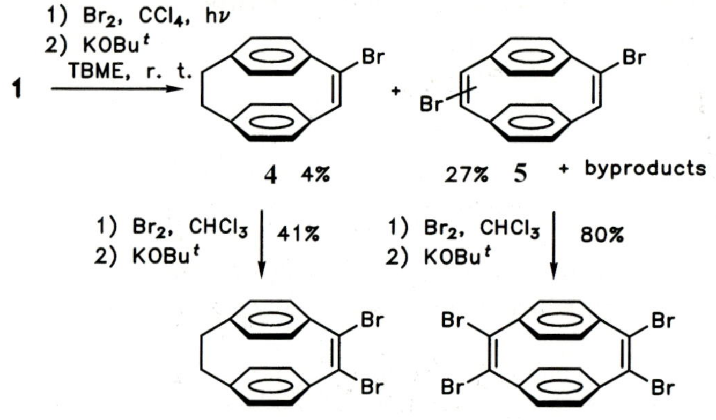

Scheme 3. Bromination – dehydrobromination sequence giving 1,2-dibromo[2.2]paracyclophane-1-ene and 1,2,9,10-tetrabromo[2.2]paracyclophane-1,9-diene [19a]

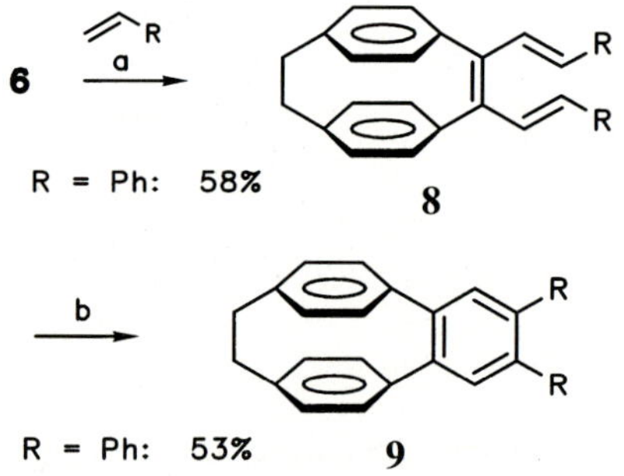

Scheme 4. Palladium-catalyzed coupling reactions yielding substituted benzo[2.2]paracyclophane-1-enes and dibenzo[2.2]paracyclophane-1,9-dienes. (a) Pd(OAc)$_2$, Bu$_4$NBr, K$_2$CO$_3$, DMF; (b) DDQ (or S), xylene, Δ. R=H, Me, SiMe$_3$, CO$_2$Me, C$_6$H$_5$, C$_6$H$_4$-4-CO$_2$Me, C$_6$H$_4$-4-F, C$_6$H$_4$-4-Bu^t, C$_6$H$_4$-4-Ph [19b]

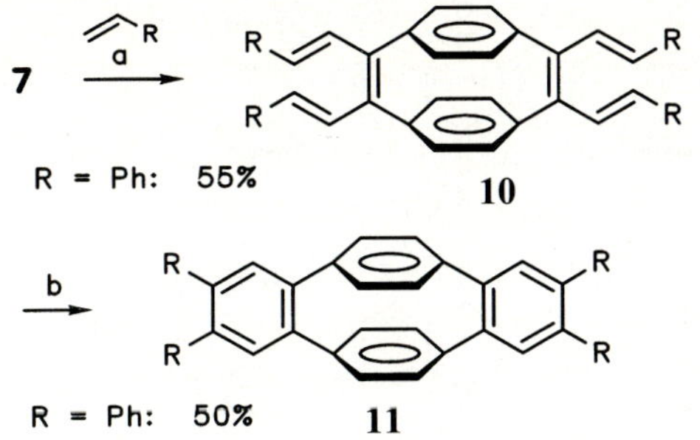

R = Ph: 55%   **10**

R = Ph: 50%   **11**

Scheme 4 (continued)

opened the way to [2.2]paracyclophanes with attached (*E,Z,E*)-1,3,5-hexatriene systems like **8** and **10**, that could be cyclized and dehydrogenated to yield substituted 1,2-benzo[2.2]paracyclophane-1-enes **9** and 1:2,9:10-dibenzo[2.2] paracyclophane-1,9-dienes **11** [19b]. Variously substituted alkenes can be employed in this procedure, however, consistently better results in the coupling step were obtained with alkenes containing electron-withdrawing groups.

A different synthetic approach to benzoannelated [2.2]paracyclophanes was realized by the Diels–Alder reaction of bis(methylene)[2.2]paracyclophane (**13**) and tetrakismethylene[2.2]paracyclophane (**15**) with *para*-benzoquinone [20]. The bis(exomethylene)paracyclophane derivatives were obtained from **6** and **7** via a copper-mediated coupling with methyl magnesium bromide, bromination and debromination. The dienes **13** and **15** react readily to the bis- and tris-phanes **16** and **17** when heated with *para*-benzoquinone in dichlorobenzene. However, both compounds are extremely poorly soluble in organic solvents and

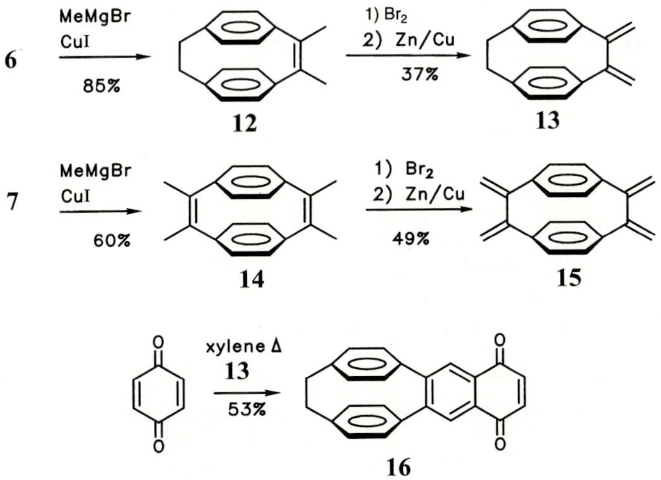

Scheme 5. Synthesis of benzoanellated [2.2]paracyclophanes via Diels–Alder reactions with *para*-benzoquinone [20]

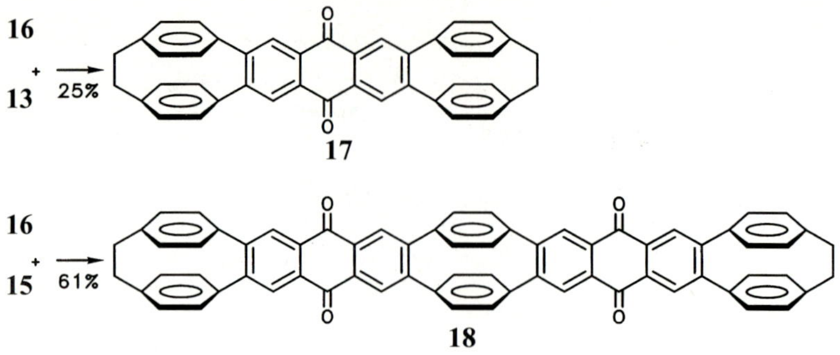

Scheme 5 (continued)

therefore not suitable for an investigation of charge distribution by electro-chemical reduction and EPR experiments.

It is well known from other rigid rod-type structures that their solubility can be significantly enhanced by substitution with flexible alkyl chains or *tert*-butyl groups [21]. A highly efficient route to soluble rigid systems with alternating orthogonal arene units was found by utilizing the mixture of 1,1,9,9- and 1,1,10,10-tetrabromo[2.2]paracyclophane (**19**), which is readily obtained from **1** in 100 g quantities by photobromination [22], as a synthetic equivalent for [2.2]paracyclophane-1,9-diyne. When the tetrabromides **19** were dehydrobro-minated with KoBu^t in the presence of excess furan, a mixture of the isomeric Diels–Alder products *anti-/syn-***20** arising from trapping of intermediately for-med [2.2]paracyclophaneynes, was obtained in moderate yield. The extrusion of acetylene from *anti-/syn-***20** in a one- or twofold retro-Diels–Alder reaction by

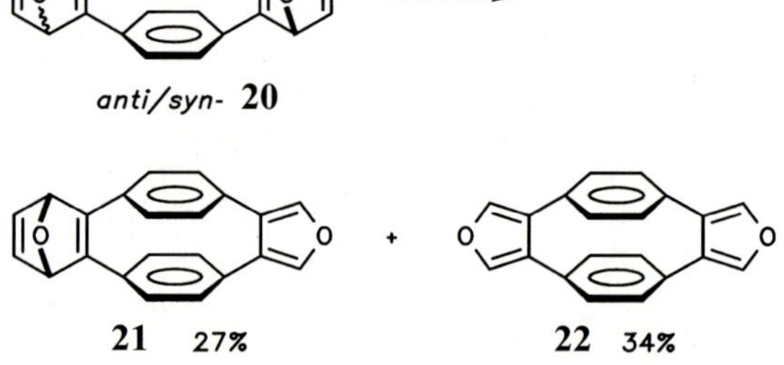

Scheme 6. Synthesis of furanoannelated [2.2]paracyclophanes; (a) KOBu^t, THF, r.t.; (b) 240 °C, 0.01 mbar [23]

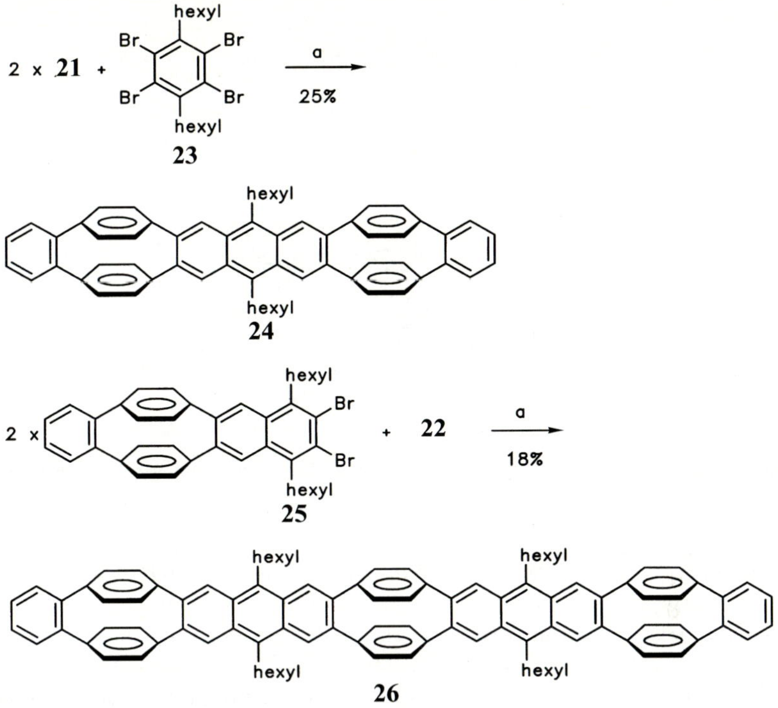

Scheme 7. Synthesis of extended alternatingly orthogonal π-systems via aryne intermediates; (a) 1) n-BuLi, toluene, $-90 \rightarrow 25\,°C$, 6 h. 2) TiCl$_4$, LiAlH$_4$, Et$_3$N, THF, 25 °C, 1 h [23]

heating under reduced pressure gave the furanoannelated phanes **21** and **22** [23]. With these compounds in hand it was possible to trap different arynes that were generated from the corresponding di- or tetrabromoarenes by treatment with butyllithium [24]. The primary addition products were reductively deoxygenated with low-valent titanium [25] to yield the desired hydrocarbons **24** and **26**. The incorporation of solubility enhancing substituents was achieved by using an appropiate aryne precursor, e.g., 3,6-dihexyl-1,2,4,5-tetrabromobenzene (**23**) [26]. This synthetic methodology does not only allow the preparation of extended alternatingly orthogonal π-systems, but also offers a very efficient access to different types of [2.2]paracyclophanes with various annelated ring systems.

Solutions of **24** and **26** exhibit an intense fluorescence. The quantum yield for **24** ($\Phi = 0.8$, excitation at 388 nm) is substantially increased in comparison with that of the parent anthracene chromophore [27]. Oxygen readily cycloadds across the 9,10-positions of **24**, when solutions are exposed to air and light [23].

The [2.2]paracyclophanediyne as generated from the precursor **19** can be trapped even more efficiently with dienes other than furan. In a single synthetic operation the twelve aromatic rings of octaphenyl-1:2,9:10-dibenzo[2.2]paracyclophane-1,9-diene (**28**) were assembled by treating the tetrabromides **19**

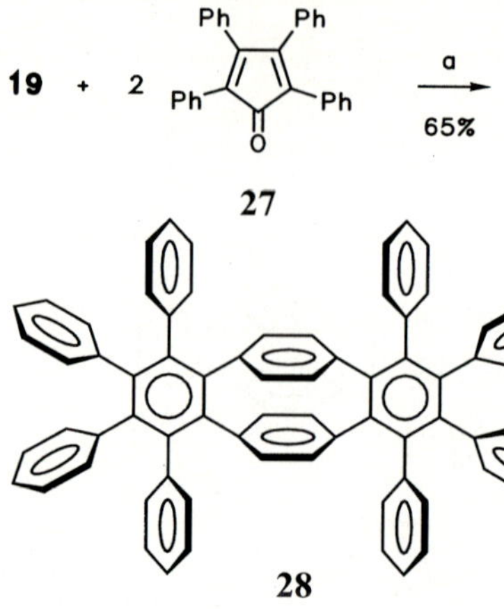

Scheme 8. Synthesis of octaphenyl-1:2,9:10-dibenzo[2.2]paracyclophane-1,9-diene; (a) KOBu^t, THF, r.t. [28]

with potassium *tert*-butoxide in the presence of tetraphenylcyclopentadienone 27 [28]. Compound 28 is apparently formed by spontaneous extrusion of carbon monoxide from the initial twofold Diels–Alder adduct.

The first cyclic array of three [2.2]paracyclophane systems linked by a common benzene ring was realized by Hopf et al. by trimerization of in situ generated [2.2]paracyclophane-1-yne 31 [29], for which the term cyclophyne was proposed. Deprotonation of 1-chloro[2.2]paracyclophane-1-ene (30) with *t*-butyllithium and subsequent elimination of LiCl by heating the THF solution under reflux gave the structurally intriguing hydrocarbon 29, which the authors named "trifoliaphane". Unfortunately, due to the rather low yield (8%), sufficient quantities of 29 to study the physical and chemical properties of this extremely interesting hydrocarbon have never been prepared.

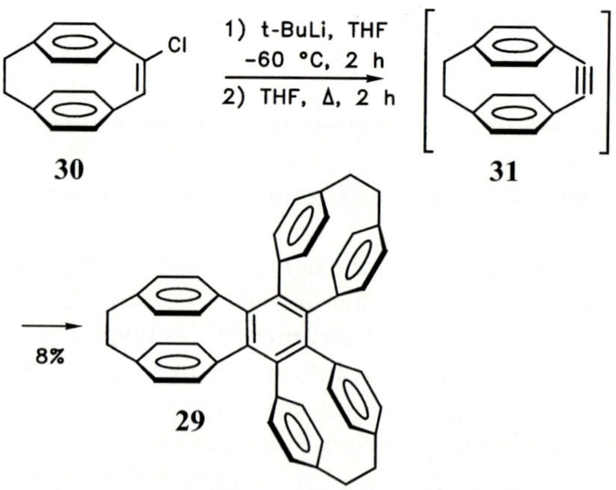

Scheme 9. Synthesis of trifoliaphane [29]

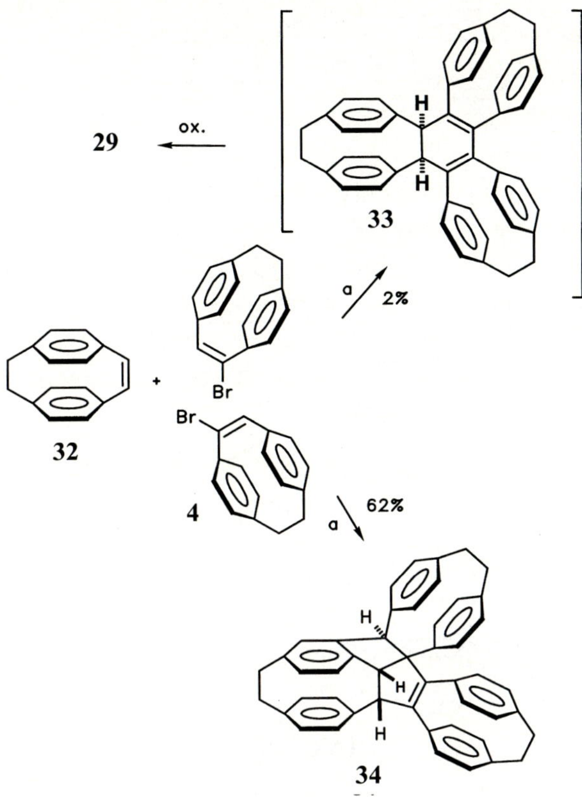

Scheme 10. Palladium-catalyzed assembly of [2.2]paracyclophane-1-ene and its 1-bromo derivative: attempted synthesis of dihydrotrifoliaphane; (a) Pd(OAc)$_2$, K$_2$CO$_3$, Bu$_4$NBr, DMF, 85 °C, 4 h [30]

In view of the successful palladium-catalyzed [2 + 2 + 2] assembly of a cyclohexa-1,3-diene derivative from one molecule of norbornene and two molecules of β-bromostyrene [30 a], a rationally designed synthesis of dihydrotrifoliaphane **33** as a precursor to **29** appeared to be plausible. However, when a 1 : 2 mixture of [2.2]paracyclophane-1-ene (**32**) and its 1-bromo derivative **4** was subjected to the reported conditions, the novel hydrocarbon **34** with a different array of three [2.2]paracyclophane units was obtained in 62 % yield [30 b]. Trifoliaphane **29**, arising from oxidation of the primarily formed dihydro derivative **33**, was obtained as a minor by-product (2 % yield) in this coupling. The constitution of **34** was established by an X-ray crystal structure analysis.

Such a rigorous proof of structure is still lacking for an equally interesting hydrocarbon with the molecular formula C$_{60}$H$_{38}$, which was isolated in 52 % yield after subjecting a 1 : 2 mixture of 1,2-benzo[2.2]paracyclophane-1,9-diene (**35**) and its bromo derivative **36** to the same conditions as applied to **32 + 4**. The NMR (^{1}H and ^{13}C) spectroscopic data, however, indicate a close structural analogy to **34** and suggest the constitution **37** for this product [30 b]. Whether

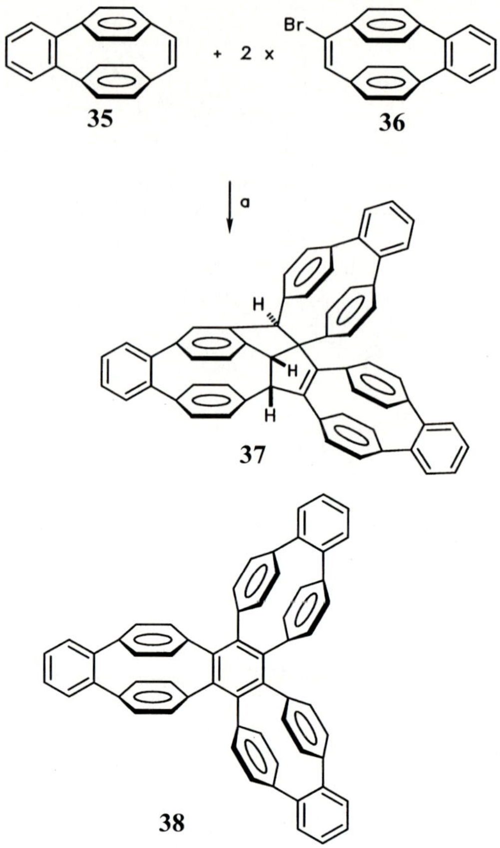

Scheme 11. Attempted synthesis of tribenzotrifoliaphane; (a) Pd(OAc)$_2$, K$_2$CO$_3$, Bu$_4$NBr, DMF, 85 °C, 17 h [30 b]

tribenzotrifoliaphane **38** or its dihydro precursor was formed as a byproduct, remains to be investigated.

The multiple reductions of a number of substituted dibenzo[2.2]paracyclophanedienes **11** were investigated by cyclic voltammetry and the reduction products characterized by their EPR (for the paramagnetic radical anions) as well as NMR spectra (for the diamagnetic oligoanions) [19]. According to these studies, the bridge-annelated π-subsystems are sufficiently isolated from each other by the central [2.2]paracyclophane unit to take up electrons one by one in consecutive steps. However, the reduction potentials as disclosed in the cyclic voltammograms are in accord with a substantial mutual electronic interaction between the formally isolated arene units transmitted by the orthogonal central π-system.

The cyclic voltammetric investigation of the bis[2.2]paracyclophane **24** showed that up to four electrons can be introduced into this molecule. The stepwise reduction of **24** with alkali metals was followed by EPR spectroscopy; this disclosed that the first electron is introduced into the central anthracene unit, as one would have expected. The anthracene is then further reduced to the dianion, but the third electron is located in one of the paracyclophane units. The coupling of the unpaired electron in this trianion with all sixteen protons of both paracyclophane units indicates a rapid exchange of the odd electron between the two lateral paracyclophane moities on the hyperfine coupling time scale (ca. 10^7 s^{-1}) [31]. The tetraanion **24**$^{4-}$ could only be detected as a fourth reduction step in the cyclic voltammogram, but not characterized by an EPR spectrum. Its nature can therefore only be speculated about. Most probably, though, the fourth electron is placed in the second paracyclophane unit, whether being paired or unpaired with the third one is unknown.

The cyclic voltammogram of octaphenyldibenzo[2.2]paracyclophanediene **28** showed three reversible two electron reduction steps, the last one of which

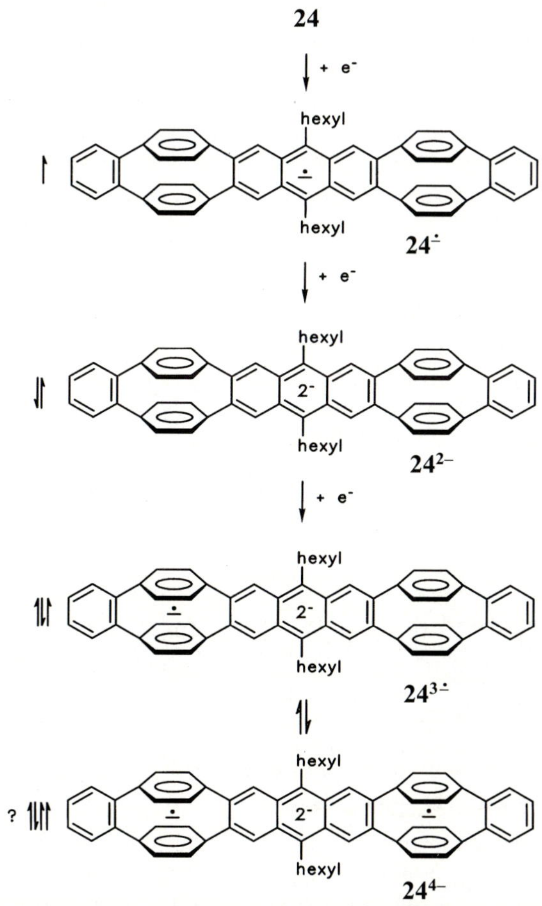

Scheme 12. Stepwise multiple reduction of a bis[2.2]paracyclophane [31]

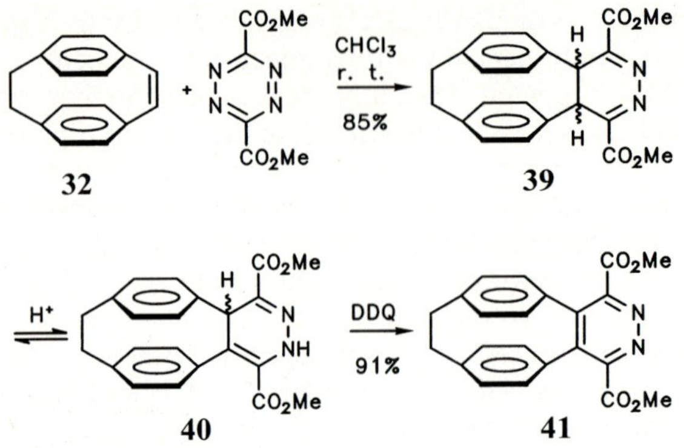

CO₂Me / CHCl₃ r. t. 85% — **32** → **39**

H⁺ ⇌ **40** — DDQ 91% → **41**

Scheme 13. Diels‑Alder reaction of [2.2]Paracyclophane‑1‑ene (**32**) [34 a]

leads to the corresponding hexaanion [28]. The Coulomb interaction between the six excess charges is minimized in this molecule by distribution throughout its unusual skeleton.

Most of the synthetic approaches towards benzoannelated [2.2]paracyclophanes use the Diels–Alder reaction to build up the annelated rings [18, 22, 29]. [2.2]Paracyclophane‑1‑ene (**32**) and [2.2]paracyclophane‑1,9‑diene (**42**) can be considered as the simplest dienophiles for this purpose. Since the first syntheses of these compounds in 1958 by Cram et al. [32], various attempts have been made to get them to react with dienes in terms of Diels–Alder reactions. However, [2 + 4] cycloadditions were never observed [33], and could not be facilitated either by the application of high pressure, or the presence of Lewis acid catalysts; Diels–Alder adducts were not even obtained with dienes such as tetrachloro‑thiophene dioxide, known for its high reactivity in [2 + 4] cycloadditions with an inverse electron demand. All the more surprising was the observation that monoene **32** reacts smoothly with dimethyl 1,2,4,5‑tetrazine‑3,6‑dicarboxylate at room temperature leading to a dihydropyridazine‑annelated paracyclophane **39** in high yield [34 a]. As reported for other tetrazine Diels–Alder reactions

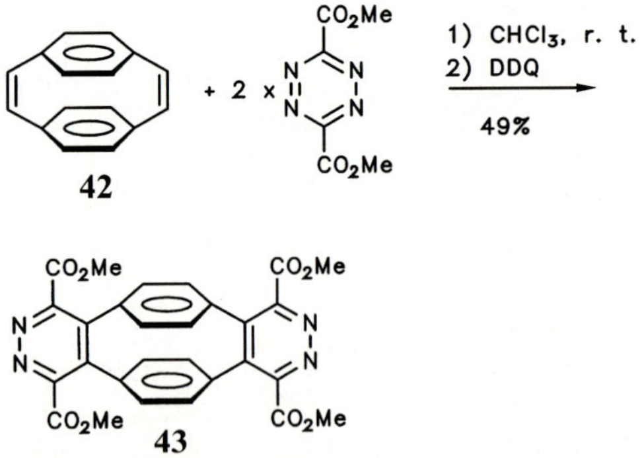

42 + 2 × [tetrazine] → 1) CHCl₃, r. t. 2) DDQ 49% → **43**

Scheme 14. Diels–Alder reaction of [2.2]paracyclophane‑1,9‑diene (**42**)

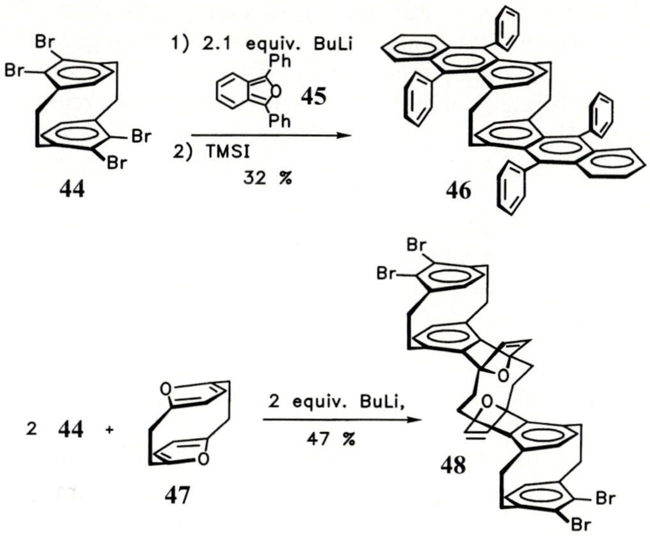

Scheme 15. Synthesis of *anti*-tetraphenyl[2.2]anthracenophane and a stair-like arrangement of cyclophanes [37]

[35a], the primary adduct of **32** and the tetrazinedicarboxylate looses N_2 instantaneously to yield **39** in which a [1,3]*H* shift occurs rapidly to give **40**. The latter is easily dehydrogenated by treatment with DDQ to give the pyridazinoannelated [2.2]paracyclophane **41** [34] [2.2]Paracyclophane-1,9-diene (**42**) smoothly reacts with two equivalents of the tetrazinecarboxylate in the same way to give the biscycloadduct which yields the bispyridazinoannelated [2.2]paracyclophanediene **43** after treatment with DDQ [34a].

Although these reactions are quite impressive, kinetic measurements have shown that **32** and **42** are less reactive dienophiles than other comparably strained cyclic olefins [35b, 35c]. It can be concluded that the inherent high reactivity of the strained double bonds in **32** and **42** is partially compensated by the steric hindrance which the *ortho*-hydrogen atoms on the phane-arene units exert on the approach of the cycloaddend.

4,5,12,13-Tetrabromo[2.2]paracyclophane (**44**) was first used by Cram as a bis-aryne equivalent. The in situ-trapping of the corresponding bis-aryne with furan gave two isomeric cycloadducts (**49**), that could be reductively deoxygenated to *anti*-[2.2]naphthophane [36]. The aryne intermediate is trapped more efficiently with 2,5-diphenylisobenzofurane (**45**) and the deoxygenation with trimethylsilyl iodide (TMSI) yields the highly fluorescent *anti*-tetraphenyl[2.2]anthracenophane **46** [37]. By treatment of **44** with 1.0 equiv. of *n*-BuLi, only one of the 1,2-dibromobenzene moieties reacts, whereas the second one remains unchanged. Thus, when two equiv. **44** were treated with 2.0 equiv. of *n*-BuLi in the presence of [2.2]furanophane (**47**) as the trapping agent, only a single product was obtained. The constitution of the product was derived from spectroscopic data to be **48**, in which three [2.2]paracyclophane units are connected in a stair-like fashion. Unfortunately, all attempts to remove the shielded oxygen atoms from **48** in order to aromatize the two central rings have failed [37].

In addition, [2.2]paracyclophane was transformed into reactive bis-dienes. By reaction of the furan cycloadducts **49** with tetraphenylcyclopentadienone and thermal retro-Diels–Alder reaction [2.2](4,7)isobenzofuranophane **51** was obtained [38]. The highly reactive molecule was trapped in situ with *para*-benzoquinone. The more stable tetraphenylisobenzofuranophane **57** was synthesized via the classical procedure to [2.2]paracyclophanes of Hopf [39]: the reaction of 1,2,4,5-hexatetraene **53** with dibenzoylacetylene **54** gave 4,5,12,13-tetrabenzoyl[2.2]paracyclophane (**56**), that was reduced and cyclized to the target molecule **57** [38].

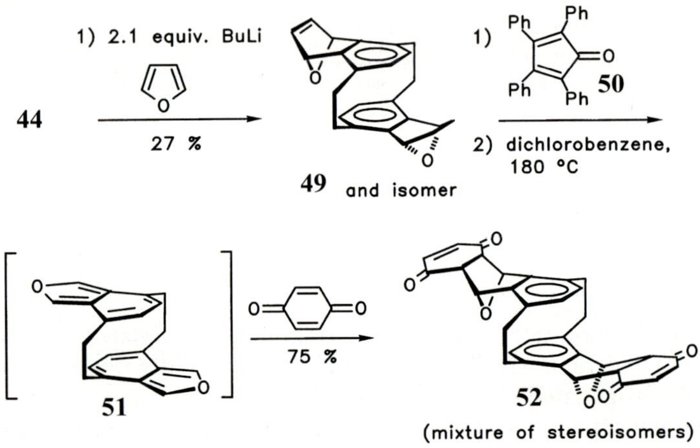

Scheme 16. Synthesis and reaction of [2.2](4,7)isobenzofuranophane; a) 2.1 equiv. *n*-BuLi, THF, –78 °C → room temperature [38]

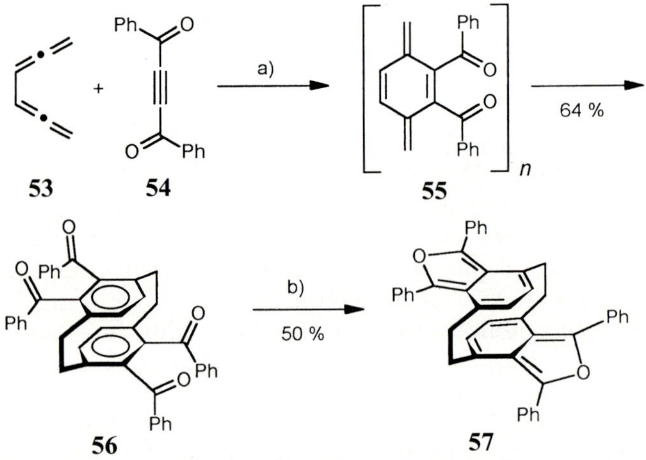

Scheme 17. Synthesis of *anti*-1,1′,3,3′-tetraphenyl[2.2](4,7)isobenzofuranophane, a) Et$_2$O/C$_6$H$_6$ (3:1), 70 °C; b) 1. LiAlH$_4$, 2. Ac$_2$O [38]

2.2
Rigid Ladder Polymers with [2.2]Paracyclophane Subunits

When extending the dimensions of the rigid-rod molecules with mutually orthogonal aromatic subunits like **26** beyond the system with three [2.2]paracyclophane subunits, the methods of purification and characterization for ordinary organic compounds soon become inadequate. Therefore a polymerization approach was applied to prepare even longer rigid-rod molecules with similar subunits as in **26**. As has been demonstrated for other combinations of appropriate diene and dienophile monomers, a completely regular connection of ladder polymers can be achieved by a repetitive Diels–Alder reaction [40]. Considering 1,2,9,10-tetramethylene[2.2]paracyclophane (**15**) and the bis(oxanorbornadiene) derivative *anti-/syn*-**20** as bifunctional diene and dienophile monomers for such repetitive Diels–Alder reactions, one simply had to apply appropriate bifunctional dienophiles and dienes, respectively. Indeed, the reaction of

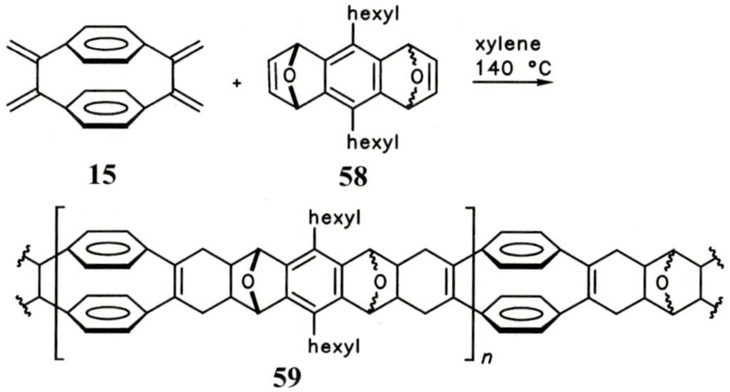

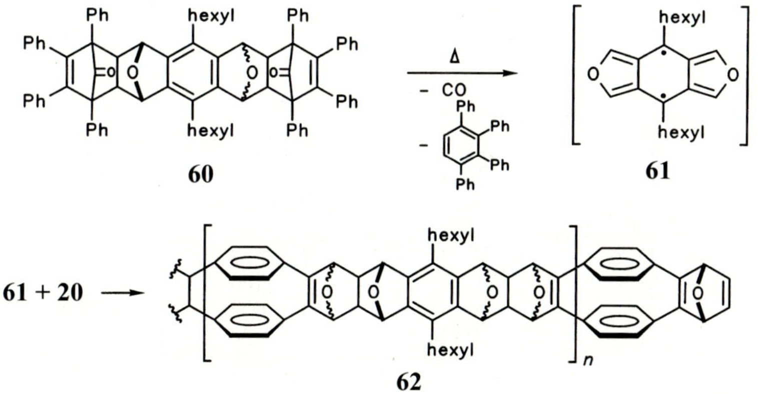

Scheme 18. Synthesis of rigid ladder oligomers with [2.2]paracyclophane units by repetitive Diels–Alder reactions [43]

15 with the known bis(dienophile) *syn-/anti*-**58** [41] gave an oligomer **59** (n = 16) with an almost monomodal molecular weight distribution (P_n = 10230, P_w = 31030), indicating that only the [4+2] cycloaddition process contributes to the product formation. The reaction of *anti-/syn*-**20** with the known highly reactive bisdiene **61**, generated from the stable precursor **62** by a retro Diels–Alder reaction, [40–42] yielded an oligomer **62** (n = 7) with a lower average molecular weight (P_n = 4400, P_w = 7150). The regular structure of the oligomeric rigid rods was confirmed by their ^{1}H and ^{13}C NMR spectra in comparison with those of model compounds containing the same repeating units [43]. Unfortunately, though, the attempted reductive removal of all oxygen bridges did not succeed, due to the very poor solubility of the partly reduced molecules.

2.3
Cyclophynes and Other Large Cyclophanes

The intermolecular dimerization of *ortho*-quinodimethanes to [2.2]paracyclophanes is an established synthetic procedure [39]. However, with extended exocyclic double bond systems, *para-* or *ortho*-quinodimethanes yield cyclophanes with alkyne groups in their bridges, such as **69**, as products of an intermolecular cyclization [44]. Such *para*-quinodimethanes with cumulated double bonds are accessible via 1,x elimination of bis-propargyl or benzyl bromides with Bu_3SnSi-Me_3/CsF [45]. Carbo- and heterocyclic unsaturated cyclophanes have been pre-

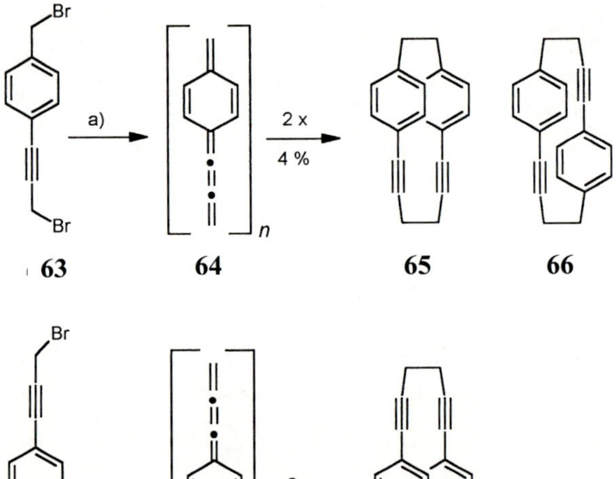

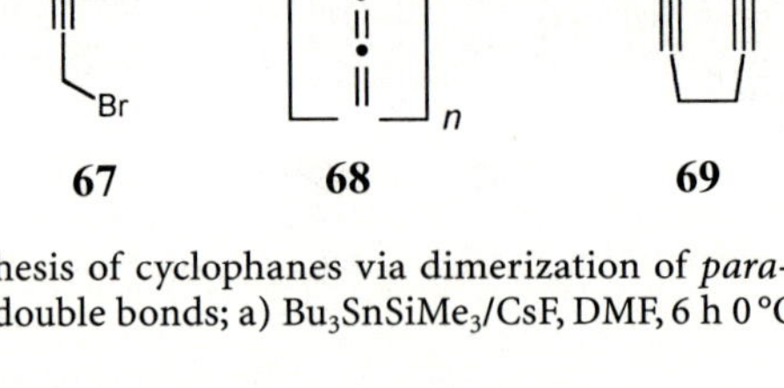

Scheme 19. Synthesis of cyclophanes via dimerization of *para*-quinodimethanes with cumulated exocyclic double bonds; a) $Bu_3SnSiMe_3$/CsF, DMF, 6 h 0 °C, 12 h 20 °C [44]

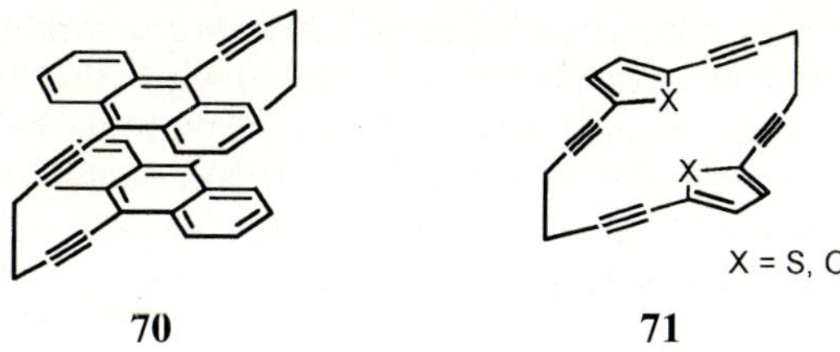

Scheme 19 (continued)

pared by this route [46]. Most of the obtained compounds are highly reactive and sensitive towards oxygen.

Only acetylene units bridge the arene rings in [8.8]paracyclophaneoctayne (**75**). The synthesis of a partially cobalt-stabilized precursor was reported by Haley et al. [47]. Palladium-catalyzed coupling of 1,4-diiodobenzene with excess triisopropylsilylbutadiyne gave a tetrayne, which reacted with 2 equiv. octacarbonyldicobalt to provide tetracobalt complex **73**. For steric reasons only the internal alkyne groups were coordinated. Ligand exchange, [47b] desilylation with fluoride ions and Eglington [48] coupling yielded the cyclophane as octacobalt complex. Iodine-promoted decomplexation of **74** to the free cyclophane **75** was not successful.

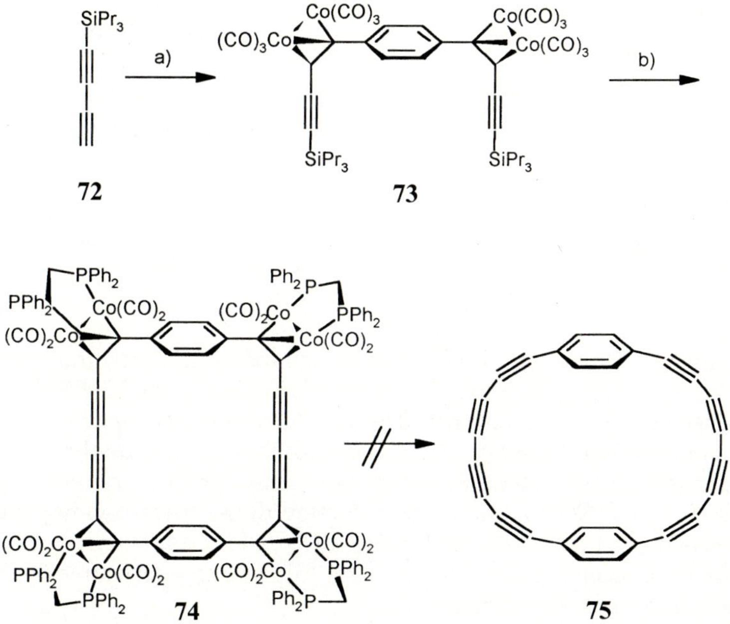

Scheme 20. Synthesis of a cobalt-stabilized [8.8]paracyclophaneoctayne; a) 1. 1,4-diiodobenzene, Pd(PPh₃)₂Cl₂, CuI, Et₃N, 73%, 2. Co₂(CO)₈, Et₂O, 66%; b) 1. dppm, toluene, 85%, 2. Bu₄NF, THF, 95%, 3. Cu(OAc)₂, py, 47% [47]

Non-benzoid phanes have always played an important role in cyclophane chemistry and many structures have been prepared with other Hückel aromatic rings such as azulene or tropone [49]. A cyclophane with two bridged annulene units was recently synthesized by Mitchell [50, 51]. Dimethyldihydropyrene (**76**), an excellent NMR probe first introduced by Boekelheide [52], was converted into the dialcohol **77** in three steps. Reaction with adipoyl chloride afforded the large [10.10]cyclophane **78**. Unfortunately the conversion of the dialcohol into the corresponding dibromide **79**, an obvious precursor to the interesting phane **80**, has failed so far.

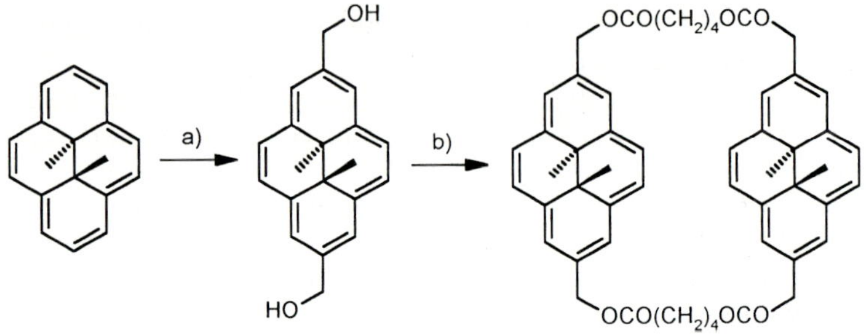

76   **77**   **78**

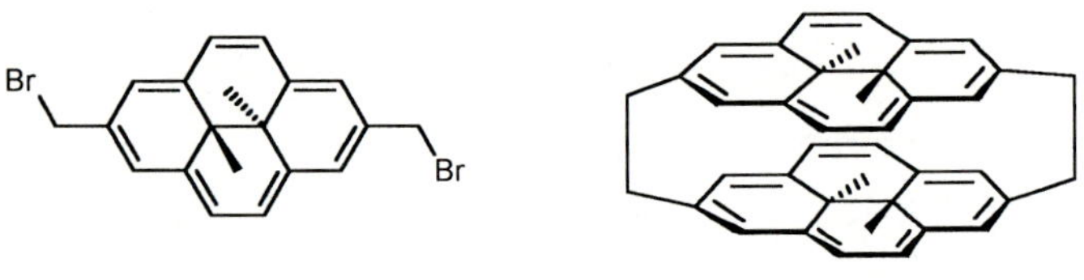

79   **80**

Scheme 21. Synthesis of a cyclophane containing two bridged annulene units; a) 1. PDC, DMF, 75%, 2. $(CH_3)_2S=CH_2$, 3. $NaBH_4$ 73%; b) adipoyl chloride, 11% [50]

Seventeen years after Boekelheide's synthesis of the first superphane $[2_6](1,2,3,4,5,6)$cyclophane [52], the higher homolog $[3_6](1,2,3,4,5,6)$cyclophane **84** was prepared [53]. The sixth and last bridge was closed by an intramolecular aldol condensation. To obtain the unsubstituted hydrocarbon the enone **81** was hydrogenated to the ketone **82**, which was resistant to most reducing agents. Only SmI_2 in 1 M KOH allowed the reduction to the corresponding alcohol **83**, which was further reduced with $LiAlH_4/AlCl_3$ yield to **84**. Theory [54] predicts a possible photoisomerization of **84** to the [6]prismane derivative **85**. However, this has not been observed jet.

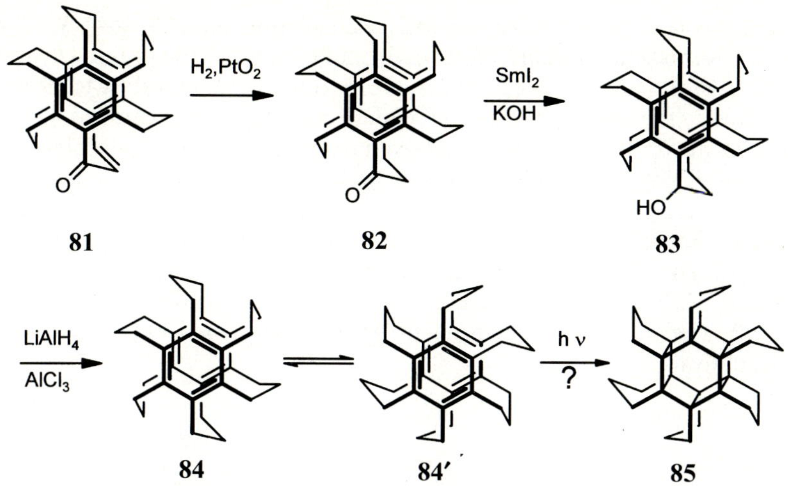

Scheme 22. Synthesis of the superphane [3$_6$](1,2,3,4,5,6)cyclophane [53]

2.4
Belt- and Cage-shaped Cyclophanes

Many calixarenes and polar cyclophanes with defined cavities have been prepared for the inclusion of guests [6, 11, 12], whereas the number of known carbon-rich, non-polar cyclophanes with a large cavity is much smaller [55]. Some new compounds of this class have been recently reported.

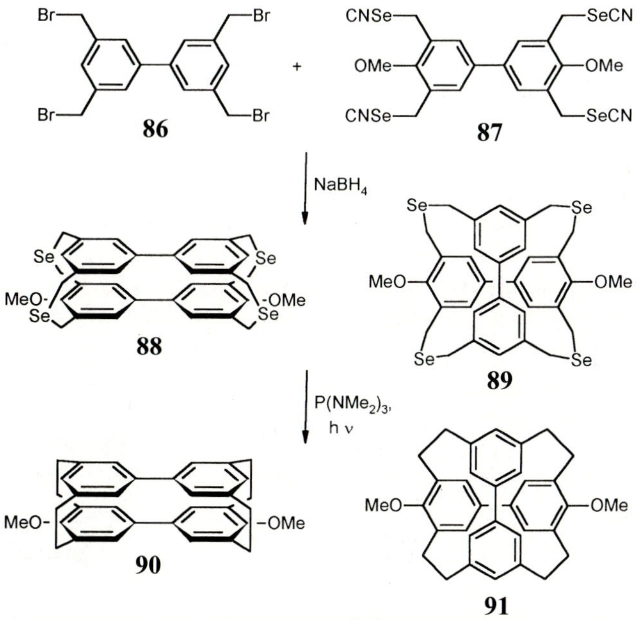

Scheme 23. Cyclophane synthesis via tetraselenabiphenylophanes [56]

The reaction of tetrabromide **86** with tetraselenocyanate **87** gave a 1:3 mixture of the tetraselenabiphenylophanes **88** and **89** [56]. In the previously reported reaction of the sulfur analog of **87** the compound corresponding to **89** was the only product, the sulfur bridges of which could not be removed [57]. The selenide atoms of **88** and **89** were removed in 40-50% yield by photolysis in the presence of P(NMe)₃ to give the hydrocarbons **90** and **91**.

New isomers of [2₄]paracyclophanetetraene and higher cyclic oligomers have been obtained by Oda et al. [58] by way of McMurry coupling of 4,4′-diformyl-(*Z*)-stilbene (**92**). When irradiated with a high-pressure Hg lamp in benzene the (*E,Z,E,Z*)-configurated compound **94** is isomerized to the (*Z,Z,Z,Z*) isomer **93**. The molecular structure of **94** was confirmed by X-ray crystallographic analysis.

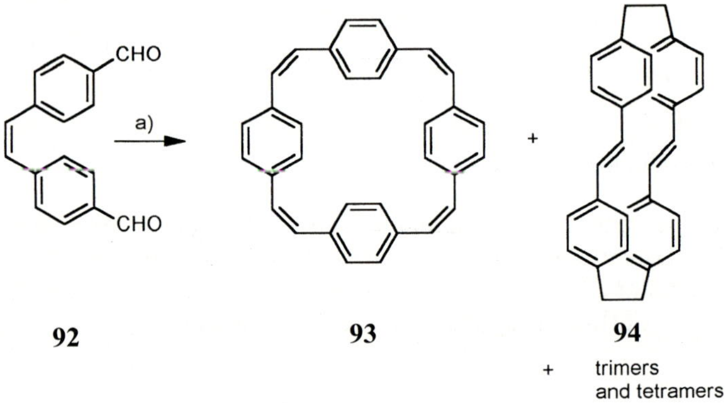

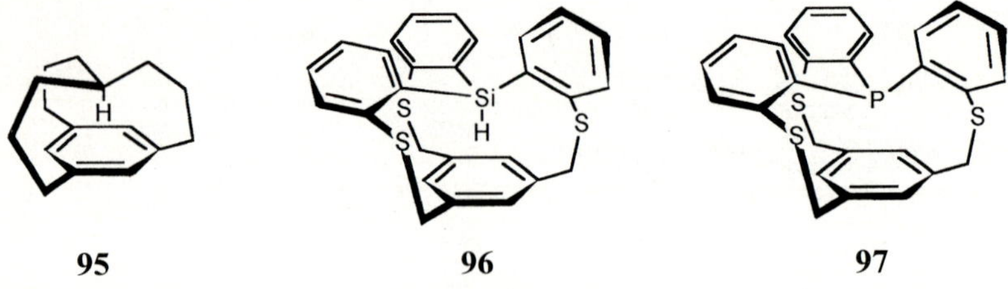

Scheme 24. Synthesis of [2₄]paracyclophanetetraene and higher cyclic oligomers; a) TiCl₄ or TiCl₃(DME)₁.₅, Zn(Cu), DME, **93** 15 – 21%, **94** 10 – 15% [58]

To study the interaction of functional groups in close proximity *in*-cyclophanes [59] are good model systems. The C-H group in **95** directs to the center of the aromatic ring leading to a proton NMR shift of $\delta = -4.03$. Heteroatoms, such as silicon or phosphorous [60], were also incorporated to study their interaction with the π-system.

Scheme 25. Examples of *in*-cyclophanes [59,60]

A molecular ribbon of nitrogen-bridged *meta*-cyclophanes (**98**) was obtained by Vögtle et al. [61]. The repetitive coupling procedure allowed the synthesis of structures in the nanometer range with up to 7 aromatic units (n = 5). Some of these fourfold functionalized ribbons were cyclized to molecular belts, such as **100, 102** or **103** [62]. In the case of **100** careful analysis of the reaction mixture revealed the formation of very large cycles **101** with up to 40 benzene rings. The higher oligomers could not be isolated in pure form.

Scheme 26. A molecular ribbon of *meta*-cyclophanes [61]

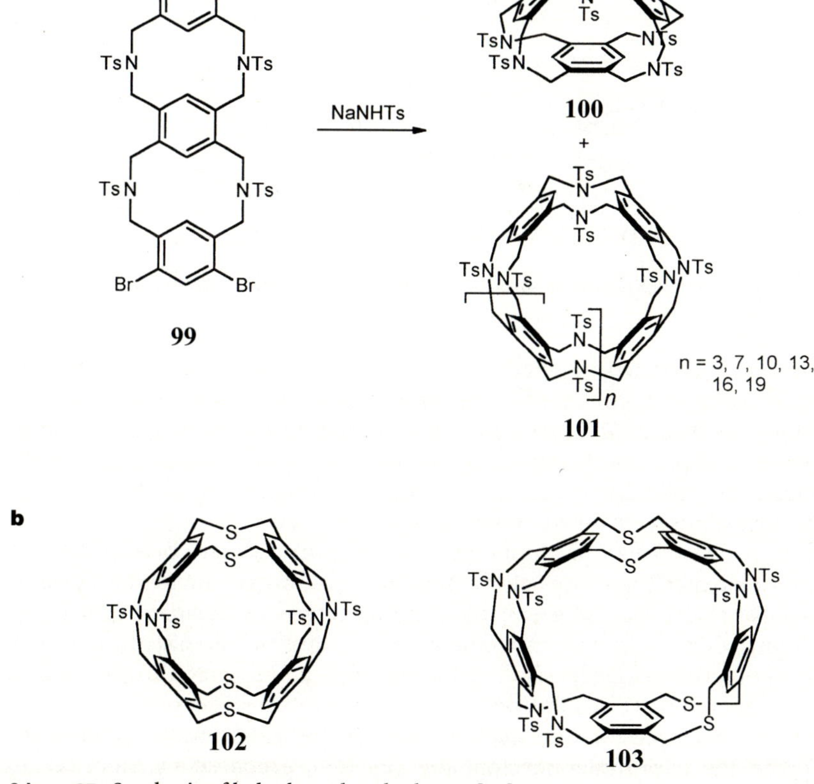

Scheme 27. Synthesis of belt-shaped cyclophanes [62]

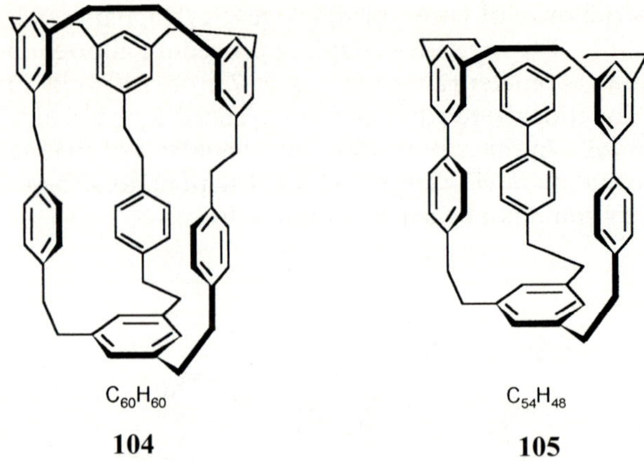

C$_{60}$H$_{60}$ C$_{54}$H$_{48}$

104   **105**

Scheme 28. Cyclophane cage compounds with spherical cavities [64]

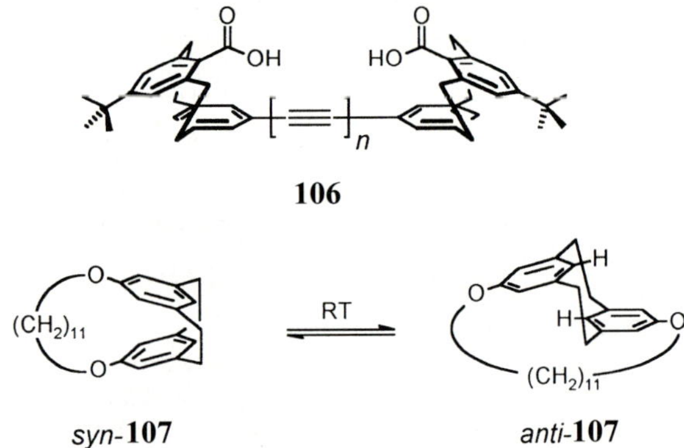

106

syn-**107**   *anti*-**107**

Scheme 29. [2.2]Metacyclophanes: as molecular tweezers and with mobile conformation [65–67]

Cage hydrocarbons with a sperical cavity, such as **104** or **105**, are fullerene models [63]. They were prepared by intramolecular sulfide cyclization of suitable hexabromo precursors, oxidation and sulfone pyrolysis [64]. The overall yields were 5% resp. 3.5%. Compound **104**, with the constitution C$_{60}$H$_{60}$, might be envisaged as a fullerene precursor. In addition, the compound shows a high affinity to silver(I) ions in solvent extraction experiments.

The linkage of two *meta*-cyclophanes by alkyne groups gave bis-cyclophane **106**. Extraction experiments revealed a moderate binding of purine- and pyrimidine by these molecular tweezers [65]. The preferred conformation of [2.2]-metacyclophanes is the *anti*-arrangement [66]. Bodwell et al. [67] showed now that a 13-membered bridge levels the energy of *syn*- and *anti*-conformer of **107**. They slowly interconvert in solution at room temperature.

Two further examples of large cyclophanes are the "Kuratowski-cyclophane" **108** [68] and the corannulenocyclophane **109** [69], prepared by Siegel et al. Compound **108** is the first achiral molecule with a K$_{3,3}$-topology.

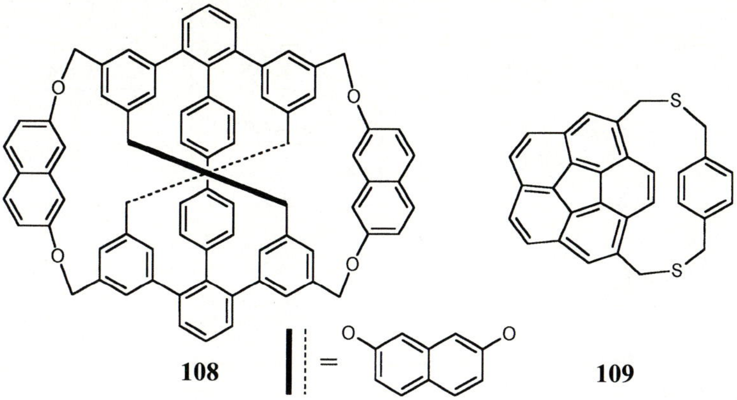

Scheme 30. "Kuratowski-cyclophane" and corannulenocyclophane [68, 69]

2.5
Short-Bridge and Strained Cyclophanes

For a long time, [2.2]para- and [2.2]metacyclophanes with two C_2 bridges were the ones with the shortest bridges in any known [*m.n*]*para-* and [*m.n*]metaparacyclophanes. Apparently, however, [2.1]paracyclo-(3,5)-naphthalinophane-1-ene (**111**) with one C_2 and one C_1 bridge is not extremely strained, as it was obtained as a stable coumpound in 54 % yield upon treating bis([2.2]paracyclophane-1,9-dienyl)methanol **110** with zinc chloride in dichloromethane. The unusual skeleton of **111** was established by an X-ray crystal structure analysis [70].

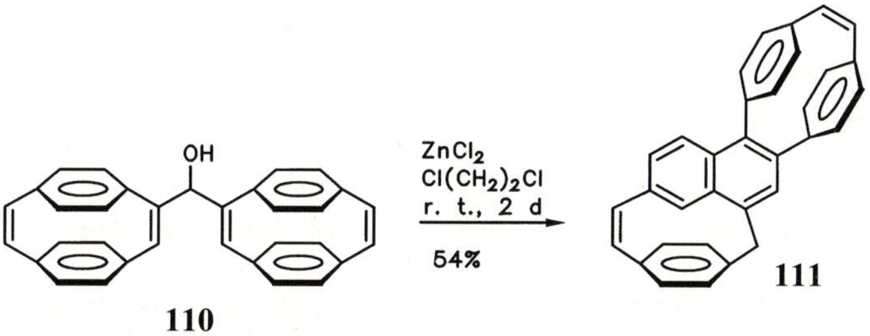

Scheme 31. A rearrangement yielding [2.1]paracyclo-(3,5)-naphthalinophane-1-ene [70]

Tsuji et al. were the first to report spectroscopic evidence for the photochemical generation of the bis(methoxycarbonyl)[1.1]paracyclophane (**113a**) from the bis(methoxycarbonyl)[1.1]Dewar-benzenophane **112a** and for its existence at low temperature [71]. The precursor **112a** was prepared along an established route from known compounds in eleven steps. The NMR spectrum and the UV absorption of **113a** persist at –20 °C, but fairly rapidly decay at room temperature, indicating that the stability of **113a** is marginal at ambient temperature. Upon irradiating a sample with longer wavelength light, the ^{1}H-NMR

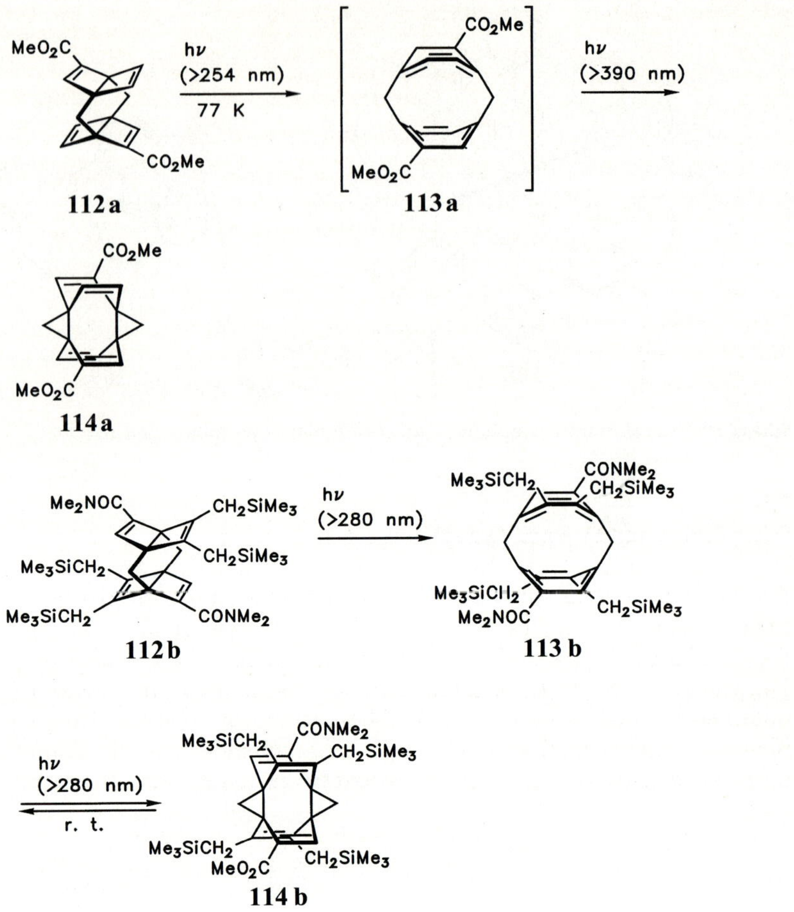

Scheme 32. Syntheses of [1.1]cyclophanes [71, 72]

spectrum of a new compound of probable structure **114a** appears at the expense of that of **113a**.

More recently, the same group was able to prepare 3,10-bis(dimethylamino-carbonyl)-6,7,13,14-tetrakis(trimethylsilylmethyl)[1.1]paracyclophane (**113b**) as a stable compound [72]. The precursor Dewar benzene **112b** was obtained essentially along the same route as **112a**. Irradiation of **112b** led to formation of a mixture of **113b** and the biscyclopropane derivative **114b**, the latter being formed by secondary photorearrangement of **113b**. The colorless crystals of **114b** are air-stable and were fully characterized by an X-ray structure analysis. Above room temperature, **114b** thermally reverts to **114b**, obtained as reddish-brown needles which are air-sensitive, but completely stable at 50°C.

One of the intriguing aspects of aromaticity is the question of how much bending a benzene ring can tolerate without losing its aromatic character. Distortions form the benzene structure can be achieved by connecting the *para* positions by a short oligomethylene bridge. Such [n]paracyclophanes have been studied in detail over the last decades [73]. The thermal stability of the compounds decreases with decreasing chain length. Thus, [7]- and [6]paracyclophane are stable compounds, [5]paracyclophane can be observed at low temperatures, [74] whereas [4]paracylophane has only been generated as a transient species via matrix isolation [75]. To further stabilize a [5]paracyclophane Bickelhaupt et al. synthesized [5](1,4)naphthalenophane [76]. The photoisomerization of the Dewar-benzene isomer **114** as starting material gave the paracyclophane **115** in 35% yield.

Polycyclic aromatic hydrocarbons can be bent by bridging in a similar fashion. Two recent examples of such [n]cyclophanes are azulenophane **116** [77] and 1,8-dioxa[8](2,7)pyrenophane **117** [78]. X-ray structure analysis revealed an angle of 3.7° between the azulene ring planes, whereas the pyrene moiety is bent by nearly 90°.

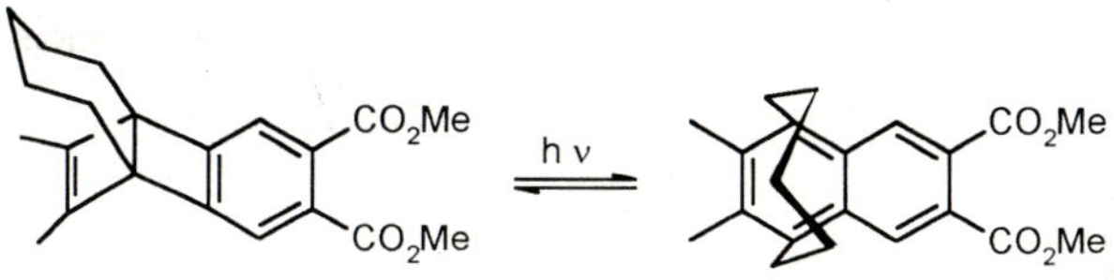

Scheme 33. Synthesis of [5](1,4)naphthalenophane from its Dewar-benzene isomer [76]

Scheme 34. [n]Cyclophanes with distorted polycyclic aromatic hydrocarbons [77, 78]

3
Cylophanes as Ligands for Metal-Ion Coordination

3.1
Platinum and Zirconium Complexes of [2.2]Paracyclophane-1-yne

The highly strained [2.2]paracyclophane-1-yne (**31**) had been proposed for the first time by Hopf et al. [29] as a reactive intermediate formed in the dehydrochlorination of 1-chloro[2.2]paracyclophane-1-ene. The same type of intermediate with a formal triple bond in one of the C_2 bridges should occur in the stepwise double dehydrobromination of **19**.

Other extremely strained cycloalkynes such as cyclohexyne and even cyclopentyne have been stabilized by the complexation with 4d and 5d transition

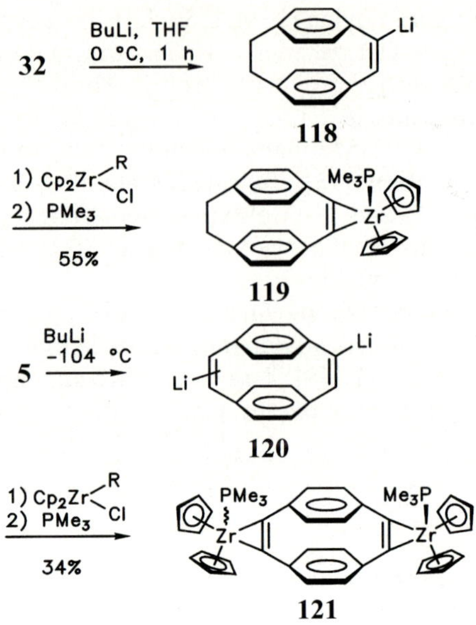

Scheme 35. [2.2]Paracyclophane-1-yne and [2.2]paracyclophane-1,9-diyne zirconium complexes [81]

metals [79]. As the triple bond in [2.2]paracyclophane-1-yne can be envisaged to be about as much bent as that in cyclohexyne, it appeared to be promising to attempt a complexation of this alkyne as well.

Treatment of [2.2]paracyclophane-1-ene (**32**) with BuLi at 0 °C yields the monolithio derivative **118** quantitatively. The reaction of **118** with $Cp_2ZrClMe$ or $Cp_2ZrClPr^i$, according to a procedure of Buchwald et al. [80] and subsequent addition of PMe_3 to stabilize the 16e intermediate gave the air-sensitive alkyne complex **119** in a yield of 55% [81]. The bis(alkyne) complex **121** was obtained in close analogy by the reaction of dilithio[2.2]paracyclophanediene **120**, generated from the corresponding 1,9(10)-dibromides **5** by bromide–lithium exchange at low temperature, with $Cp_2ZrClMe$ and excess PMe_3. The dinuclear complex **121** was formed as a mixture of diastereomeric *syn*- and *anti*-isomers. Both complexes readily undergo insertion reactions with 2-butyne to form zirconacyclopentadienes.

Stable platinum(0) complexes of a number of strained cycloalkyne and even aryne intermediates have also been reported [79]. Upon generating the [2.2]paracyclophane-1-yne by debromination of 1,2-dibromo[2.2]para-cyclophane-1-ene (**6**) with sodium amalgam in the presence of $Pt(PPh_3)_4$ the platinum complex **122** was obtained. The compound is stable in air and its constitution has been confirmed by an X-ray crystal structure analysis [82].

The isolation and characterization of the complexes **119**, **121** and **122** makes the [2.2]paracyclophane-1-yne a likely intermediate in elimination reactions, but direct spectroscopic evidence for the free cycloalkyne would certainly be even more convincing.

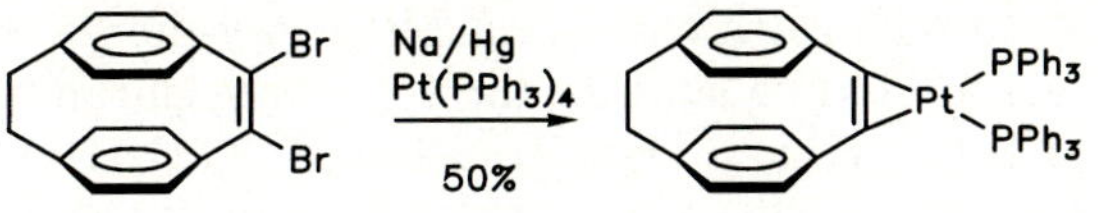

6 → **122** 50%

Scheme 36. Synthesis of a platinum complex of [2.2]paracyclophane-1-yne [82]

3.2
Oligonuclear Metal Complexes with Cyclophane Ligands

The [2.2]paracyclophanediene **42** with an enhanced electron density outside the aromatic rings has been used as a bidentate ligand for tricarbonylarenechromium complexes. The mono- and bis(tricarbonyl)chromium complexes **123** and **124** were obtained from the reaction of tris(propanenitrile)tricarbonylchromium with **42** and characterized by X-ray crystal structure analyses [83].

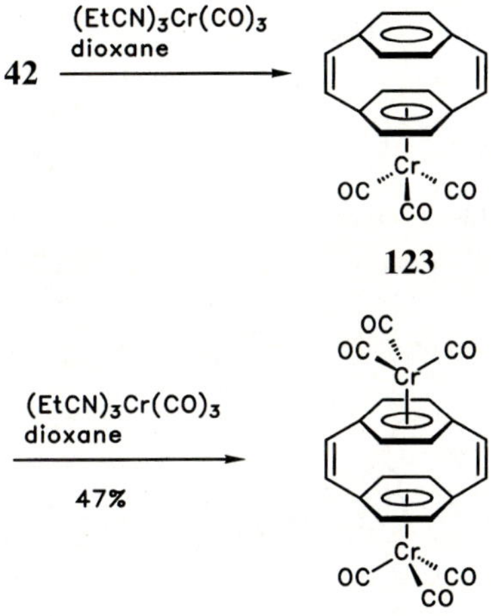

42 → (EtCN)$_3$Cr(CO)$_3$, dioxane → **123**

(EtCN)$_3$Cr(CO)$_3$, dioxane, 47% → **124**

Scheme 37. Tricarbonylarenechromium complexes of [2.2]paracylophane-1,9-diene [83]

11-R → (EtCN)$_3$Cr(CO)$_3$, dioxane, 36% → **125**

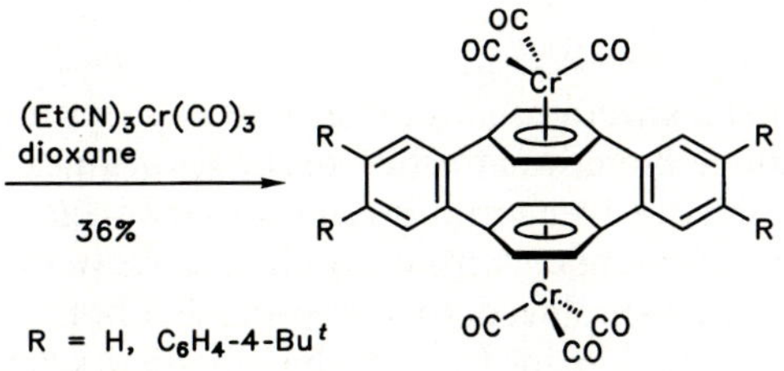

R = H, C$_6$H$_4$-4-Bu^t

Scheme 38. Tricarbonylarenechromium complexes of dibenzo[2.2]paracyclophane-1,9-dienes [83]

Substituted 1 : 2,9 : 10-dibenzo[2.2]paracyclophane-1,9-dienes **11** offer electronically different arene binding sites for the tricarbonylchromium moieties. Experiments with **11-H** (R = H) and **11-C$_6$H$_4$-4-Bu^t** (R = C$_6$H$_4$-4-Bu^t) were good for a surprise in that not only the first chromium coordinates to the central [2.2]paracyclophane, but also the second to yield the dinuclear complex **125** [83].

A variety of novel cyclopentadienyl complexes was obtained from [2.2]-paracyclophanes with annelated five-membered rings. *anti*-[2.2]Indenophane [84] was used to obtain multi-layered metallocenophanes. The reactions of mono- and dianions [85] of **127** with FeCl$_2$ · THF yielded di- and trinuclear ferrocenes respectively [86].

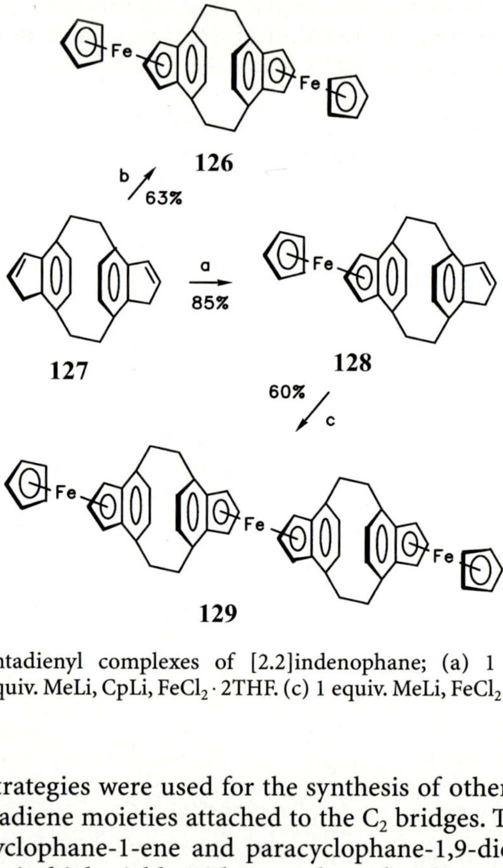

Scheme 39. Cyclopentadienyl complexes of [2.2]indenophane; (a) 1 equiv. MeLi, CpLi, FeCl$_2$ · 2THF. (b) 2 equiv. MeLi, CpLi, FeCl$_2$ · 2THF. (c) 1 equiv. MeLi, FeCl$_2$ · 2THF [84–86]

Two general strategies were used for the synthesis of other types of ligands having cyclopentadiene moieties attached to the C$_2$ bridges. The lithiated derivatives of paracyclophane-1-ene and paracyclophane-1,9-diene, **118** and **120** respectively, react in high yields with a number of carbonyl compounds [87]. When benzaldehyde was employed, the resulting alcohols could be intramolecularly cyclized with zinc chloride to form the indenoannelated [2.2]paracyclophanes **130** and *syn-/anti*-**131**. All indeno[2.2]paracyclophanes were easily deprotonated to the corresponding indenyl anions which formed ferrocene complexes like **132** with iron(II) chloride tetrahydrofuran complex.

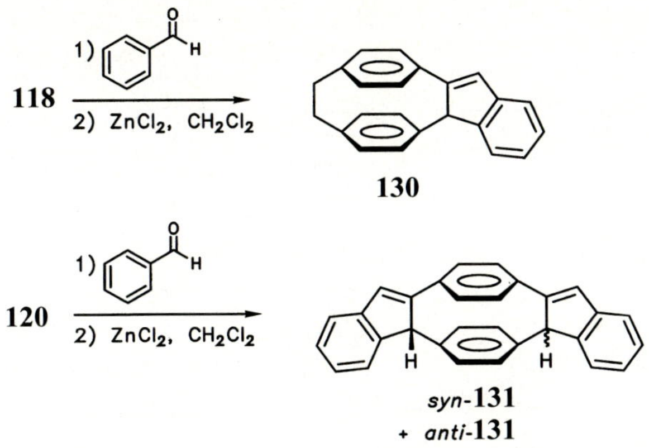

Scheme 40. Synthesis of indenoannelated [2.2]paracylophanes [87]

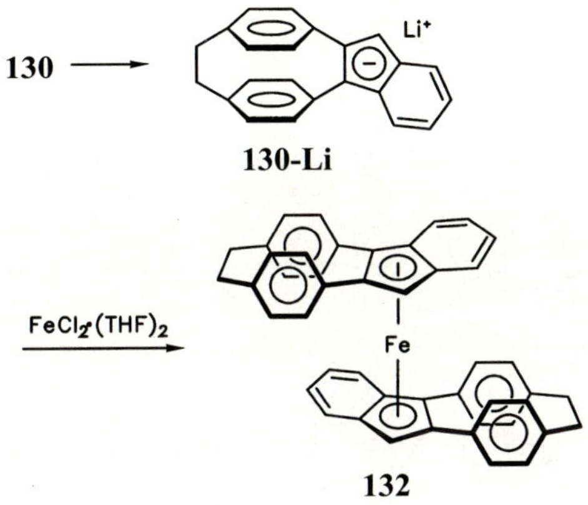

Scheme 41. Ferrocene formation form an indenoannelated [2.2]paracylophane [88]

The interesting tris([2.2]paracyclophane-1,9-dieno)triindene **134** was obtained from benzene-1,3,5-tricarboxaldehyde and monodeprotonated paracyclophane-1,9-diene **42**, followed by cyclization with zinc chloride. The trianion **134-Li₃** could be generated by treatment with MeLi in THF/Et₂O and characterized by its ¹H NMR spectrum as well as by quenching it with deuterium oxide to give the trideuterio derivative of **134** [87 c].

A bis([2.2]paracyclophane)-annelated cyclopentadiene could be synthesized via a Nazarov cyclization of the bis([2.2]paracyclophane-1,9-dienyl) ketone **136** prepared by addition of monolithiodiene **133** to ethyl formate and subsequent oxidation of the carbinol **110**. The synthesis was completed by diisobutylaluminum hydride (DIBAl-H) reduction of the resulting cyclopentenone, and dehydration with *p*-toluenesulfonic acid. Treatment with methyllithium gave the corresponding cyclopentadienyllithium **136-Li**, which was identified by NMR spec-

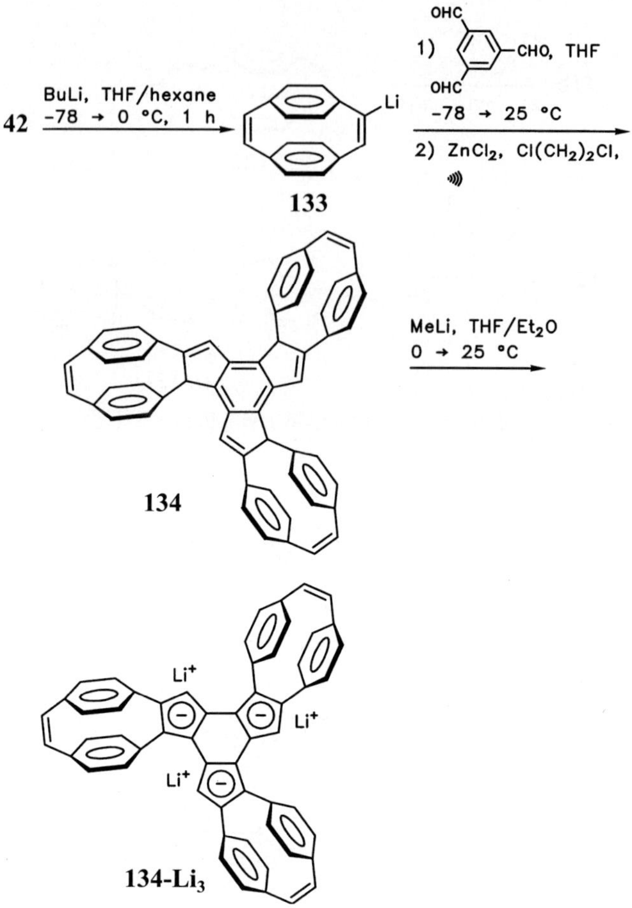

Scheme 42. Synthesis and trifold deprotonation of tris([2.2]paracyclophane-1,9-dieno)triindene [87]

troscopy and by quenching with D_2O to give a monodeuterated analog of bis([2.2]paracyclphanedieno)cyclopentadiene [88].

The meanwhile well established so-called Pauson–Khand reaction for the construction of five-membered rings from alkenes which is known to work particularly well with strained alkenes [89], could favorably be applied to [2.2]paracyclophane-1-ene (**32**) and -1,9-diene (**42**). Cyclopentenone-annelated [2.2]paracyclophanes **138** and (*Z/E*)-**139** were thus obtained by reacting **42** with trimethylsilylacetylene (**137**) and $Co_2(CO)_8$. Reduction of the carbonyl groups with DIBAl·H, dehydration of the alcohols and concomitant desilylation with trimethylsilyl chloride/lithium bromide in acetonetrile gave the parent cyclopentadiene-annelated [2.2]paracyclophanes **140** and **141** [90].

Coordination compounds with up to three [2.2]paracyclophane units and two metal centers, like **142**, were obtained from the deprotonated ligands in the

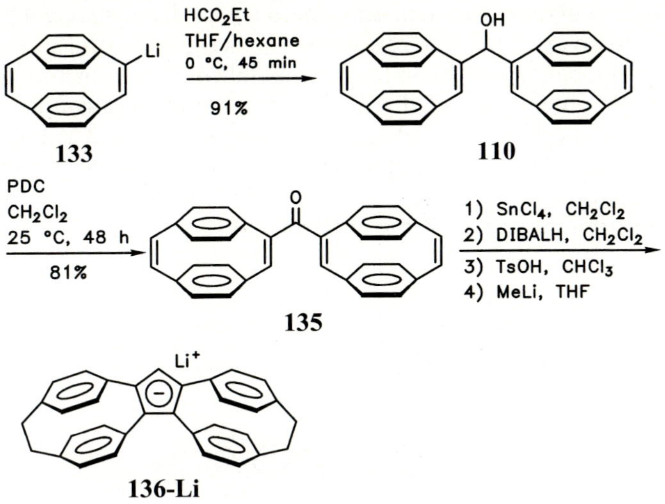

Scheme 43. Synthesis of bis([2.2]paracyclphanedieno)cyclopentadiene [88]

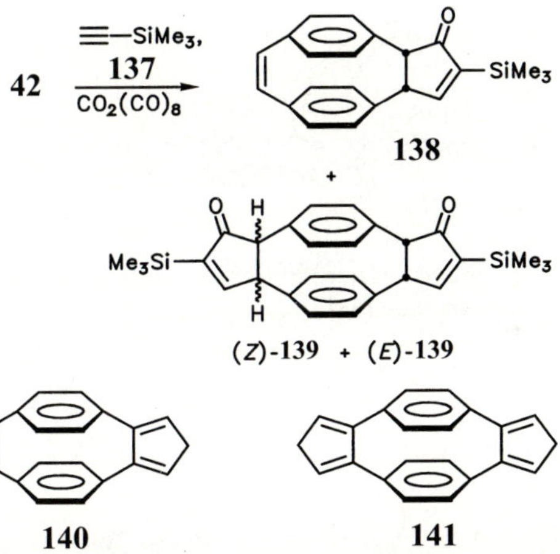

Scheme 44. Pauson–Khand reaction of [2.2]paracyclophane-1-ene and -1,9-diene [90]

reaction with appropiate metal salts as e.g. iron(II) chloride tetrahydrofuran complex. While such coordination compounds with more than one ferrocene unit are interesting model systems to study electron transfer phenomena between redox centers, zirconocenes of type **143** with bulky groups attached to the cyclopentadienyl units also bear on practically applicable metallocenes, which have gained considerable importance in recent years as polymerization catalysts [91]. The bis([2.2]paracyclophanenocyclopentadienyl)zirconium dichloride (**143**) could be prepared from [2.2]paracyclophanenocyclopentadienylanion and zirconium tetrachloride. When **143** was applied in an alumoxane-derived catalyst for the polymerization of ethylene, a low molecular weight polymer was obtained with an extremely narrow molecular weight distribution [88].

Larger cyclophanes have been used for metal coordination, too [92]. A recent example is the preparation of the disilver(I) complex of a conjugated polyene with [2.2.2]paracyclophane end groups (144) [93]. The NMR spectroscopic studies and X-ray structure analysis showed that the sites of complexation are exclusively the cyclophane groups and the olefinic spacers do not overly perturb the complexation properties of the cyclophanes.

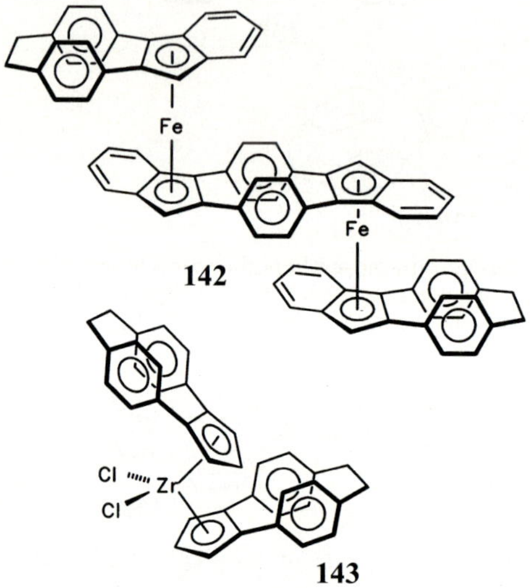

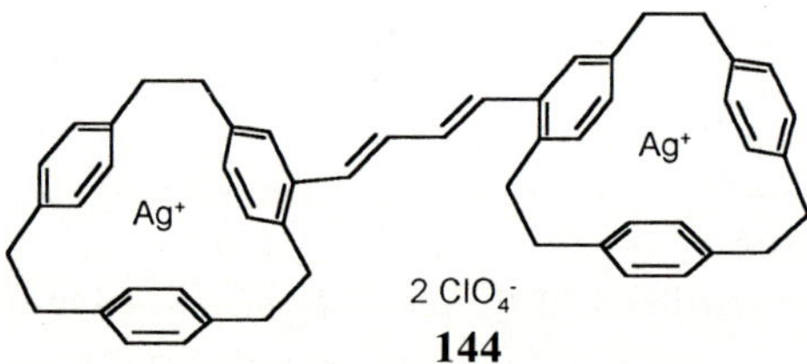

Scheme 45. Oligonuclear cyclophane ferrocenes and an active [2.2]paracyclophane zirconium catalyst for polymerization of ethylene [88]

Scheme 46. Disilver(I) complex of a conjugated polyene with [2.2.2]paracyclophanyl end groups [93]

An interesting application of metal-containing cyclophanes is the preparation of doped polymer films. Such metal–polymer systems are promising as new catalysts or magnetic or non-linear optical materials. The technologically most promising method of obtaining such polymer composites is vapor deposition of metal-monomer systems followed by polymerization of the monomer and inclusion of the metal into the matrix. Dimethylgermanium-bridged bis-cyclophane **145** has been used to prepare a poly(p-xylylene) film that contains Ge crystals [94]. At 550 °C the compound is converted into highly reactive p-xylylenes **146** and **147**. The pyrolysis products are deposited onto a silicon substrate cooled to 280 K. Polymerization proceeds during condensation leading to a Ge-containing polymer (**148**). Thermal treatment (20 min at 300 °C) leads to a destruction of the Ge-organic units in the polymer. Ge-crystals are formed that cause a decrease of the electrical resistance of the polymer film. In addition the film becomes sensitive to air humidity.

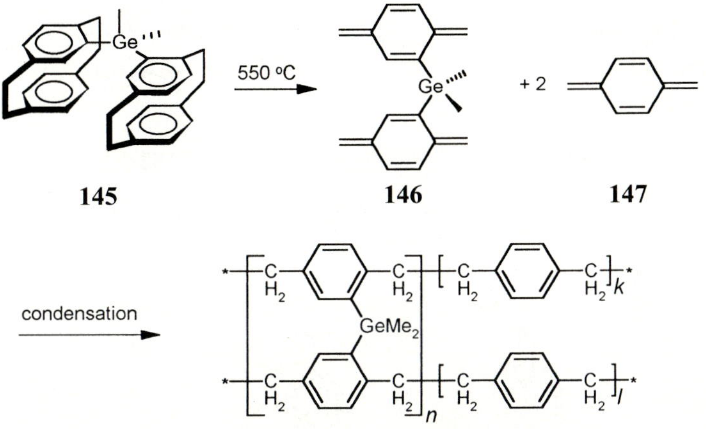

Scheme 47. Formation of a Ge-containing poly(p-xylylene) film by chemical vapor deposition [94]

4
Chiral Cyclophanes as Auxiliaries and Ligands for Asymmetric Catalysis

Chiral derivatives of [2.2]paracyclophanes [95] are chemically stable and do not racemize under standard reaction conditions. Although these are ideal prerequisites for their use as auxiliaries or ligands in asymmetric synthesis only very few examples have been reported in recent years. A [2.2]paracyclophane-substituted selenide **149** was used for chirality transfer in a [2.3]sigmatropic rearrangement to give the optically active linalool **150** [96]. The cyclophane **151** derivative of desazaflavine showed high diastereoselectivity in oxidation and reduction of the desazaflavine moiety [97] NaBD$_4$ reduction gave the deuterated compound **152** as the only product.

4-Formyl-5-hydroxy[2.2]paracyclophane (**153**) was used as a chiral auxiliary in the synthesis of α-amino acids [98]. The reported enantiomeric excess was in the range of 90 – 98 %. Racemic **153** was first prepared by Hopf and Barrett [99]. To separate the enantiomers, their Schiff bases with the dipeptide (S)valyl-(S)valine was prepared. The diastereomeric copper(II) complexes of this compound show different solubility in 2-propanol. Alternatively they can be separ-

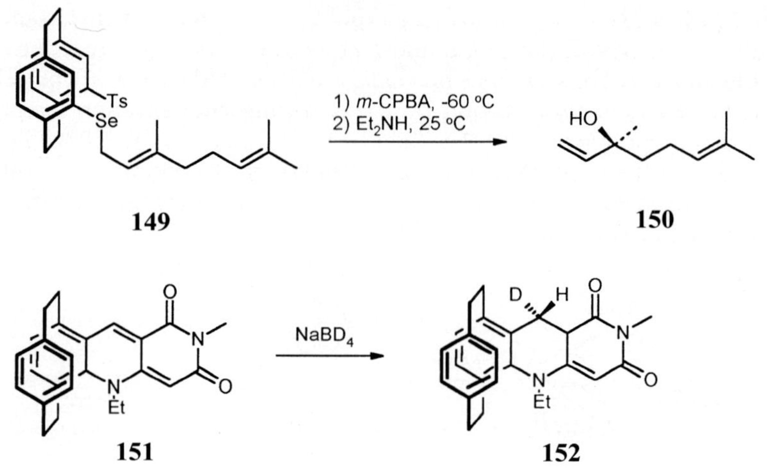

1) *m*-CPBA, -60 °C
2) Et$_2$NH, 25 °C

149   **150**

NaBD$_4$

151   **152**

Scheme 48. Chirality transfer in [2.2]paracyclophane derivatives [96, 97]

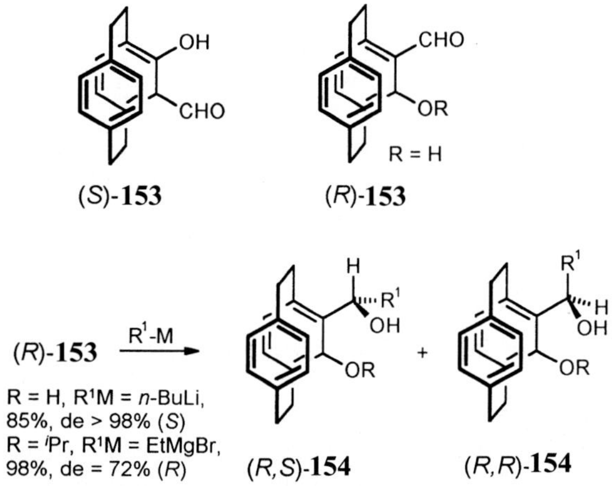

(*S*)-**153**   (*R*)-**153**

R = H

(*R*)-**153** $\xrightarrow{\text{R}^1\text{-M}}$

R = H, R^{1}M = *n*-BuLi,
85%, de > 98% (*S*)
R = ^{i}Pr, R^{1}M = EtMgBr,
98%, de = 72% (*R*)   (*R,S*)-**154**   (*R,R*)-**154**

Scheme 49. 4-Formyl-5-hydroxy[2.2]paracyclophane and diastereoselective nucleophile additions [100]

ated by chromatography on silica gel. An even simpler procedure is the Schiff-base formation of **153** with (*S*)-α-phenylethylamine and recrystallization from hexane [98b]. The nucleophilic additions of organometallic reagents to formyl[2.2]paracyclophane derivatives which are ortho substituted by hydroxy-, alkoxy- or trimethylsiloxy-groups proceed with high diastereoselectivity. The magnitude of the asymmetric induction and the sense of the chirality of the newly formed asymmetric carbon atom of the alcohols **154** depends on the nature of the *ortho*-substituents [100].

Although planar chirality has not been found in nature so far, biological tools can be used for resolution of planar chiral molecules. The synthesis of enantiomerically pure (*S*)-4-formyl[2.2]paracyclohane (>99% ee) (*S*)-**155** and (*R*)-4-hydroxymethyl[2.2]paracyclophane (*R*)-**156** (>78% ee) was achieved by bioreduction in 49 and 34% yield respectively. From several mircoorganisms screened only one strain of the yeast *Saccharomyces cerevisiae* showed a stereospecific reduction of the planar chiral substrate. Despite the high enantiomeric ratio it was necessary to maintain the conversion of the process at almost 50% in order to obtain high enantiomeric excesses of both substrate and reduction product [101].

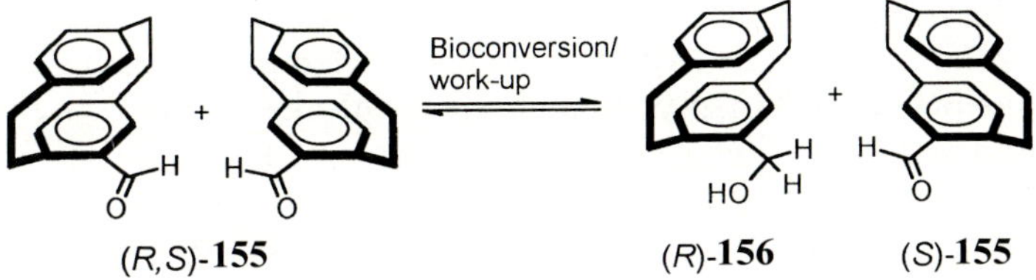

Scheme 50. Synthesis of planar chiral [2.2]paracyclophanes by biotransformation [101]

Recently the first use of the paracyclophane backbone for the placement of two diphenylphosphano groups to give a planar chiral C_2-symmetric bisphosphane was reported [102]. The compound **159** abbreviated as [2.2]PHANEPHOS was used as a ligand in Rh-catalyzed hydrogenations. The catalytic system is exceptionally active and works highly enantioselective [103]. The preparation of [2.2]PHANEPHOS starts with *rac*-4,12-dibromo[2.2]paracyclophane (*rac*-**157**), which was metalated, transmetalated and reacted with diphenylphosphoryl chloride to give racemic bisphosphane oxide (*rac*-**158**). Resolution with dibenzoyltartaric acid and subsequent reduction of the phosphine oxides led to the enantiomerically pure ligand **159**.

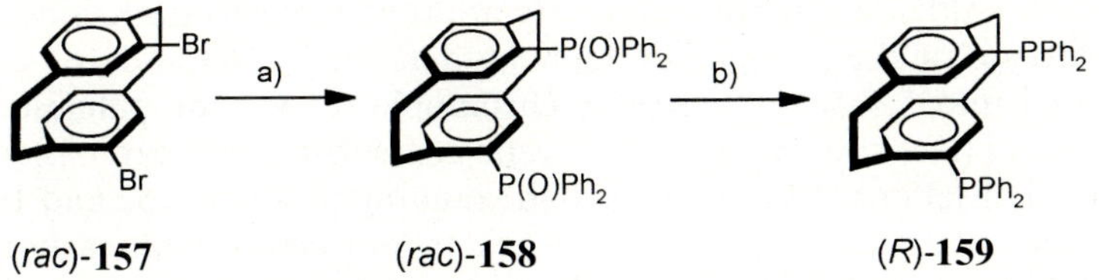

Scheme 51. Synthesis of [2.2]PHANEPHOS; a) 1. 4.2 equiv. ᵗBuLi; 2. 2.3 equiv. MgBr₂·Et₂O; 3. Ph₂P(O)Cl; b) 1. Dibenzoyl-d-tartaric acid; 2. NaOH; 3. HSiCl₃ [102]

5
Cyclophanes in Photochemistry

5.1
Double-Layered Chromophores Derived from [2.2]Paracyclophane

The UV and fluorescence spectra of aromatic systems can change dramatically with any change in the size of the conjugated system and the extent of participation of polar groups. However, besides such intramolecular effects, intermolecular interactions can lead to phenomena such as fluorescence of excimers [104] or intense charge-transfer absorptions. The response of a composite chromophore to light is strongly influenced by the relative orientation of its subunits. In view of the strong π,π-interaction between the two "decks" of aromatic rings (as two chromophores) in [2.2]paracyclophane "clamped" together at a short distance, one may regard them as a single electronically coupled system.

A fourfold arene functionalization of [2.2]paracyclophane was achieved in a single operation by treatment of **1** with liquid bromine and a catalytic amount of iodine [105]. Equal amounts of only two isomeric tetrabromides **160** and **161**

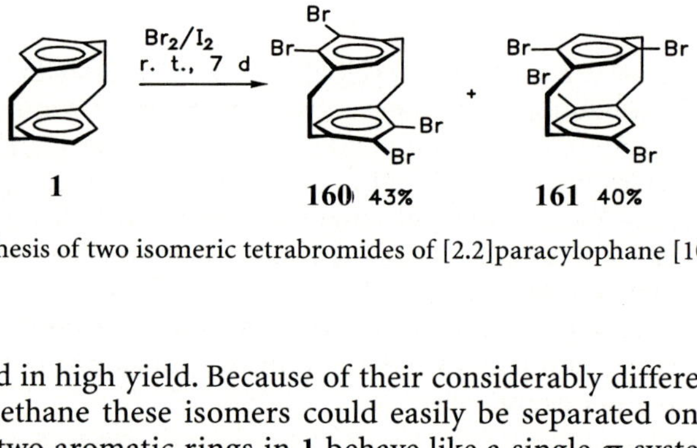

Br₂/I₂ r. t., 7 d

1   160 43%   161 40%

Scheme 52. Synthesis of two isomeric tetrabromides of [2.2]paracylophane [105]

were obtained in high yield. Because of their considerably different solubilities in dichloromethane these isomers could easily be separated on a multigram scale. As the two aromatic rings in **1** behave like a single π-system, this transformation corresponds to a perbromination. Further bromination is at least drastically slowed down, if not completely prohibited due to steric hindrance.

Both "decks" of the [2.2]paracyclophane derivatives **160** and **161** could be transformed into 1,4-distyrylbenzene chromophores via the palladium-catalyzed coupling (Heck reaction) with styrene and substituted styrenes adopting the protocol of Jeffery [106]. The fourfold coupling products **162** and **163** were thus obtained with yields up to 70% [107a]. As four new carbon-carbon bonds are formed in this transformation, each single coupling step must proceed with excellent yield. All these reactions lead stereoselectively to products with (E)-configurated double bonds.

These phanes consisting of 1,4-distyrylbenzene units could also be reduced with alkali metals (Li, Na, K) in THF solution to give multiply charged species. The tetraanion of **163** (R = C_6H_5) is produced in four steps, its charge distribu-

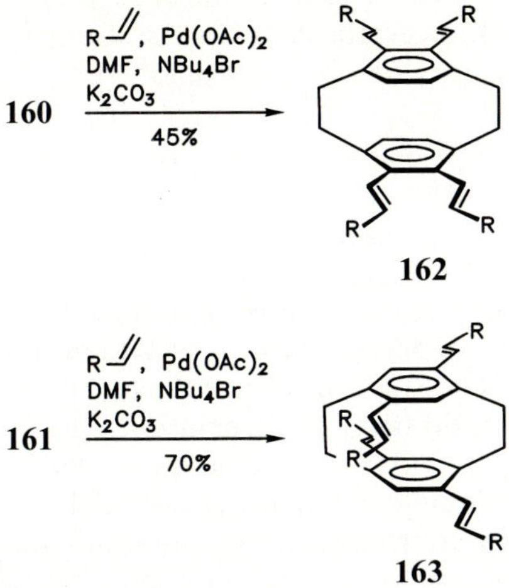

Scheme 53. Synthesis of double-layered distyrylbenzene chromophors via Pd-catalyzed coupling reactions; R = CO$_2$Me, C$_6$H$_5$, C$_6$H$_4$-4-CO$_2$Me, C$_6$H$_4$-4-OMe [107]

tion is determined by the close vicinity of the two decks in the [2.2]paracyclophane system [107b].

An analogous coupling reaction of **161** with phenylacetylene under tetrakis(triphenylphosphane)palladium and copper iodide co-catalysis [108] yielded the double-layered 1,4-bis(phenylethynyl)benzene system **164** [107b].

Solutions of both the tetrastyryl[2.2]paracyclophanes **162** and **163**, respectively, show intense blue-green fluorescence when exposed to sunlight. In comparison to that of 2,5-dimethyl-1,4-distyrylbenzene the UV/vis spectrum of **163** (R=Ph) shows a bathochromic shift of 41 nm for the longest wavelength absorption λ_{max}=394 nm) and a significant increase in the extinction coefficients. Excitation of **163** (R=Ph) at 394, 355 or 339 nm leads to fluorescence with a broad, unstructured emission band at 465 nm, which is shifted by 61 nm towards longer wavelengths in comparison to that of 2,5-dimethyl-1,4-distyrylbenzene with a well structured band at 404 nm. The relative quantum yield Φ of **163** (R=Ph) has been determined to be 0.6 (1,4-distyrylbenzene: Φ=0.78 [109]). It is

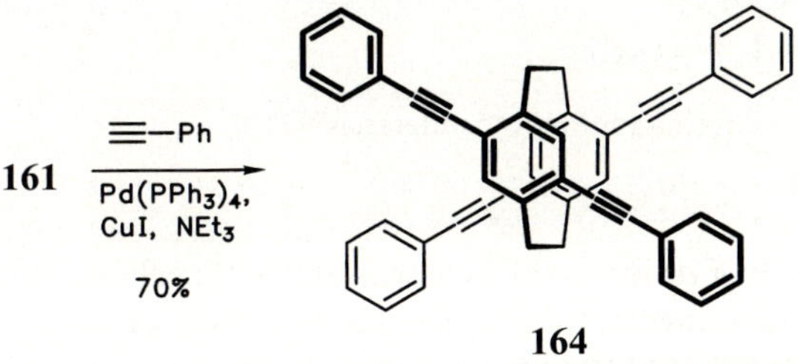

Scheme 54. Synthesis of a double-layered 1,4-bis(phenylethynyl)benzene system [107b]

most likely that the bathochromic shifts of the absorption and emission bands of **163** (R=Ph) are caused to a certain extent by the strong electronic interaction between the two 1,4-distyrylbenzene chromophores through the central [2.2]paracyclophane unit.

5.2
Topological Reaction Control in Solution

The [2 + 2]-photocycloaddition reaction is important for the synthesis of four-membered rings. However, in dilute solutions cyclodimerizations are inefficient due to rapid *cis,trans*-photoisomerization. Efficient photodimerizations are, however, observed in the solid state, and in solution when high local concentrations are achieved. With the aid of a suitable molecular framework, the topological reaction control principle, originally established by Schmidt et al. for the solid state [110], can be transferred to dilute solution. The bis[2.2]cinnamatophane **165** yields a single photoproduct **166** upon radiation with a quantum yield of 0.8 [111]. The reaction can be extended up to a bistetraene **167** that gives a single stereoisomer of the "ladderane" **168** [112]. Irradiation of the colorless product with light of shorter wavelength causes the reverse reaction. This unusual "molecular zipper" is stable for many cycles.

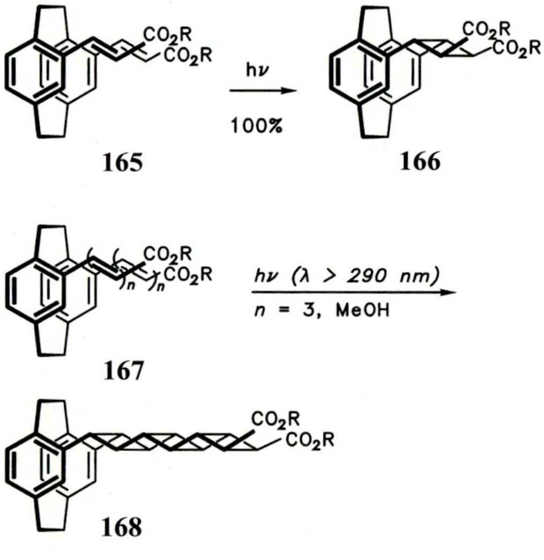

Scheme 55. Multiple [2 + 2]cycloaddition to "ladderanes" [112]

An analogous trick of doing a photochemical [2+2] cycloaddition intramolecularly [113] was also successful for the construction of [3.2]para- **170** and [3.2.]metacyclophanes **172** with one and two cyclobutane rings annelated to the C_2 bridges [114]. These intramolecular reactions gave good to very good yields,

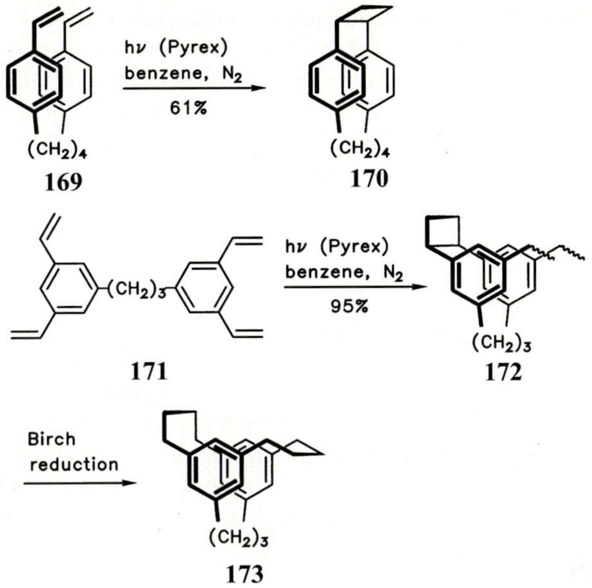

Scheme 56. Cyclophane synthesis via intramolecular photocycloaddition [114]

while the intermolecular photodimerization of styrene can only be achieved in poor yield. The intramolecular version of vinylarene dimerization was successfully applied to prepare a variety of para-, meta-, orthocyclophanes as well as naphthalenophanes and phanes with even larger aromatic systems. A problem with oligobridged systems like **172** arises from the possible formation of different isomers, which may be difficult to separate. However, Birch reduction in such cases yields $[4_n.3]$arenophanes as single isomers. Multi-bridged cyclophanes that contain only cyclobutane rings as bridges are called "paddlanes". Substituted derivatives of 1,3- and 1,3,5-"paddlanes" have been prepared previously. Now Nishimura et al. [115] reported the synthesis of the parent compounds 1,3,5-"paddlane" (**174**) and 1,2,4,5-"paddlane" (**175**) by intermolecular photocycloaddition of vinyl-benzenes in 1 % and 0.6 % yield.

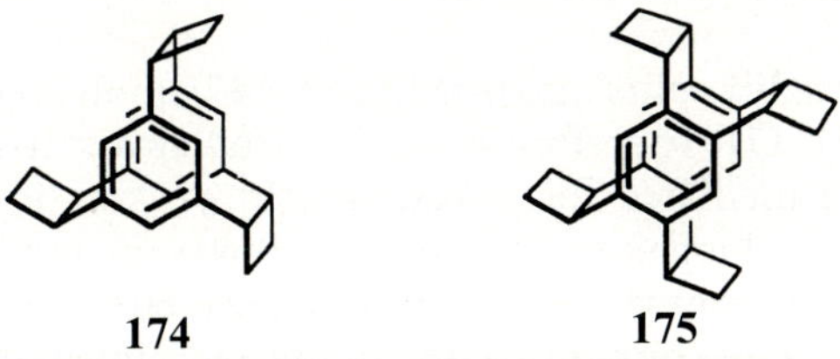

Scheme 57. 1,3,5-"Paddlane" and 1,2,4,5-"paddlane" [115]

A surprisingly efficient intermolecular photochemical [2+2] dimerization was achieved in the synthesis of belt-shaped cyclophanes by irradiation of arenoannelated [18]annulenes like **176** [116]. The highly selective threefold [2π+2π] cycloaddition yields dimers **177** in up to 68%. The efficiency of this dimerization process even at high dilution can be explained with the perfect preorganization of the discotic mesogenes **176**.

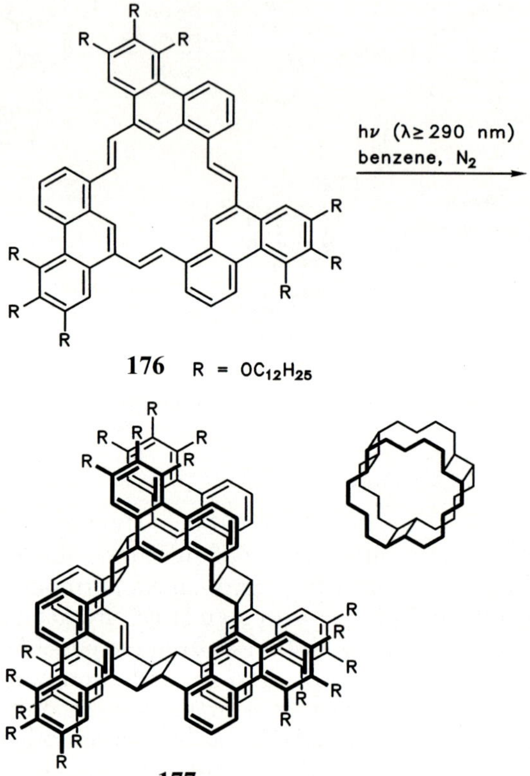

Scheme 58. Photocyclization of preorganized arenoannelated [18] annulenes [116]

5.3
Photolysis of [2.2]Paracyclophane-1,10-dione

[2.2]Paracyclophane readily splits into *p*-xylylene (**147**) upon thermolysis due to its strained structure [117]. Therefore it can be used as a starting material to generate reactive intermediates for spectroscopic or kinetic studies [118]. Photolysis of [2.2]paracyclophane-1,10-dione (**178**) was the way to generate the reactive bisketene 1,4-dicarbonyl-2,5-cyclohexadiene (**180**) in argon at 10 K. The compound was characterized by comparison of the experimental and calculated IR spectrum [119].

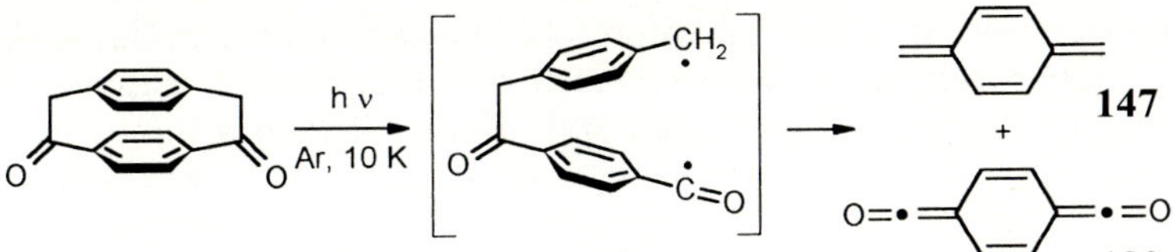

Scheme 59. Photolysis of [2.2]paracyclophane-1,10-dione in argon at 10 K yields *p*-xylylene and 1,4-dicarbonyl-2,5-cyclohexadiene [119]

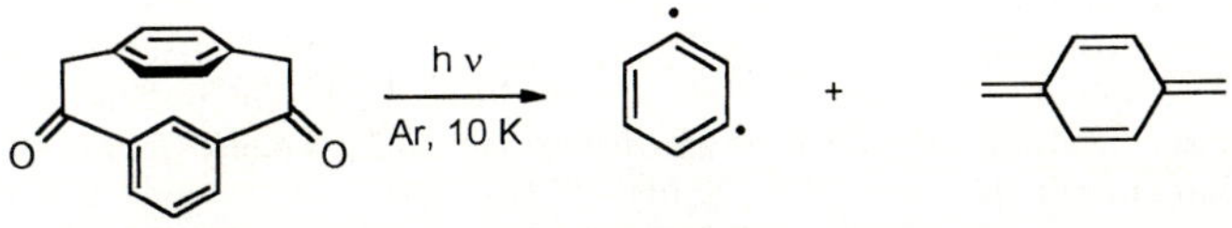

Scheme 60. Photolysis of [2.2]metacyclophane-2,9-dione in Ar at 10 K yields *p*-xylylene, carbon monoxide and 1,3-didehydrobenzene [120]

Irradiation of [2.2]metacyclophane-2,9-dione (**181**) in Ar at 10 K yielded carbon monoxide, *p*-xylylene (**147**) and a new compound, to which structure **182** was assigned by comparison of experimental and calculated IR spectra. Additional structure proof was obtained by the independent generation of **182** by pyrolysis of isophthaloyldiacetyl peroxide and matrix-isolation of the products [120].

6
Perspectives

The objective of research on cyclophanes has changed over the past two decades. While in the beginning the physical properties of the cyclophanes and *ansa*-compounds were investigated, most of the recent work has been focused on the function of molecules containing cyclophane units. Although the cavity of small cyclophanes, such as [2.2]paracyclophane (**1**), is not useful for inclusion – only a "naked" chromium atom has been incorporated into **1** [121] and **37** [122] with very low efficiency by metal vapour-ligand co-condensation – the rigid skeletons make these molecules ideal building blocks for model systems to study intramolecular electron transfer phenomena. Much deeper going investigations would be desirable to fully understand even the simplest examples presented here. The high π-electron density on the two faces of such molecules as **1** could be useful in the development of non-linear optical materials, whereas chemical vapor deposition techniques of metal-containing cyclophanes allows the preparation of polymer films doped with metals.

The use of cyclophanes as chiral auxiliaries for stereoselective synthesis or as chiral ligands in asymmetric catalysis are even more promising future applications. The structure of chiral substituted [2.2]paracyclophanes that effectively prohibits their racemization, renders such compounds most suitable for this task, and it is likely that many more applications will follow the first reported examples.

7
References

1. Pellegrin MM (1899) Rec Trav Chim Pays-Bas 18:457
2. Brown CJ, Farthing AC (1949) Nature 164:915
3. a) Keehn PM, Rosenfeld SM (1983) Cyclophanes, I, II. Academic Press, New York; b) Vögtle F (1993) Cyclophane Chemistry, Wiley, New York; c) for recent reviews on cyclophane chemistry see: Bodwell, GJ (1996) Angew Chem 108:2221; Angew Chem Int Ed Engl 35:2085; d) deMeijere A, König B (1997) Synlett 1221
4. Atwood JL, Davies JED, Macnicol DD, Vögtle F, Lehn JM (eds) (1996) Comprehensive Supramolecular Chemistry Vol. 1, 2 and 10, Elsevier Science, Oxford
5. a) Diederich F, Carcanague DR (1994) Helv Chim Acta 77: 800; b) examples of small cyclophanes with heteroatom-containing bridges: Müller D, Vögtle F (1995) Synthesis 759; c) Irie S, Yamamoto M, Iida I, Nishio T, Kishikawa K, Kohmoto S, Yamada K (1994) J Org Chem 59:935
6. a) Diederich F (1991) In: Cyclophanes; Monographs in Supramolecular Chemistry, Stoddart JF (ed) The Royal Society of Chemistry, London; b) for the effect of cavity size, see: Whitlock BJ, Whitlock HW (1994) J Am Chem Soc 116:2301
7. Cram DJ, Cram JM (1994) In: Container molecules and their guests; Monographs in supramolecular chemistry, Stoddart JF (ed) The Royal Society of Chemistry, London
8. a) Mogck O, Pons M, Böhmer V, Vogt W (1997) J Am Chem Soc 119:5706; b) Shimuzu KD, Rebek Jr J (1995) Proc Natl Acad Sci USA 92:12403
9. Cram DJ, Tanner ME, Thomas R (1991) Angew Chem; Angew Chem Int Ed Engl 30:1024
10. Warmuth R (1997) Angew Chem109:1406; Angew Chem Int Ed Engl 36:1347
11. a) Gutsche CD (1989) In: Calixarenes; monographs in supramolecular chemistry, Stoddart JF (ed) The Royal Society of Chemistry, London; b) Vögtle F (1992) Supramolecular chemistry Wiley, New York
12. a) Böhmer V (1995) Angew Chem 107:785; Angew Chem Int Ed Engl 34:713; b) Kubo Y, Maeda S, Tokita S, Kubo M (1996) Nature 382:522
13. a) Cram DJ, Steinberg H (1951) J Am Chem Soc 73:5691; b) Cram DJ, Hornby RB, Truesdale EA, Reich HJ, Delton MH, Cram JM (1974) Tetrahedron 30:1757
14. a) Lüttringhaus A, Gralheer H (1942) Justus Liebigs Ann Chem 550: 67; b) (1947) ibid 557:108
15. A series of difluoro-substituted cyclophanes with systematically varying F-F distance has been used to determine an equation that relates ^{19}F-^{19}F coupling with through space distance: Ernst L, Ibrom K (1995) Angew Chem 107:2010; Angew Chem Int Ed Engl 34:1881
16. Cohen MD, Schmidt GM, Sonntag FI (1964) J Chem Soc 2000
17. a) Wegner G (1981) Angew Chem 93: 352; Angew Chem Int Ed Engl 20: 361; b) Hanack M (1980) Nachr Chem Tech Lab 28:632
18. a) Chan CW, Wong HNC (1988) J Am Chem Soc 110: 462; b) Chan CW, Wong HNC (1985) J Am Chem Soc 107:4790
19. a) Reiser O, Reichow S, de Meijere A (1987) Angew Chem 99: 1285; Angew Chem Int Ed Engl 26:1277; b) Reiser O, König B, Meerholz K, Heinze J, Wellauer T, Gerson F, Frim R, Rabinovitz M, de Meijere A (1993) J Am Chem Soc 115:3511
20. König B, de Meijere A (1992) Chem Ber 125:1895
21. Blatter K, Schlüter AD, Wegner G (1989) J Org Chem 54:2396
22. Stöbbe M, Reiser O, Näder R, de Meijere A (1987) Chem Ber 120:1667

23. König B, Heinze J, Meerholz K, de Meijere A (1991) Angew Chem 103:1350; Angew Chem Int Ed Engl 30:1361
24. a) Hart H, Lai C, Nwokogu GC, Shamouilian S (1987) Tetrahedron 43: 5203; b) Hart H, Bashir-Hashemi A, Luo J, Meador MA (1986) Tetrahedron 42:1641
25. König B (1995) J Prakt Chem 337:250
26. Blatter K, Schlüter AD (1989) Chem Ber 122:1351
27. Krasovitskii BM, Bolotin BM (1988) Organic luminescent materials, VCH, Weinheim
28. de Meijere A, Heinze J, Meerholz K, Reiser O, König B (1990) Angew Chem 102:1443; Angew Chem Int Ed Engl 29:1418.
29. Psiorz M, Hopf H (1982) Angew Chem 94: 639; Angew Chem Int Ed Engl 21:623
30. a) Albrecht K, de Meijere A (1994) Chem Ber 127: 2539; b) de Meijere A, Rauch K, Noltemeyer M, Knieriem B, Albrecht K, Hopf H, unpublished results.
31. Gerson F, Scholz M, de Meijere A, König B, Heinze J, Meerholz K (1992) Helv Chim Acta 75: 2307
32. a) Dewhirst KC, Cram DJ (1958) J Am Chem Soc 80:3115
33. a) Hopf H, Psiorz M (1986) Chem Ber 119:1836; b) Raasch MS (1980) J Org Chem 45:856
34. a) König B, de Meijere A (1992) Helv Chim Acta 75:901; b) Taylor EC, Ehrhart WA (1960) J Am Chem Soc 82:3138
35. a) Beck K, Höhn A, Hünig S, Prokschy F (1984) Chem Ber 117:517; b) Sauer J, Meier A (1990) Tetrahedron Lett 31:6855; c) Sauer J, Thalhammer F, Wallfahrer U (1990) Tetrahedron Lett 31:6851
36. Reich HJ, Cram DJ (1969) J Am Chem Soc 91:3527
37. König B, Knieriem B, Rauch K, de Meijere A (1993) Chem Ber 126:2531
38. König B, Ramm S, Bubenitschek P, Jones PG, Hopf H, Knieriem B, de Meijere A (1994) Chem Ber 127:2263
39. a) Böhm I, Herrmann H, Menke K, Hopf H (1978) Chem Ber 111:523; b) Hopf H, Böhm I, Kleinschroth J (1980) Org Synth 60:41
40. a) Ashton PR, Isaacs NS, Kohnke FH, Mathias JP, Stoddart JF (1989) Angew Chem 101:1266; Angew Chem Int Ed Engl 28:1258; b) Ashton PR, Isaacs NS, Kohnke FH, Stagno d'Alcontres G, Stoddart JF (1989) Angew Chem 101:1269; Angew Chem Int Ed Engl 28:1261
41. Blatter K, Schlüter AD (1989) Macromolecules 22:3506
42. Blatter K, Schlüter AD (1989) Chem Ber 122:1351
43. König B, de Meijere A, Schlüter AD, unpublished results
44. Hopf H, Jones PG, Bubenitschek P, Werner C (1995) Angew Chem 107:2592; Angew Chem Int Ed Engl 34:2367
45. Sato H, Isono N, Okamura K, Date T, Mori M (1994) Tetrahedron Lett 35:2035
46. a) Werner C (1997) Dissertation, Technische Universtität Braunschweig; b) Rubin Y (1997) Chem Eur J 3:1009
47. a) Haley MM, Langsdorf BL (1997) Chem Commun 1121; b) The ligand exchange was necessary to make the complex stable to fluoride ions
48. Eglington G, McRae W (1963) Adv Org Chem 4:225
49. For examples, see: ref. 3b, p 131
50. Mitchell RH, Jin X (1995) Tetrahedron Lett 36:4357
51. Mitchell RH (1989) Adv Theoret Interesting Mol 1:135
52. Sekine Y, Brown M, Boekelheide V (1979) J Am Chem Soc 101:3126
53. a) Sakamoto Y, Miyoshi N, Shinmyozu T (1996) Angew Chem 108:585, Angew Chem Int Ed Engl 35: 549; b) Bettinger HF, Schleyer PvR, Schaefer III HF (1998) J Am Chem Soc 120:1074
54. Cha OJ, Osawa E, Park S (1993) J Mol Struct 300: 73
55. a) Example: Olsson T, Tanner D, Thulin B, Wennerström O (1981) Tetrahedron 37:3485; b) Cyclophanes have been used as precursors for saturated and unsaturated small cage hydrocarbons. For recent work on this topic, see: Hopf H, Savinsky R, Jones PG, Dix I, Ahrens B (1997) Liebigs Ann/Recueil 1499; c) Ernst L, Hopf H, Savinsky R (1997) Liebigs Ann/Recueil 1915; d) Hopf H, Savinsky R, Diesselkämper B, Daniels RG, de Meisere A (1997) J Org Chem 62:8941

56. Tani K, Seo H, Maeda M, Imagawa K, Nishiwaki N, Ariga M, Tohda Y, Higuchi H, Kuma H (1995) Tetradedron Lett 36:1883
57. a) Vögtle F, Hohner G, Weber E (1973) J Chem Soc Chem Commun 366; b) Matsumoto K, Kugimiya M (1975) Z Kristallogr 141:260
58. a) Darabi, HR, Kawase T, Oda M (1995) Tetrahedron Lett 36:9525; b) Müllen K, Unterberg H, Huber W, Wennerström O, Norinder U, Tanner D, Thulin S (1984) J Am Chem Soc 106:7514
59. a) Pascal Jr RA, Grossman RB, Van Engen D (1987) J Am Chem Soc 109:6878; b) Pascal Jr RA, Winans CG, Van Engen D (1989) ibid 111:3007; c) Pascal Jr RA, Grossman RB (1987) J Org Chem 52:4616
60. a) L'Esperance RP, West AP, Van Engen D, Pascal Jr RA (1991) J Am Chem Soc 113:2672; b) West AP, Smyth N, Craml CM, Ho DM, Pascal Jr RA (1993) J Org Chem 58:3502
61. a) Breidenbach S, Ohren S, Nieger M, Vögtle F (1995) J Chem Soc Chem Commun 1237; b) Breidenbach S, Ohren S, Vögtle F (1996) Chem Eur J 2:832
62. a) Josten W, Karbach D, Nieger M, Vögtle F, Hägele K, Svoboda M, Przybylsk M (1994) Chem Ber 127:767; b) Josten W, Neumann S, Vögtle F, Nieger M, Hägele K, Przybylsk M, Beer F, Müllen K (1994) Chem Ber 127:2089; c) Schröder A, Karbach D, Güther R, Vögtle F (1992) Chem Ber 125:1881
63. Faust R (1995) Angew Chem 107:1559; Angew Chem Int Ed Engl 34:1429
64. Groß J, Harder G, Vögtle F, Stephan H, Gloe K (1995) Angew Chem 107:523; Angew Chem Int Ed Engl 34:481
65. Güter R, Nieger M, Rissanen K, Vögtle F (1994) Chem Ber 127:743
66. Brown CJ (1953) J Chem Soc 3278
67. Bodwell GJ, Houghton TJ, Kennedy JWJ, Mannion MR (1996) Angew Chem 108:2280; Angew Chem Int Ed Engl 35:2121
68. Chen CT, Ganzel P, Siegel JS, Baldridge KK, English RB, Ho DM (1995) Angew Chem 107:2870; Angew Chem Int Ed Engl 34:2657
69. Seiders TJ, Baldridge KK, Siegel JS (1996) J Am Chem Soc 118:2754
70. Buchholz H, de Meijere A (1993) Synlett 253
71. Tsuji T, Ohkita M, Nishida S (1993) J Am Chem Soc 115:5284
72. a) Tsuji T, Ohkta M, Konno T, Nishida S (1997) J Am Chem Soc 119:8425; b) Kawai H, Suzuki T, Ohkita M, Tsuji T (1998) Angew Chem 110:827; Angew Chem Int Ed Engl 37:817
73. Ma B, Sulzbach MH, Remington RB, Schaefer HF (1995) J Am Chem Soc 117:8392
74. Jenneskens LW, deKanter FJJ, Kraakman PA, Turkenburg LAM, Koolhaas WE, deWolf WH, Bickelhaupt F (1985) J Am Chem Soc 107:3716
75. a) Bickelhaupt F (1990) Pure Appl Chem 62:373; b) Okuyama M, Tsuji T (1997) Angew Chem 109:1157; Angew Chem Int Ed Engl 36:1085
76. van Es DS, deKanter FJJ, deWolf WH, Bickelhaupt F (1995) Angew Chem 107:2728; Angew Chem Int Ed Engl 34:2553
77. Schuchmann P, Hafner K (1995) Tetrahedron Lett 36:2603
78. Bodwell GJ, Bridson JN, Houghton TJ, Kennedy JWJ, Mannion MR (1996) Angew Chem 108:1418; Angew Chem Int Ed Engl 35:1320
79. a) Bennett MA, Schwemlein HP (1989) Angew Chem 101:1349; Angew Chem Int Ed Engl 28:1296; b) Bennett, MA, Wenger E (1997) Chem Ber/Recueil 130:1029
80. a) Buchwald SL, Lum RT, Dewan JC (1986) J Am Chem Soc 108:7441; b) Barr KJ, Watson BT, Buchwald SL (1991) Tetrahedron Lett 32:5465
81. König B, Bennett MA, de Meijere A (1994) Synlett 653
82. Albrecht K, Hockless DCR, König B, Neumann H, Bennett MA, de Meijere A (1996) J Chem Soc Chem Commun 543
83. a) de Meijere A, Reiser O, Stöbbe M, Kopf J, Adiwidjaja G, Sinnwell V, Khan SI (1988) Acta Chem Scand A42:611; b) A detailed study of the site selectivity of tricarbonylchromium coordination on substituted [2.2]paracyclophane derivatives has been performed by Rozenberg V, Academy of Science, Moscow. Rozenberg L personal communication
84. a) Frim R, Raulfs FW, Hopf H, Rabinovitz M (1986) Angew Chem 98:160; Angew Chem Int Ed Engl 25:174; b) for the synthesis of [2]metacyclo[2]indenophanes, see: Bodwell GJ, Frim R, Hopf H, Rabinovitz M (1993) Chem Ber 126:167

85. Frim R, Rabinovitz M, Bodwell G, Raulfs FW, Hopf H (1989) Chem Ber 122:737
86. Hopf H, Dannheim J (1988) Angew Chem 100:724; Angew Chem Int Ed Engl 27:701
87. a) Buchholz HA, Höfer J, Noltemeyer M, de Meijere A (1998) Eur J Org Chem, submitted; b) Höfer J (1989) Dissertation, Universität Hamburg; c) Buchholz H (1992) Dissertation, Universität Hamburg; d) König B (1991) Dissertation, Universität Hamburg
88. Buchholz HA, de Meijere A (1998) Eur J Org Chem, submitted; Buchholz H (1992) Dissertation Universität Hamburg
89. Reviews see: a) Pauson PL (1985) Tetrahedron 41:5855; b) Schore NE (1988) Chem Rev 88:1081; c) Schore NE (1991) In: Comprehensive organic synthesis, Trost BM, Fleming I (Eds) Pergamon Press, Oxford, Vol. 5, p 1037
90. Buchholz H, Reiser O, de Meijere A (1991) Synlett 1:20
91. Brintzinger HH, Fischer D, Mülhaupt R, Rieger B, Waymouth R (1995) Angew Chem 107:1255; Angew Chem Int Ed Engl 34:1143 and references cited therein
92. Vögtle F (1991) Supramolecular chemistry, Wiley, New York, p 107
93. Heirtzler F, Hopf H, Jones PG, Bubenitschek P (1995) Chem Ber 128:1079
94. Hopf H, Gerasimov GN, Chvalun SN, Rozenberg V, Popova E, Nikolaeva E, Grigoriev E, Zavjalov S, Trakhtenberg L (1997) Chem Vap Deposition 3:197
95. a) Molecules with planar chirality: Eliel EL, Wilen SH (1994) Stereochemistry of organic compounds, Wiley, New York, p 1166; b) for the preparation of chiral (R)-4-hydroxy- and (R)-4-halogeno[2.2]paracyclophanes and the assignment of their absolute configuration by chemical correlation, see: Cipiciani A, Fringuelli F, Mancini V, Piermatti O, Pizzo F (1997) J Org Chem 62:3744; c) for the preparation of chiral paracyclophane amino acids, see: Pelter A, Crump RANC, Kidwell H (1996) Tetrahderon Lett 37:1273
96. Reich HJ, Yelm YE (1991) J Org Chem 56:5672
97. Yanada R, Higashikawa H, Mura Y, Taga T, Yoneda F (1992) Tetrahedron: Asymmetry 3:1387
98. a) Rozenberg V, Kharitonov V, Antonov D, Sergeeva E, Aleshkin A, Ikonnikov N, Orlova S, Bekolon Y (1994) Angew Chem 106:106; Angew Chem Int Ed Engl 33:91; b) Antonov D, Belokon YN, Ikonnikov N, Orlova S, Pisarevsky AP, Raevski NI, Rozenberg V, Sergeeva E, Struchkov YT, Tararov VI, Vorontsov EV (1995) J Chem Soc Perkin Trans 1:1873
99. a) Barrett DG (1988) Diplomarbeit, Technische Universität Braunschweig; b) Hopf H, Barrett, DG (1995) Liebigs Ann 449
100. Sergeeva, EV, Rozenberg, V, Vorontsov E, Danilova T, Starikova Z, Yanovsky A, Belokon Y, Hopf H (1996) Tetrahedron: Asymmetry 3445
101. Pamperin D, Hopf H, Syldatk C, Pietzsch M (1997) Tetrahedron: Asymmetry 8:319
102. a) Pye PJ, Rossen K, Reamer RA, Tsou NN, Volante RP, Reider PJ (1997) J Am Chem Soc 119:6207; b) for the use of chiral ferrocenes, see: Togni A (1996) Angew Chem 108:1581; Angew Chem Int Ed Engl 35:1475
103. More recently the catalytic system Pd/[2.2]PHANEPHOS was successfully used in the Buchwald-Hartwig amination: Rossen K, Pye PJ, Maliakal A, Volante RP (1997) J Org Chem 62:6462
104. Förster T (1969) Angew Chem 81:364; Angew Chem Int Ed Engl 8:333
105. a) Method adopted from: Fieser LF, Haddadin MJ (1965) Can J Chem 43:1599; b) the synthesis of the tetrabromo derivatives of [2.2]paracyclophane with Fe/Br_2 has been previously reported. Reich HJ, Cram DJ (1969) J Am Chem Soc 91:3527
106. a) Jeffery T (1985) Tetrahedron Lett 26:2667; b) Jeffery T (1996) Tetrahedron 52:10113
107. a) König B, Knieriem B, de Meijere A (1993) Chem Ber 126:1643; b) Shabtai E, Rabinovitz M, König B, Knieriem B, de Meijere A (1996) J Chem Soc Perkin Trans II:2589
108. Method adopted from: Sanechika K, Yamamoto T, Yamamoto A (1984) Bull Chem Soc Jpn 57:752
109. Krasovitskii BM, Bolotin BM (1988) Organic luminescent materials, VCH, Weinheim, p 41
110. Cohen MD, Schmidt GMJ, Sonntag FI (1964) J Chem Soc 2000
111. Greiving H, Hopf H, Jones PG, Bubenitschek P, Desvergne JP, Bouas-Laurent H (1994) J Chem Soc Chem Commun 1075

112. a) Hopf H, Greiving H, Jones PG, Bubenitschek P (1995) Angew Chem 107:742; Angew Chem Int Ed Engl 34:685; b) Greiving H, Hopf H, Jones PG, Bubenitschek P, Desvergne JP, Bouas-Laurent H (1995) Liebigs Ann 1949

113. For an example of an intramolecular cyclophane [4+4] photocycloaddition, see: Hua DH, Dantoing, Robinson PD, Tran-Cong Q, Meyers CY (1994) Acta Cryst C50:1090

114. a) Nishimura J (1992) In Osawa E, Yonemitsu O (Eds) Carbocyclic cage compounds: chemistry and applications, VCH, New York, p 155; b) Nishimura J, Okada Y, Inokuma S, Nakamura Y, Gao SR (1994) Synlett 884; c) Nishimura J, Doi H, Ueda E, Ohbayashi A, Oku A (1987) J Am Chem Soc 109:5293

115. a) Nakamura Y, Hayashida Y, Wada Y, Nishimura J (1997) Tetrahedron 53:4593; b) Wada Y, Ishimura T, Nishimura J (1992) Chem Ber 125:2155

116. a) Meier H, Müller K (1995) Angew Chem 107: 1598; Angew Chem Int Ed Engl 34:1437; b) Meier H (1992) Angew Chem 104:1425; Angew Chem Int Ed Engl 31:1399; c) Müller K, Meier H, Bouas-Laurent H, Desvergne JP (1996) J Org Chem 61:5474

117. Pearson JM, Six HA, Williams DJ, Levy M (1971) J Am Chem Soc 93:5034

118. a) Cleavage of [2.2]paracyclophane radical cation: Adam W, Miranda MA, Mojarrad F, Sheikh H (1994) Chem Ber 127:875; b) cyclophane formation form *m*-xylylenes: Gajewski JJ, Paul GC, Chang MJ, Gortva AM (1994) J Am Chem Soc 116:5150

119. Marquardt R, Sander W, Laue T, Hopf H (1995) Liebigs Ann 1643

120. Marquardt R, Sander W, Kraka E (1996) Angew Chem 108:825; Angew Chem Int Ed Engl 35:746

121. Elschenbroich C, Möckel R, Zenneck U (1978) Angew Chem 90:560; Angew Chem Int Ed Engl 17:531

122. Stöbbe M (1986) Dissertation, Universität Hamburg

Unsaturated Oligoquinanes and Related Systems

Rainer Haag · Armin de Meijere

Institut für Organische Chemie, Georg-August-Universität Göttingen, Tammannstrasse 2,
D-37077 Göttingen, Germany. *E-mail: ameijer1@uni-goettingen.de*

Fully unsaturated oligoquinanes comprise a class of fused five-membered ring compounds which contain extended π-systems with interesting electronic properties. This class of rigid molecules contains planar, bowl- and even ball-shaped structures, and their strain energy increases for each additionally fused five-membered ring. The first three members of this family, fulvene, pentalene and acepentalene have been synthesized or at least generated, whereas all higher members were only approached or studied by computational methods. In this review the class of fully unsaturated oligoquinanes – ranging from fulvene to C_{20}-fullerene – is presented with respect to their syntheses and their properties. Also, related molecules with similar structural features will be discussed, which have not been highlighted in previous volumes of this series [1, 2].

Keywords: Polyquinanes (unsaturated), fulvene, pentalene, acepentalene, C_{20}-fullerene.

Topics in Current Chemistry, Vol. 196
© Springer Verlag Berlin Heidelberg 1998

1
Introduction

The chemistry of oligoquinanes, the class of fused-ring compounds consisting of five-membered rings only, has experienced an exceptional growth since the 1960s [1, 2]. Saturated cyclopentanoid structures are important motifs in many natural products, whereas the fully unsaturated oligoquinanes (Fig. 1) are among the most highly strained π-systems conceivable and are therefore unlikely to be found in natural sources. In addition to their high strain these polyenes possess interesting electronic properties [3]. The continuous advances in analytical and computational techniques have made these π-systems persistent targets for both structurally and theoretically oriented organic chemists. The challenges are manifold: to learn about the electronic and structural properties of these highly strained π-systems; to generate non-natural products and explore the frontiers of modern synthetic methods. Finally, chemists can be driven by purely aesthetic reasons in trying to synthesize some of these molecules which are highly symmetrical.

Although several groups have been and still are engaged in the synthesis of some of these fully unsaturated oligoquinanes [1, 2, 4], only three members in the series, **1**, **2**, and **3**, have actually been prepared or at least characterized as short-lived species so far. Higher members of the series have been approached synthetically but the increasing ring strain from fulvene **1** to C$_{20}$-fullerene **7** makes them even more unlikely to be observed. Nevertheless, ab initio calculations have been performed for many of these strained π-systems predicting their structural and electronic properties.

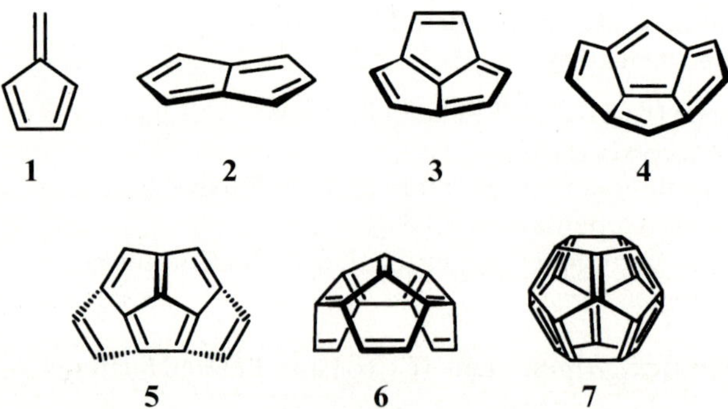

Fig. 1. A series representative of fully unsaturated oligoquinanes

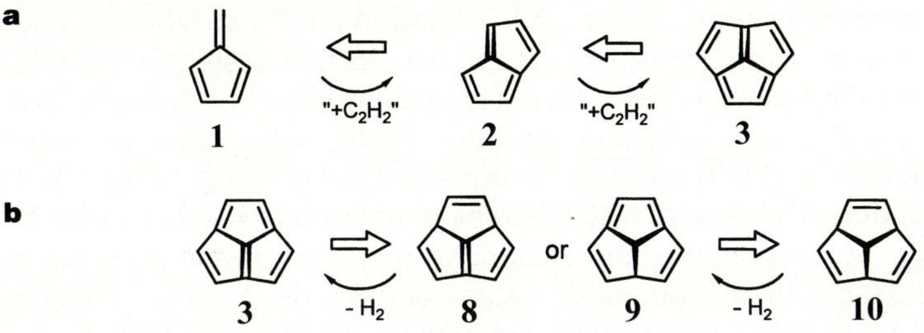

a

1 "+C$_2$H$_2$" **2** "+C$_2$H$_2$" **3**

b

3 – H$_2$ **8** or **9** – H$_2$ **10**

Scheme 1

Polycyclic aromatic hydrocarbons (PAHs) such as naphthalene, anthracene, pyrene etc. which might be termed "oligobenzenes" in this context are mostly planar as long as they do not contain any five-membered rings, the higher members of the series of fully unsaturated oligoquinanes 3–7 are bowl- or even ball-shaped. This topological feature puts additional strain into these skeletons 3–7 consisting of sp^2-hybridized carbons, far more than in the similarly ball-shaped C$_{60}$-, C$_{70}$-fullerenes etc. In order to form a stable fullerene each fully unsaturated five-membered ring must be exclusively surrounded by six-membered rings [5]. According to this empirical rule C$_{20}$-fullerene 7 has to be the least stable spherical fullerene.

Formally, higher oligoquinanes can be formed from fulvene 1 by consecutive addition of etheno bridges (Scheme 1a). This concept has actually been successfully applied in the synthesis of pentalene 2 but it is unlikely to be practicable for higher members of this family due to the drastical increase in ring strain on going from, e.g., 2 to 3. Therefore all realistic approaches towards compounds 3–7 start from a less unsaturated skeleton and try to introduce the additional double bonds in a stepwise fashion (e.g. 3 ⇒ {8 or 9} ⇒ 10, Scheme 1b).

In this report several new approaches towards the generation of highly unsaturated oligoquinanes, the chemistry of their precursors and related systems will be covered. In general, unsaturated oligoquinanes and related systems will be presented, which are in line with the general formula C$_x$H$_y$ with y ≤ x. Also, in this account any unnecessary duplication of material covered in previous reviews on polyquinane chemistry in this series is avoided [1, 2].

2
Fulvene (C$_6$H$_6$) and Related Systems

2.1
Fulvene and Substituted Fulvene Hydrocarbons

Fulvene 1 (Fig. 1) [6, 7] may be considered as the monocyclic parent member in the series of fully unsaturated oligoquinanes, as it exhibits a cross-conjugated π-system which is a key feature in all higher members of the family 2–7 (Fig. 1). Unsubstituted fulvene, in fact, is the only compound in this family which is stable at room temperature for a short while. Detailed NMR experiments have demonstrated the non-aromatic property of this 6π-electron system [8]. In contrast to the isomeric benzene, fulvene 1 has localized double bonds [9]. Therefore, it behaves like a diene or like a dienophile, and it readily oligomerizes at temperatures above 0 °C [7]. In contrast, terminally disubstituted fulvene derivatives such as dimethylfulvene 11 which had been obtained as early as 1906 by Thiele et al. from cyclopentadiene and acetone in methanol in the presence of potassium hydroxide [10], are perfectly stable compounds. A rather efficient new synthesis of 11 and the monosubstituted fulvene derivative 12 by the reaction of cyclopentadiene with unsaturated carbonyl compounds in the presence of pyrrolidine (Scheme 2), has been described by Griesbeck [11]. Dimethylfulvene 11 in this transformation arises via Michael addition of the cyclopentadienide ion to the α,β-unsaturated ketone mesityloxide and a subsequent retro-

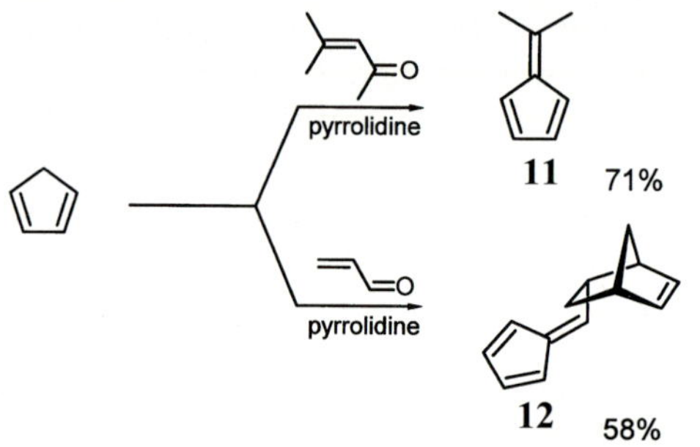

Scheme 2

aldol type reaction. The norbornenylfulvene 12, a formal Diels–Alder adduct of 6-ethenylfulvene to cyclopentadiene, can be transformed into 1,2-dihydropentalenes by flash vacuum thermolysis (see Scheme 10). The syntheses of many other fulvene derivatives have been covered in reviews previously [1, 2].

A species with an even higher degree of unsaturation is the cyclopentadienylidenecarbene 15. Ab initio calculations predict that 15 represents a shallow minimum on the C_6H_4 energy surface and should readily convert into the far more stable dehydrobenzene [12]. Indeed, carbene 15 has been generated and characterized by Wiersum et al. [13]. Flash vacuum thermolysis of Meldrum's acid 13 yields, upon loss of ethene, acetone and carbon dioxide, the cyclopentadienylideneketene 14 (Scheme 3). Subsequent thermolysis of 14 in the presence

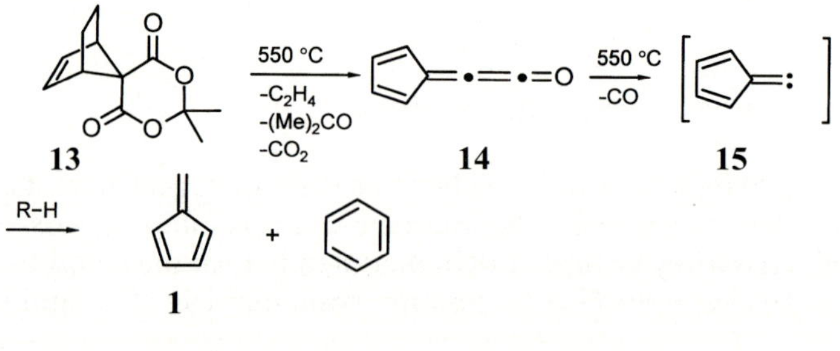

Scheme 3

of a hydrogen-donor source yields fulvene **1** and benzene. The formation of **1** can only be explained by the intermediate occurrence of the cyclopentadienyli- denecarbene **15**.

2.2
Heteroanalogous Fulvene Derivatives

Heteroanalogous fulvene derivatives, such as diazocyclopentadiene **16** and tetra-*tert*-butylcyclopentadienone **19** have also been of great interest as reactive precursors to other theoretically interesting molecules. Irradiation of diazocy- clopentadiene **16** readily generates the cyclopentadienylidene **17** which has been trapped with alkynes to form spiro-fused cyclopentadiene derivatives such as **18** (Scheme 4) [14]. It has been demonstrated by UV spectroscopy [15] and calcu- lations [16] that these spiro[2.4]heptatrienes (so-called [1.2]spirenes) **18** are spiroconjugated across their central sp^3 carbon.

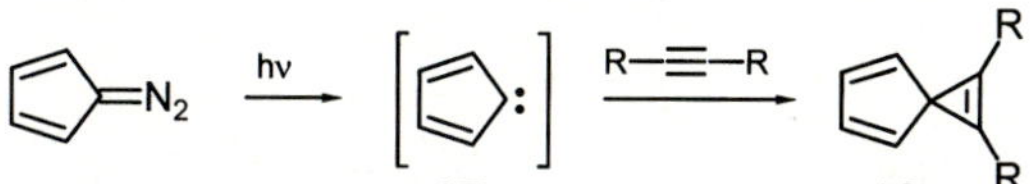

Scheme 4   **16**   **17**   **18**

The tetra-*tert*-butylcyclopentadienone **19** [17] is another prominent example of heterosubstituted fulvene derivatives. It was used in the generation of the highly strained, yet kinetically stabilized tetra-*tert*-butyltetrahedrane **21**, a deri- vative of the smallest of the so-called 'Platonic hydrocarbons' which contains the most highly bent single bonds (deviation of the interorbitae angle from the regular volume 26°) known for organic compounds [18]. As reported by Maier et al. (Scheme 5) [19], irradiation of the cyclopentadienone derivative **19** initial- ly forms the intermediate **20** by an intramolecular [2 + 2] cycloaddition. Upon further irradiation the extrusion of CO is observed, and the tetrahedrane deri- vative **21** can be isolated.

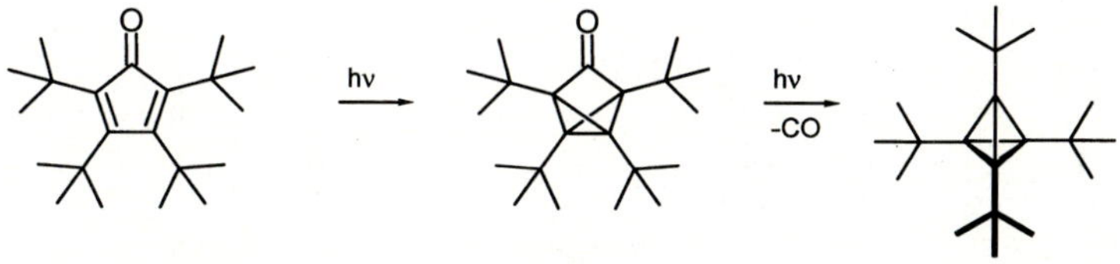

Scheme 5   **19**   **20**   **21**

The tetracyclopropyl-substituted fulvene derivative **23** and the tetracyclopro- pylcyclopentadienone **24** have both been envisaged as logical precursors to the pentacyclopropylcyclopentadienyl cation **22** (Scheme 6), which according to Becke 3LYP/3–21G and HF/3–21G computations [20] should have a singlet ground state unlike the antiaromatic parent cyclopentadienyl cation [21]. The fulvene derivative **23** and the pentacyclopropylcyclopentadienol **25**, both

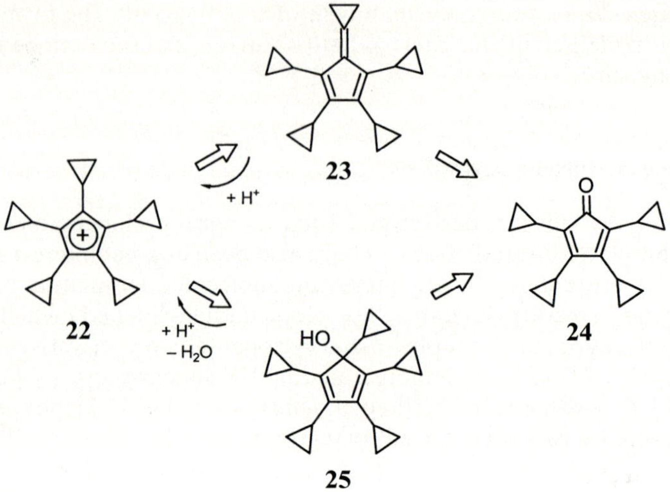

Scheme 6

derived from tetracyclopropylcyclopentadienone **24** have therefore been
defined as desirable targets for synthesis [22].

A number of symmetrically substituted cyclopentadienone metal complexes
have been prepared by metalcarbonyl-mediated dimerization of alkynes [23a,
24]. The (tetracyclopropylcyclopentadienone)tricarbonyliron complex **27** can
easily be obtained as the major product by ironcarbonyl-mediated dimerization
with CO insertion of dicyclopropylacetylene [25]. Upon treatment of the
complex **27** with triethylamine *N*-oxide, the uncomplexed tetracyclopropylcyclopentadienone **28** apparently is liberated; however, in contrast to the kinetically sufficiently stabilized tetra-*tert*-butylcyclopentadienone **19** (see Scheme 5)

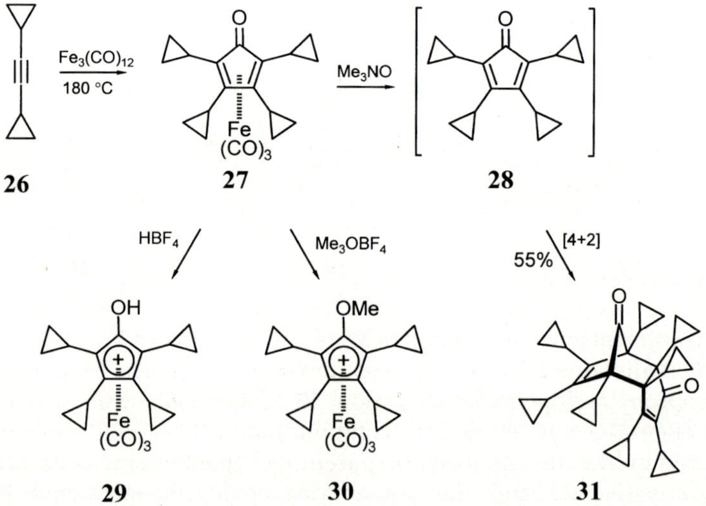

Scheme 7

and the tetrakis(trimethylsilyl)cyclopentadienone [23b], the steric stabilization by the four cyclopropyl substituents in **28** is not sufficient to prevent it from undergoing [4 + 2] dimerization to give **31** which was characterized by an X-ray crystal structure analysis (Scheme 7) [20, 22]. On the other hand, the cyclopentadienone complex **27** could be protonated with HBF_4 and methylated with trimethyloxonium tetrafluoroborate to yield the stable cationic cyclopentadienyltricarbonyliron complexes **29** and **30**, respectively.

Yet another rather general approach towards highly substituted cyclopentadienones has been developed via β-aminosubstituted α,β-unsaturated chromiumcarbene complexes such as **32**. Their cycloaddition to alkynes proceeds without carbonyl insertion to yield 1,2,3-trisubstituted 5-dimethylamino-3-ethoxy-cyclopentadienes which are readily hydrolyzed to the correspondingly substituted cyclopentenones [26]. After quaternization with methyl iodide the ammonium salt **33** is obtained. Treatment of the latter with a base such as NaOMe or NaOH results in deprotonation and elimination of trimethylamine to yield a trisubstituted cyclopentadienone which immediately dimerizes by [2 + 2] cycloaddition probably via a 1,4-diradical intermediate like **35** to yield the bis(cyclopentenone)-annelated cyclobutane derivative **34** (Scheme 8) [27].

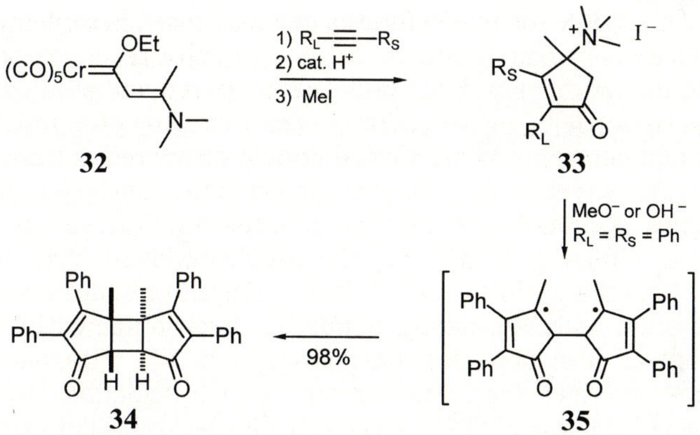

Scheme 8   **34**   **35**

3
Pentalene (C$_8$H$_6$) and Related Systems

3.1
Pentalene and Substituted Pentalene Hydrocarbons

Bicyclo[3.3.0]octa-1,3,5,7-tetraene, trivially called pentalene (**2**) (Fig. 1) [28] is the second member in the series of fully unsaturated oligoquinanes. Hückel-MO theory predicts that this planar hydrocarbon with its 8π-electron system should be an antiaromatic species [3]. The 2-methylpentalene **36** could be generated by a retro-Diels-Alder reaction and deposited as a film at $-196\,°C$ on an NaCl or quartz plate to be characterized spectroscopically. It rapidly dimerized upon warming the cold plates to temperatures above –140 °C [29]. So far only two stable derivatives of pentalene, the hexaphenyl- 37 [30], and 2,4,7-tri-*tert*-butyl-pentalene 38 [31], not complexed to a metal have been reported (Fig. 2). Many other pentalene derivatives and their syntheses have been mentioned in previous reviews [1, 2], and therefore only some recent developments are highlighted in this article.

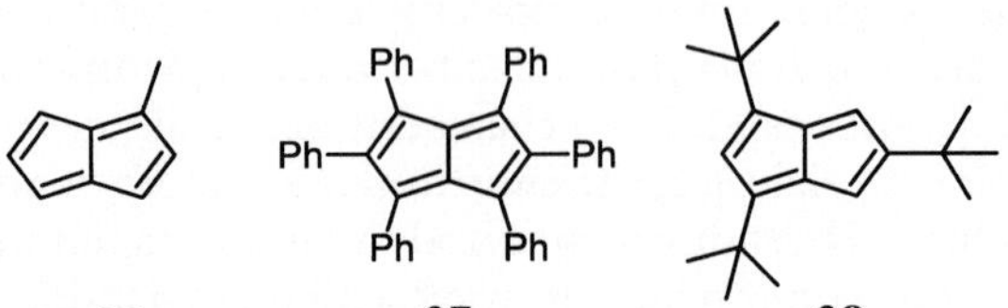

Fig. 2. An unstable and two stable pentalene derivatives

The question of aromaticity vs antiaromaticity and delocalized vs localized double bonds in pentalene 2 dates back to 1922, when Armit and Robinson compared it with naphthalene and postulated that the former might be similarly aromatic [32, 33]. While the first synthesis of a non-fused hexaphenylpentalene 37 [30] provided only some clues as to the non-aromatic reactivity of the pentalene skeleton, the tri-*tert*-butyl derivative 38, prepared and studied by Hafner and Süss in great detail [31], gave a better insight. The ring proton signals of this alkyl-substituted pentalene 38 are shifted upfield compared to those of fulvene 1 and other cyclic polyenes. This observation led to the conclusion that the pentalene derivative 38 should be an antiaromatic species. However, the results did not permit one to distinguish whether the double bonds in 38 are localized or undergoing a rapid bond shift between the two Kekulé structures. An indication that 2 is a system with alternating bonds has been obtained by comparing the photoelectron (PE) and electronic absorption (UV-VIS) data with calculated orbital energies [34]. More recent DFT calculations at the Becke 3LYP/6–311+G* // Becke 3LYP/6–31G*+ZPVE (Becke 3LYP/6–31G*) level of theory have revealed that pentalene 2 has a localized antiaromatic π system with a singlet ground state lying 9.7 kcal/mol below the triplet state [35]. The calculated structural parameters of 2 and its dication 2^{2+} (see Fig. 3) appear to be quite reliable as the agreement between calculated and experimental bond lenghts for pentalene dianion 2^{2-} and its dilithium derivative 45 is excellent.

Brown et al. [36] attempted to approach pentalene 2 by a thermolytic ring contraction-cyclization of 3-ethenylbenzyne 40 which in turn was generated by flash vacuum thermolysis of 3-ethenylphthalic anhydride 39. Indeed, the pentalene dimer 42 (~ 50 %) was formed along with phenylacetylene 43 (~ 50 %) in 80 % overall yield (Scheme 9). The failure to detect monomeric pentalene 2 is in accord with the observation of Bloch et al. that 1-methylpentalene dimerizes above –140 °C [29]. The formation of phenylacetylene 43 was unexpected, and it is as yet unclear whether it arises by migration of two hydrogens in the aryne 40 or the intermediate carbene 41, or whether it is a secondary product formed from pentalene 2.

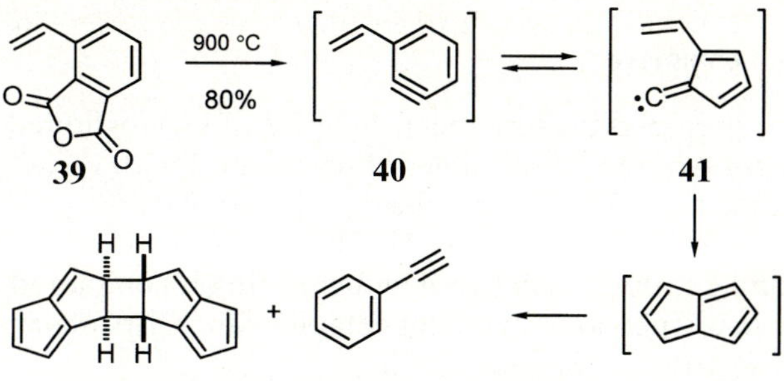

Scheme 9 42 ~1:1 43 2

3.2 Dilithium Pentalenediide

Both, the synthesis [37] and X-ray crystal structure [38] of dilithium pentalenediide **45** have been reported and covered in a previous review (Scheme 10, Fig. 3) [2]. Reaction of the dihydropentalene **44** with *n*-butyllithium yielded the crystalline dilithium pentalenediide **45** (Scheme 10). The more recently reported flash vacuum thermolysis of 6-norbornenylfulvene **12**, a precursor to 6-ethenylfulvene, allows one to prepare the dihydropentalene **44** in gram quantities [11] (see also Scheme 2).

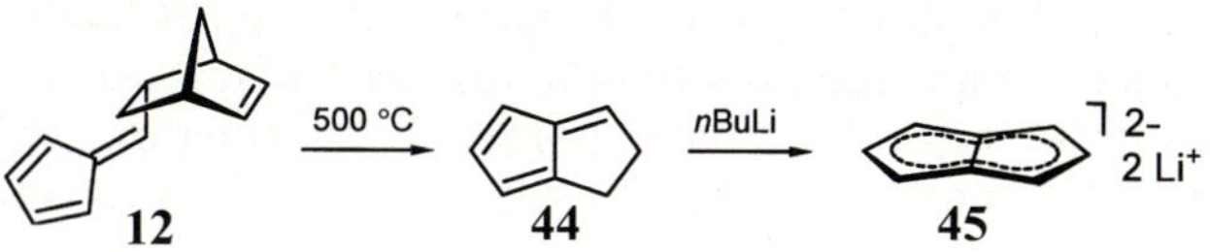

Scheme 10 12 44 45

The structure analysis of the dilithium pentalenediide **45** reveals a C_2-symmetric ion triplet with the two lithium cations located on opposite sides of the two different rings. The structural parameters were extremely well reproduced by ab initio calculations [35] (see Fig. 3). Both the experimental structural parameters and calculated magnetic susceptibility exaltation classify the 10π-electron species **45** as an aromatic compound. Apparently, the lithium counterions in **45** do not exert any significant effect on the bond lenghts of the dianion $\mathbf{2}^{2-}$. On the other hand, the antiaromatic pentalene **2** and its aromatic dication $\mathbf{2}^{2+}$ show the characteristic bond length alternation (Fig. 3) [35].

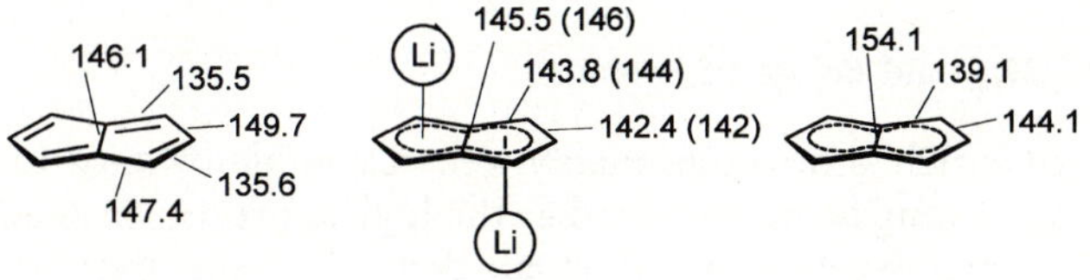

2 (S_0, C_{2h}) **45** (D_{2h}) $\mathbf{2}^{2+}$ (D_{2h})

Fig. 3. Calculated structural parameters of pentalene **2**, pentalene dianion $\mathbf{2}^{2-}$ (experimental values for dilithium pentalenediide **45** in parentheses [38]) and pentalene dication $\mathbf{2}^{2+}$

3.3
Pentalene Metal Complexes

In 1973 Hafner prepared the first metal complex of unsubstituted pentalene **46** (Fig. 4) by reaction of the [2 + 2] dimer of pentalene **2** with Fe_2CO_9 [39]. Several other dinuclear complexes of **2** have been prepared from the pentalenediide **45** as reviewed previously by Paquette [1]. All these structures contain a planar pentalene ligand, in which each five-membered ring is complexed to a different metal atom as in bis(pentamethylcyclopentadienylnickel)pentalene **47** obtained by reaction of **45** with Cp*Ni(acac) [40].

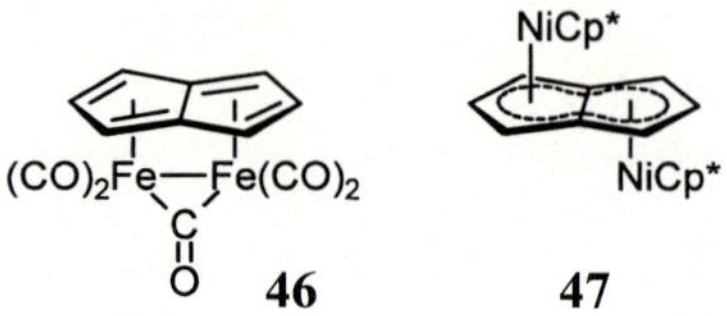

Fig. 4. Dinuclear pentalene complexes

Recently, Jonas et al. presented a new mode of complexation for the pentalene ligand in which all eight carbon atoms are coordinated to only one metal atom (Scheme 11). Dilithium pentalenediide reacts with Cp_2VCl to yield the vanadium pentalene-Cp-complex **48** [41, 42]. Reaction with Cp_2ZrCl_2 or Cp_2HfCl_2 gives the bispentalene metal complexes **49** [42]. In these complexes all carbon-metal bonds show similar lengths and the out of plane deformation angle of the pentalene skeleton (the four substituents on the central carbon-carbon bond) is ca. 43°.

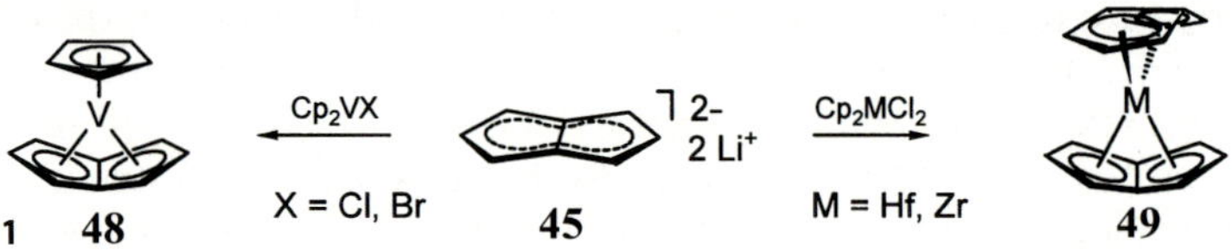

Scheme 11

4
Acepentalene ($C_{10}H_6$) and Related Unsaturated Triquinane Derivatives

4.1
Triquinacene ($C_{10}H_{10}$) and Related Systems

Triquinacene **10** which is now reasonably well accessible by a six to seven step synthesis [1, 2], has long been envisaged as the logical precursor to acepentalene **3** [43] (see Scheme 1b). Woodward et al. developed the first synthesis for triquinacene **10** [43] and suggested that it should be a potential precursor to both the fully unsaturated tricycle acepentalene **3** and dodecahedrane **88** [1, 44]. Although many attempts have been made to cyclodimerize triquinacene **10**, none has been successful to even approach dodecahedrane [1]. However, the proposed introduction of additional double bonds into the tricyclic ring system **10** to eventually form acepentalene **3** and its dianion has been achieved (see below) [4].

10   **50**   **51**

Fig. 5. Triquinacene **10** and related hydrocarbons

Based on its heat of hydrogenation, triquinacene **10** has been proposed to be stabilized by 4.5 kcal/mol in terms of neutral homoaromaticity [45]. Such homoaromaticity [46] has also been claimed for *cis,cis,cis*-1,4,7-cyclononatriene **50** (Fig. 5) [47]. In this context it is interesting to note that the most highly symmetrical valence tautomer of triquinacene **10**, the $(CH)_{10}$ hydrocarbon **51** has also been prepared, and shown to readily rearrange to triquinacene **10** [48, 49]. The suggestion that the double bonds in triquinacene **10** are similarly homoconjugated through space like those in **50**, and that **10** with its 6π-system therefore should be homoaromatic, has provoked a vigorous discussion going on for over ten years [45, 50, 51]. An examination of the bond pattern for triquinacene **10** applying the AIM (Atoms In Molecules) theory [50] could not locate any bond critical points between the centers supposed to exhibit homoaromatic interaction. Inspired by these contradictive results the heat of formation ΔH^f for triquinacene **10** has been determined experimentally for the first time by measuring its heat of combustion [52]. The experimental value of $\Delta H^f =$ 57.5 kcal/mol is about 4 kcal/mol higher than that calculated from heats of hydrogenation and in good agreement with the theoretically derived results, especially those reported by Schulman et al. [51]; this shows that there is no additional stabilization by homoaromaticity at least in triquinacene **10**.

The facile thermal [2 + 2 + 2] cycloreversion of diademane **51** (R = H, Scheme 12) is an interesting formation, but not a viable synthesis of triquinacene **10** [48, 49]. This rearrangement in which three cyclopropyl σ-bonds of **51** open to form the three π-bonds of **10** is a concerted process and occurs at only 80 °C

X = H, Br   **52-R**   **51-R**   **53-R**

R = H, Me, CH$_2$OH, SiMe$_3$

Scheme 12   **51-R**   **10-R**

with an Arrhenius activation energy of 31 kcal/mol [49, 52]. Diademane derivatives **51-R** with an electron donating substituent at the central bridgehead (e. g. R = TMS, Me, CH$_2$OH) undergo this cycloreversion even faster than the unsubstituted parent compound **51-H** [53]. This reaction cleanly leads to centro-substituted triquinacene derivatives **10-R** (R = Me, CH$_2$OH, TMS) which are otherwise inaccessible or at best much more difficult to make [1, 2]. The diademane derivatives **51-R** (R ≠ H) are obtained along with the isomeric ones **53-R** (R ≠ H) by photochemical intramolecular [2 + 2] cycloaddition of the corresponding snoutene derivatives **52-R** (R ≠ H) which can be prepared from bromocycloooctatetraene in an eight-step sequence analogous to the preparation of snoutene **52-H** from the unsubstituted cyclooctatetraene.

Although diademane **51-H** itself crystallizes very well, several attempts to determine its structure by X-ray crystallography have failed because the crystals were plastic and completely disordered. It was conceived that the intermolecular hydrogen bonding in 1-hydroxymethyldiademane **51-R** (R = CH$_2$OH) would overcome this packing disorder. Indeed, the centrally substituted isomer **51-R** (R = CH$_2$OH) crystallized well and could be separated from the isomer **53-R** (R = CH$_2$OH). In the crystal, the molecules of **51-R** (R = CH$_2$OH) are aggregated due to intermolecular hydrogen bonding between the hydroxymethyl groups in such a way that the almost spherical diademane residues cannot rotate around the bond to their substituents [53].

4.2
Azatriquinacene (C$_9$H$_9$N)

A remarkably straightforward access to perchloro-10-azatriquinacene **59** has been reported by Mascal et al. [54]. Towards this end, the disubstituted pyrrole **54** was hydrogenated selectively to give the *cis*-2,5-disubstituted pyrrolidine **55** (Scheme 13). This could be cyclized in refluxing xylene yielding the pyrrolizidinone **56**. After hydrolysis of the ester functionality the final cyclization was performed by dry distillation of the carboxylic acid from sodalime to yield the stable hemiaminal **57**. The parent azatriquinane **58** can be obtained by reduction with lithium aluminum hydride in refluxing tetrahydrofuran. Attempts to dehydrogenate **58** have so far only been successful by the photochemical chlorina-

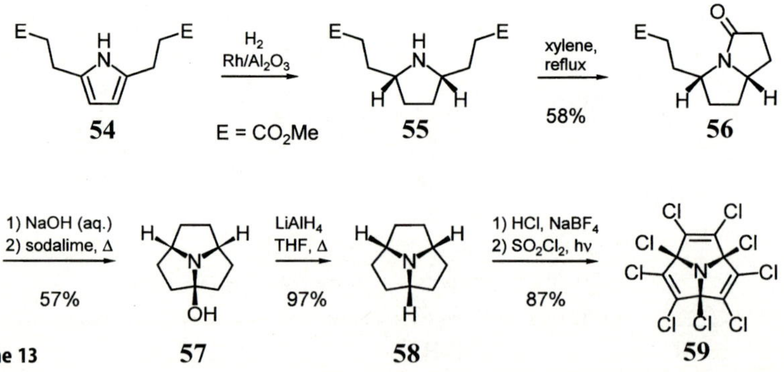

Scheme 13

tion-dehydrochlorination sequence to give the perchlorinated azatriquinacene **59**. The corresponding perchlorinated triquinacene had previously been prepared by this method from the saturated hydrocarbon triquinane by Jacobson [55].

4.3
Potential Acepentalene Precursors and the Elusive Acepentalene

In the series of fully unsaturated oligoquinanes the tricyclic acepentalene **3** is the first member to exhibit a curved molecular surface. According to Hückel MO theory, **3** should have a triplet ground state [3] but more recent ab initio calculations have disclosed that the singlet ground state of acepentalene **3** is more favorable by 3.9 kcal [35]. However, the prohibitively large strain in the molecule which by far exceeds that of pentalene **2**, renders **3** impossible to isolate at ambient temperature. While pentalene **2** was successfully generated from monocyclic precursors (compare Schemes 1a and 9) [36] analogous attempts to approach acepentalene **3** from 7,7-diethynylfulvene have not been successful [56]. When introducing two additional double bonds one by one into the triquinacene skeleton **10**, the two different isomeric tetraene isomers **8** and **9** (see Scheme 1b) can be envisaged as potential intermediates. According to force field calculations, however, the latter is more strained by 18 kcal/mol [57]. It was therefore more reasonable to aim at the tetraene **8** as a promising precursor for acepentalene **3**, and consequently all efforts were concentrated on the synthesis of an isolable derivative of 4,7-dihydroacepentalene **8**.

One kind of 4,7-disubstituted tetraene of type **8**, namely a series of 4,7-bis(dialkylamino)dihydroacepentalenes **61**, can be obtained from triquinacene **10** via its tribromide **60** and a subsequent sequence of a twofold elimination-addition followed by an elimination brought about by treatment with secondary amines (Scheme 14) [58]. An attempted quaternization of one dialkylamino group in **61** and elimination of both (analogous to the generation of pentalene [28]), did not succeed [59]. Complexation of these bis(dialkylamino)tetraenes **61** with transition metals gave stable metal complexes **62** and **63** in good yields [60, 61]. The X-ray crystal structure of the tricarbonyliron complex **62** clearly indicates that the tricarbonyliron unit is attached to only one of the two five-membered rings with a slightly predominating coordination to the central

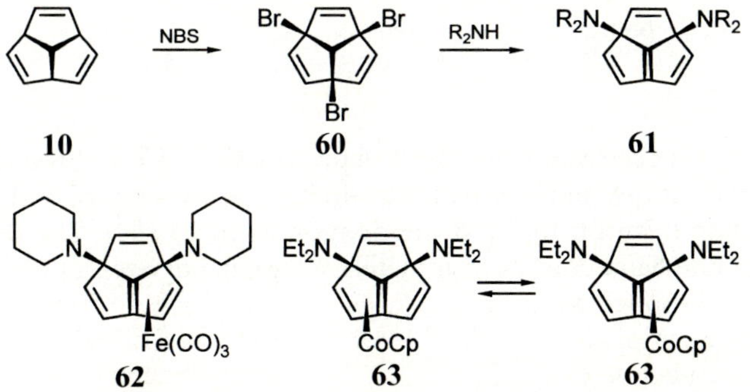

Scheme 14

double bond. It also revealed a considerable out of plane deformation for the
central double bond with an angle of 35° (see Fig. 13) [60]. No metallotropic
mobility of the $(CO)_3Fe$ unit was detectable up to a temperature of 150 °C for the
tricarbonyliron complex **62**, but the cyclopentadienylcobalt complex **63** showed
fluxional behaviour even at room temperature. The barrier for the metallotropic
rearrangement of **63** was determined from the coalescence temperature observed
by measuring the temperature dependent ^1H NMR spectra to be 15 kcal/mol [61].

A more versatile approach to 4,7-disubstituted dihydroacepentalenes **65** is via
the stable acepentalene dianion **64** as an easily accessible intermediate. Dipotassium acepentalenediide **64** can be obtained in virtually quantitative yield by
treatment of triquinacene **10** with the superbasic mixture of potassium *t*-pent-
oxide and butyllithium [or even better potassium *t*-butoxide, butyllithium and
tetramethylethylenediamine (TMEDA)] in hexane, the so-called Lochmann-
Schlosser base (Scheme 15) [62, 63]. Mechanistically this transformation has

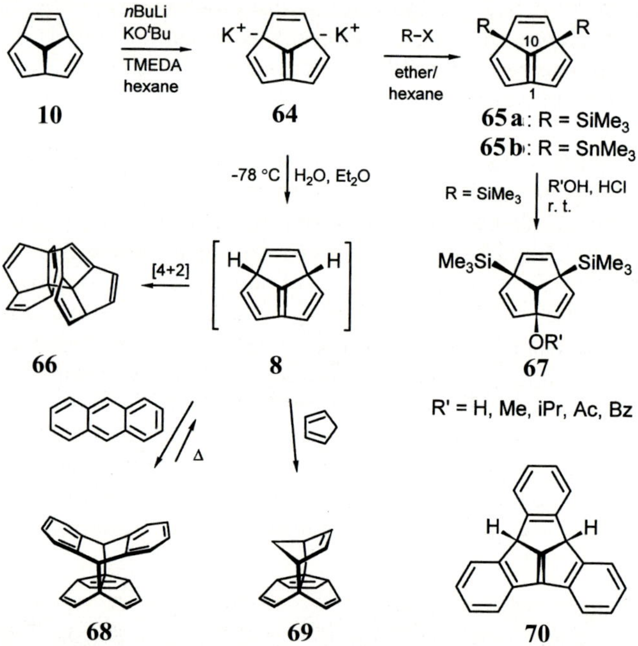

Scheme 15

been shown to proceed via a threefold deprotonation at the three allylic positions in **10** and a subsequent hydride elimination of the central hydrogen from
the intermediate trianion to form the dianion as in **64** [64]. The dipotassium
acepentalenediide **64** cannot be completely purified, but even in its crude form
very efficiently trapped with various electrophiles such as trimethylsilyl and
trimethylstannyl chloride to yield the 4,7-disubstituted dihydroacepentalenes
65a and **65b**, respectively.

The sp^2-hybridized carbon atoms of the central double bond in these dihydroacepentalene derivatives **65** are highly pyramidalized with an angle between
the two adjacent five-membered rings of 143° (see Fig. 13) [64, 65]. Therefore,
the central double bond in compounds of type **65** is highly reactive towards electrophiles, e.g., **65a** adds water and alcohols under acid catalysis, and organic
acids readily at room temperature to yield the 1-substituted 4,7-bis(trimethylsilyl)triquinacenes **67**. When the dipotassium acepentalendiide **64** is trapped
with water (by addition of moist ether at –78°C), the 4,7-dihydroacepentalene **8**
is formed, but it immediately dimerizes. This dimerization is remarkable on its
own since only a single product, the [4 + 2] cycloadduct **66**, is isolated in very
high yield. When generated in the presence of a reasonably reactive 1,3-diene
the highly reactive dihydroacepentalene **8** can be trapped as the corresponding
[4 + 2] cycloadduct of that particular diene, e.g. **68** and **69**, from anthracene and
cyclopentadiene, respectively (Scheme 15). The highly reactive tribenzoanalog
of **8**, tribenzodihydroacepentalene **70** could be observed by ^{1}H NMR spectroscopy at –80 °C. Upon warming to 0 °C, the hydrocarbon **70** dimerized to yield
the head to tail type [2 + 2] cycloadduct across its most highly strained central
double bond [64].

4.4
Generation of Acepentalene (C$_{10}$H$_6$)

Formally, acepentalene **3** can be generated by homolytic cleavage of both
bridgehead carbon to substituent bonds in any dihydroacepentalene derivative
of type **65**. In the case of the bis(trimethylstannyl) derivative **65b** this cleavage
could be brought about in the mass spectrometer by chemical ionization with
N$_2$O to generate the anion radical **3**$^{\cdot-}$. A subsequent neutralization-reionization
sequence (NRMS) of the selected anion radical **3**$^{\cdot-}$ yielded the acepentalene cati-
on-radical **3**$^{\cdot+}$ and thus proved the intermediate existence of the neutral acepentalene **3** (Scheme 16) [4]. Ab initio calculations at the B3LYP/6 – 311 + G* //
B3LYP/6 – 31G* + ZPVE (B3LYP/6 – 31G*) level of theory predict the molecule of
3 to be bowl shaped with a typical bond length alternation (Fig. 6b).

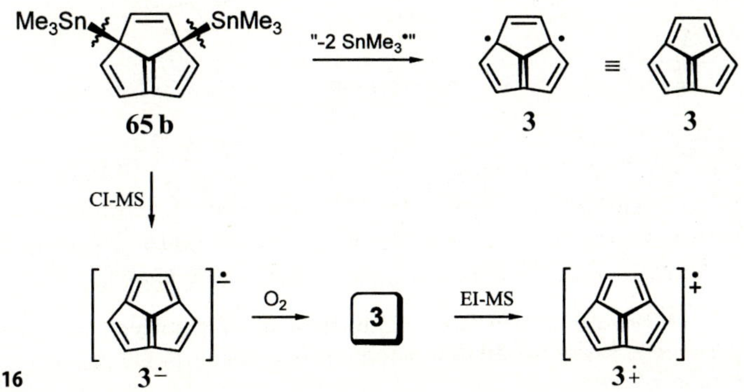

Scheme 16

4.5
Dilithium Acepentalenediide and Other Metal Complexes of Acepentalene

The 4,7-bis(trimethylstannyl)dihydroacepentalene **65b** also turned out to be perfectly suited for the generation of pure dilithium acepentalenediide **71** by transmetallation (Scheme 17) [63, 65]. Treatment of **65b** with salt-free methyllithium in dimethoxyethane yielded a crystalline compound which according to an X-ray crystal structure analysis turned out to be an interesting dimeric aggregate of two ion triplets held together by dimethoxyethane ligands (Fig. 6a) [63]. Two of the lithium cations are sandwiched between the convex surfaces of the acepentalene units and two lithium ions are located on the outer surfaces, all four located off the threefold axes of the carbon skeleton. The tetralithium bis(acepentalenediide) can thus be regarded as a lithium sandwich complex of the acepentalene dianion. A similar dimeric arrangement has been concluded from NMR spectral phenomena for the bowl-shaped corannulene tetraanion in **72** [66]. Both the pentalene dianion in **45** (see Fig. 3) and the acepentalene dianion in **71** are mesomerically stabilized aromatic systems as revealed by NMR spectroscopy, X-ray crystallography [38, 63], and ab initio calculations [35]. For the bowl-shaped acepentalene dianion in **71**, a facile bowl-to-bowl inversion with a calculated energy barrier of 7.9 kcal/mol has been predicted (Fig. 6a) [67]. The bowl-to-bowl inversion barrier for the corannulene tetraanion in **72** has been experimentally determined to be 13.4 kcal/mol [66]. The calculated optimized (B3LYP/6–31G*) geometry of acepentalene dication **3**$^{2+}$

Scheme 17

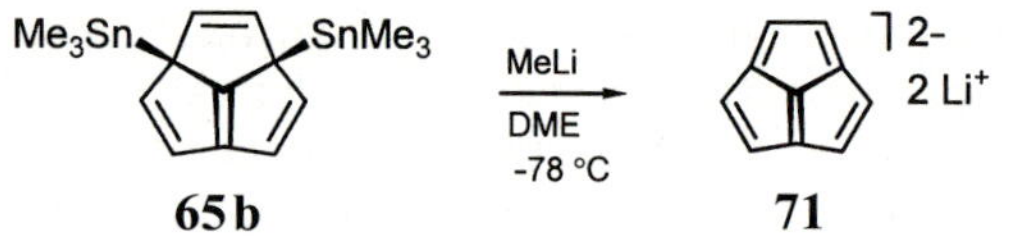

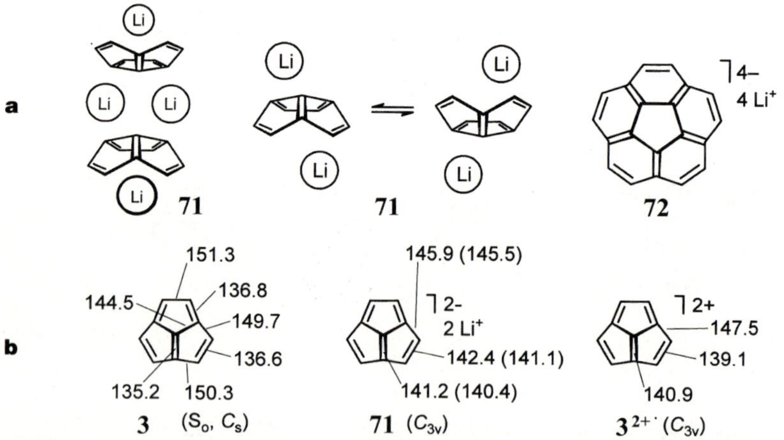

Fig. 6. Structures and calculated structural parameters of acepentalene **3**, acepentalene dianion **3**$^{2-}$ (experimental values for dilithium acepentalendiide **71** in parentheses [63]) and acepentalene dication **3**$^{2+}$

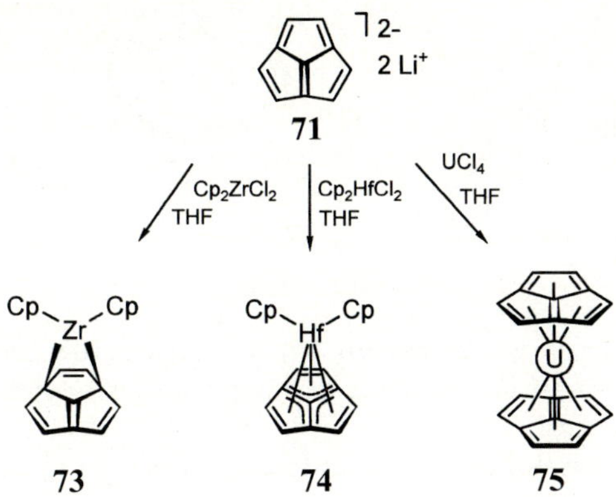

Scheme 18   **73**   **74**   **75**

is C_{3v} symmetric like that of $\mathbf{3}^{2-}$, but has a significantly alternating bond length pattern (Fig. 6b) [35].

The dilithium acepentalenediide **71** is also a good precursor for other metal and transition metal complexes, as it is easily obtained with high purity, in contrast to the dipotassium salt **64** which could not be fully separated from admixed butyllithium, potassium t-pentoxide and other impurities. The reaction of **71** with Cp_2ZrCl_2 in tetrahydrofuran gave a complex which, according to its ^1H NMR spectrum, has a C_s-symmetric carbon skeleton and therefore, in analogy to the known bis(pentamethylcyclopentadienylzirkonium)trimethylenemethane complex [68], is best formulated as **73** [69]. The dilithium derivative **71** also reacted with Cp_2HfCl_2 to give an oxygen-sensitive complex, the ^1H NMR spectrum showing a single line for the six protons on the acepentalene residue in accord with its C_3-symmetric structure in **74** [69].

In analogy to the cyclooctatetraene dianion, **71** also forms a very stable sandwich complex **75** upon reaction with UCl_4 in THF (Scheme 18) [69]. The ^1H NMR spectrum of this complex **75** revealed a very characteristic singlet at $\delta = -21$ ppm (!) which is in good agreement with the chemical shift of uranocene, the bis(cyclooctatetraenediyl)uranium complex [70].

5
Dicyclopenta[*cd,gh*]pentalene ($C_{12}H_6$) and Related Systems

5.1
Dicyclopenta[*cd,gh*]pentalene and Other Unsaturated Tetraquinanes

The fully unsaturated dicyclopenta[*cd,gh*]pentalene **4** (Fig. 7) remains elusive. Several attempts at generating **4** have been made, but so far all available knowledge about the system rests on computational studies which predict the structure and properties of this highly strained π-system [35]. In contrast to computational results for pentalene **2** and acepentalene **3**, calculations for **4** predict an aromatic singlet ground state with a delocalized π-system. While the dianion of **4** is calculated to also be aromatic, the dication $\mathbf{4}^{2+}$ should be antiaromatic [35].

Fig. 7. Unsaturated C_{12}-tetraquinanes

In this context the as yet unknown triply bridged 10π annulene **76** is of interest, as it appears to be able to serve as a logical intermediate en route to the highly strained hexaene **4**.

The Cook-Weiss reaction, i.e. the condensation of two molecules of 3-oxo-glutarate with glyoxal, has proved to be a very powerful method to build up highly functionalized polyquinane frameworks and has been successfully utilized for several oligoquinane syntheses [2]. It has also been applied to access the dicyclopenta[*cd,gh*]tetraene **83** as reported by Cook et al. (Scheme 19) [71]. The important steps in this synthesis of **83** were the bisalkylation of the bisenol ether **77** to provide the 2,8- and 2,6-diallylbicyclo[3.3.0]octane-3,7-diones **78** and **79** in a 2:3 ratio, and the twofold aldol addition-type cyclization of the bisaldehyde **80** to generate the tetracyclic dihydroxydione **81**. The required 2,6-bisaldehyde **80** was prepared by ozonolytic cleavage of the two olefinic bonds of **79**. Reduction of the diketone **81** was executed with diborane in THF solution and the resulting tetraalcohol was transformed into the tetraxanthate **82**. Thermolysis of **82** in the mass spectrometer indicated the formation of the desired tetraene **83**. However, in the laboratory, generation of **83** by thermolysis of **82** with or without the addition of diphenylisobenzofuran was unsuccessful [72].

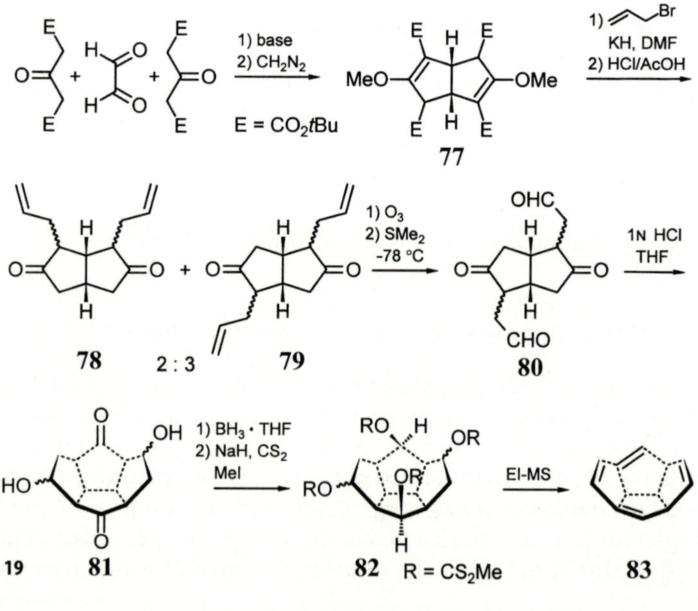

Scheme 19

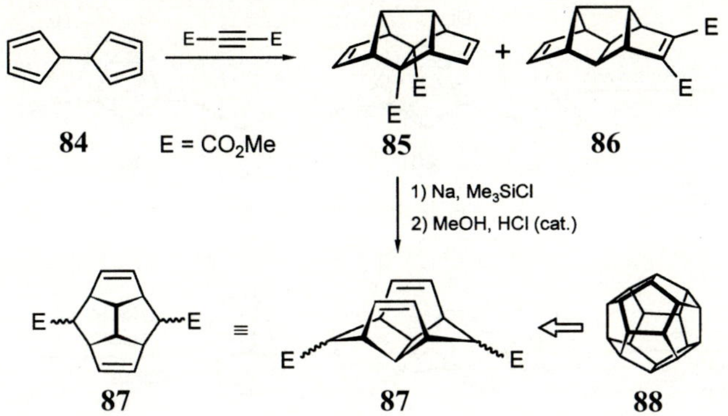

Scheme 20

The diester **87** with the same tetracyclic skeleton as **83** had previously been prepared by Paquette et al. via a domino Diels–Alder reaction of 5,5′-bicyclopentadienyl **84** with dimethyl acetylenedicarboxylate (Scheme 20) [73]. The key precursor **84** was obtained by iodine-induced oxidative coupling of the copper cyclopentadienide derived from the sodium derivative. The diester **85** formed along with **86** was transformed into a bissilyl bisenol ether by reductive cleavage of the central bond in the succinate moiety with sodium in the presence of trimethylsilyl chloride. Subsequent hydrolysis of the bisenol ether – actually a bisketene acetal – gave the dienic tetraquinacenedicarboxylate **87**. This compound served as the key intermediate in the first synthesis of dodecahedrane **88** [74].

5.2
Octahedrane ($C_{12}H_{12}$)

The tetracyclic diester **87** (Scheme 20) was also used as the starting material for the longtime elusive D_{3d}-symmetric $(CH)_{12}$ hydrocarbon octahedrane **93** (Scheme 21) [75]. Prior to that, only the unsymmetric 2,9-dimethyloctahedrane **94** had been reported [1, 76]. Octahedrane **93** can be regarded as the second congener of hydrocarbons consisting of a central n-membered ring capped with two $n/2$-membered rings. In this respect, dodecahedrane **88** consists of a central ten-membered ring capped with two five-membered while **93** has a central six- capped with two three-membered rings [77].

The key step in the synthesis of the parent hydrocarbon **93** was the transmissive bromination of the two closely spaced double bonds in **87** in which the neighboring double bond is attacked by the first intermediate bromonium ion. Among the two isomeric dibromides **89** and **90,** only the latter, minor product can be transformed into the dimethyl octahedranedicarboxylate **91** upon treatment with sodium methoxide in methanol solution. After ester hydrolysis and Barton decarboxylation of the diacid, the highly symmetric octahedrane **93** was obtained in good overall yield. The ^{13}C–H coupling constant $^{1}J_{CH}$ for the six protons on the two three-membered rings is slightly larger ($^{1}J_{CH} = 170$ Hz) than for normal cyclopropane C–H bonds which indicates an increased s-character of

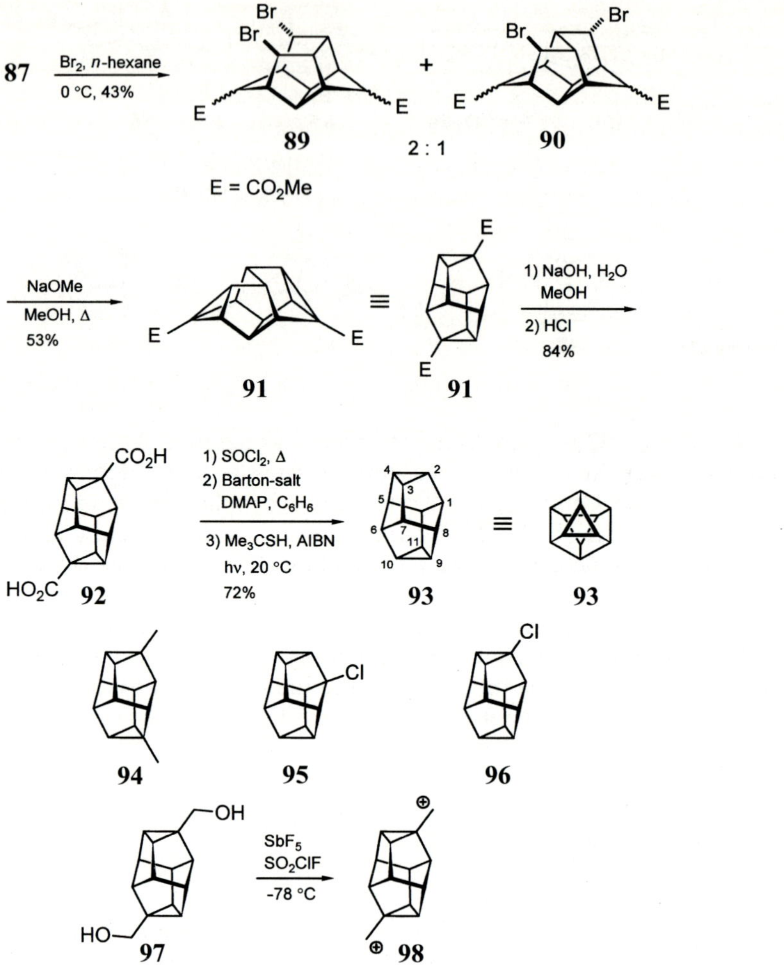

Scheme 21

34% for these bonds in **93** [75]. The photochlorination of **93** with *t*-butyl hypo-chlorite, accordingly, gives the 1-chlorooctahedrane **95** as the major product along with the 2-chloro isomer **96**. On the other hand, the higher *s*-character of the cyclopropane C–H bonds in **93** implies an enhanced C–H acidity which might be utilized for further functionalization.

Reduction of the diester **91** yielded the bis(hydroxymethyl)octahedrane **97**. Treatment of **97** with superacid led to the interesting cage dication **98** which according to its NMR data must be regarded as a bis(bicyclobutonium) ion (Scheme 21) [78]. The positive charge in the dication **98** is significantly delocalized into the cyclopropane rings.

6
Higher Unsaturated Oligoquinanes

The fully unsaturated linear and angular tetraquinanes **99**–**103**, the C_{13}-fenestranes **104**, **105** and the bisangular tetraquinanes **107**, **108** with two quaternary carbon atoms each (Fig. 8), resemble hydrocarbon skeletons consisting of four five-membered rings. While saturated congeners of **99** (see Note added in proof, p. *165*) and **100** are known as skeletons of natural products [1, 2], none of the fully unsaturated compounds has even been tackled. While **99** and **100** consist of a pentalene and an added fulvene unit each, and **102** is a cyclopentene-annelated acepentalene, they have no chance to be stable at ambient temperature. Compounds **101**, **107** and **108**, however, should be less strained.

The fully unsaturated [5.5.5.5]fenestrahexaene **104** (Fig. 8) has also not yet been synthesized. MINDO/3 calculations by Gleiter et al. predict that this hydrocarbon should be nonplanar and show pronounced bond alternation in its ground state [79]. In the context of discussions about the strain involved in fully planarizing a tetracoordinated carbon, a number of fenestrane derivatives have been prepared and investigated [80, 81]. The saturated *all-cis*-fenestrane **106** has been synthesized rather efficiently and its structure analyzed by gas-phase electron diffraction [80]. X-ray crystal structure analyses of unsaturated fenestrane derivatives, such as that of the tetrabenzofenestrane **109** indicate an almost normal tetrahedral arrangement around the central carbon atom with bond angles of 117° [82].

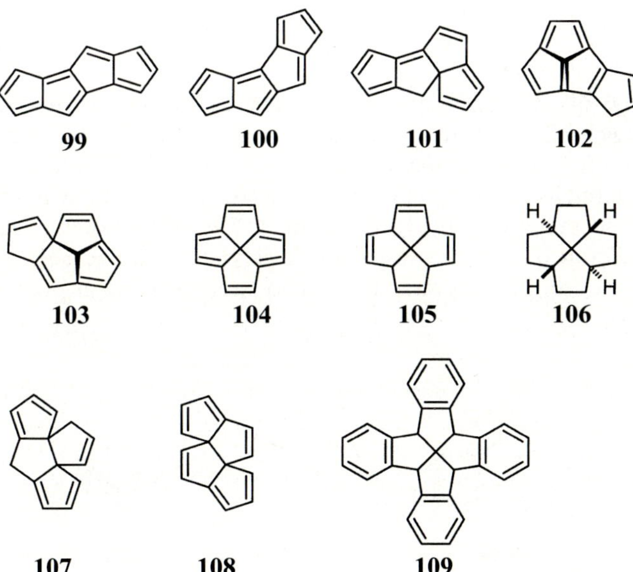

Fig. 8. Unsaturated linear and angular tetraquinanes as well as C_{13}-fenestranes

The most highly unsaturated fenestrane known so far is the [5.5.5.5]fenestra-tetraene **105** ($C_{13}H_{12}$) which has been synthesized by Cook et al. utilizing the Weiss-Cook reaction (see Scheme 19) [2, 72]. More highly unsaturated fenestranes with their expected extremely high strain such as **104** therefore remain a synthetic challenge.

An even larger variety of fully unsaturated pentaquinanes consisting of five five-membered rings can be constructed by attaching an additional cyclopentene ring to each of the tetraquinanes **99–105, 107, 108** in all conceivable modes, resulting in, e. g., **110–115** and others (Fig. 9), but none of them have so far been approached. Several saturated pentaquinanes of different constitutions have been reported, and their chemistry has been reviewed extensively by Paquette [1, 2].

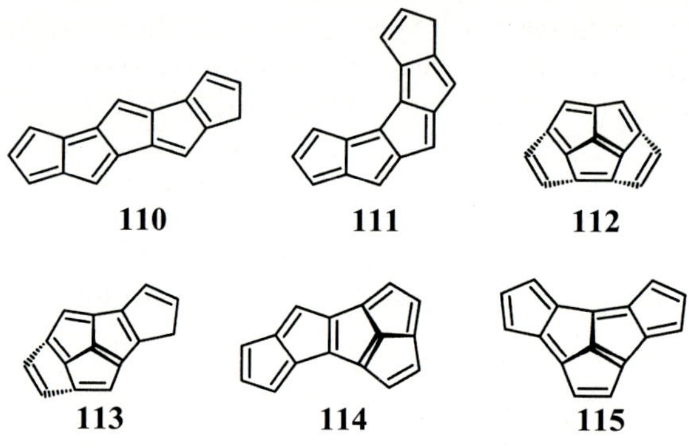

Fig. 9. Fully unsaturated pentaquinanes (incomplete listing)

A particular situation arises, when the linear C_{17}-pentaquinane **111** is bridged between its two ends to result in the skeleton of the so-called [5]peristylene **117** which essentially is a hexaquinane (Fig. 10). Decamethyl[5]radialene **116** which can be considered as the inner core of **117**, has been reported by Iyoda et al. [83], but the fivefold methylene-bridged analog **117** is still unkown. Only the fully saturated skeleton of **117**, the so-called peristylane **118** has been prepared by Eaton et al. [84], and its chemistry has also been covered in previous reviews [1, 2].

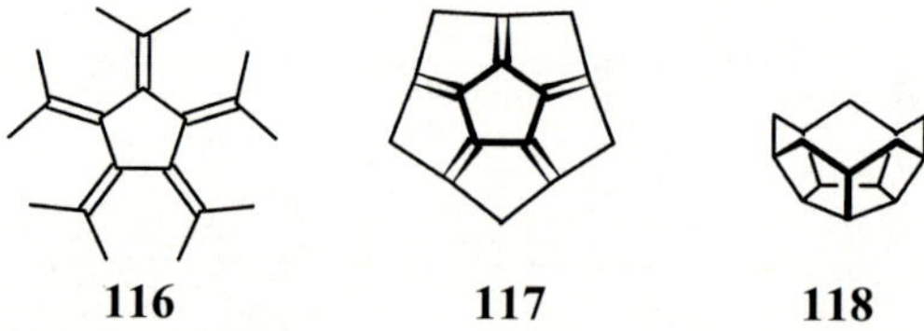

Fig. 10. Peristylane **118** and congeners

More recently, an elegant access to a pentaoxa[5]peristylane **123** has been reported by Wu et al. (Scheme 22) [85]. The Diels–Alder reaction of 5-(bismethoxy)methylcyclopentadiene **119** and *cis*-enediones **120** furnished the *anti-endo* adduct **121**. After cleavage of the dimethyl acetal moiety and ozonolysis of the double bond in **122**, the pentaoxa[5]peristylanes **123** (R = Me, *n*-Bu, *n*-Oct) were obtained in high yields. Compounds **123** resemble an interesting novel type of crown ether, the cation binding properties of which are being studied [85].

In the large family of conceivable fully unsaturated hexaquinanes (Fig. 11), only the partially unsaturated skeletons of **126** in form of the dione **127** [86] and of the C_3-symmetric C_{16}-hexaquinanoctaene **6** have been synthesized. The hexaquinatriene **130** has actually been synthesized by Paquette et al. and covered in previous reviews [1]. In analogy to triquinacene **10** (cf. Sect. 4.1) no homocon-

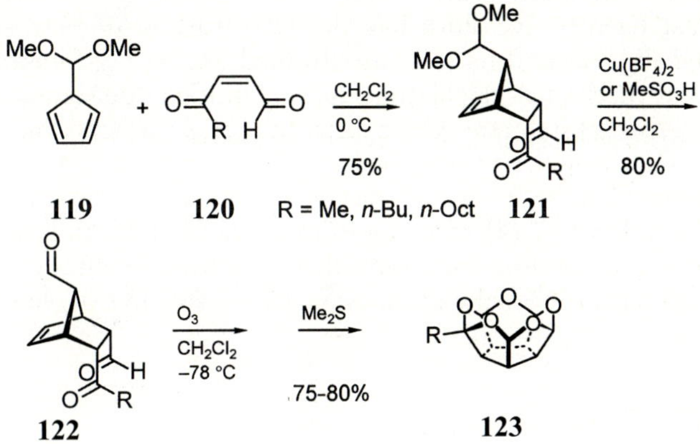

Scheme 22

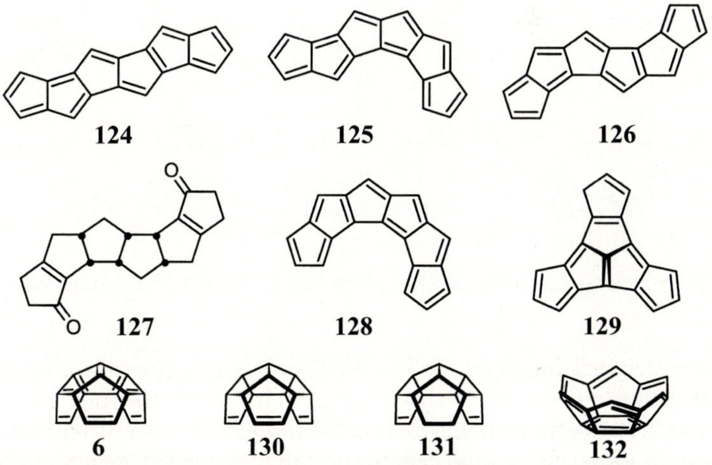

Fig. 11. Fully unsaturated hexaquinanes (incomplete listing)

jugative interaction between the three double bonds in **130** has been found [87]. An alternative approach to the fully saturated skeleton of **6** and to the diene **131** has been reported by Eaton et al. [88].

The fully unsaturated linear hexaquinanes **124–126, 128** and the angular derivative **129** as well as other isomers thereof have no chance ever to be prepared in view of the instability of pentalene **2** and acepentalene **3**. The fully unsaturated hexaquinane **132**, formally arising by bridging the two ends of the bent linear hexaquinane system **128**, is a particular case in that it is a corannulene (compare Fig. 6a) with an inner 6π and an outer 12π system.

7
Unsaturated C$_{20}$-Dodecahedranes

According to Euler's theorem, a spherical structure consisting of five-membered rings only, must have twelve sides like the hydrocarbon dodecahedrane **88** (Scheme 20 and Fig. 12) and its fully unsaturated analog C$_{20}$-fullerene **7** (see Fig. 1). Molecular mechanics calculations have identified dodecahedrane **88** as the lowest energy structure in the whole series of (CH)$_{20}$ isomers (Fig. 12) [5, 44, 89]. Among the C$_{20}$ clusters, however, the bowl-shaped isomer **134** (B), reminescent of corannulene (see **72** in Fig. 6a), has been calculated to be comparable in energy with the fullerene **7** (F) and ring isomer **133** (R) [89]. In laser ablation experiments of graphite under conditions that produce fullerenes from carbon clusters higher than C$_{30}$, only the monocyclic C$_{20}$ cluster **133** (R) has been observed [90].

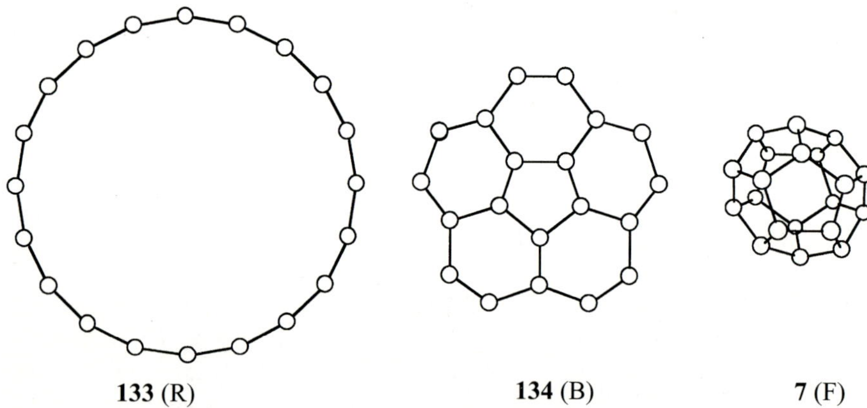

133 (R)   **134** (B)   **7** (F)

Fig. 12. Three isomeric C$_{20}$ clusters, **133** (R = Ring), **134** (B = Bowl), **7** (F = Fullerene)

Therefore stepwise introduction of double bonds into the saturated dodecahedrane **88** is the only logical alternative approach towards the fully unsaturated C$_{20}$-fullerene **7**. Paquette et al. have reported a β-elimination of trifluoroacetic acid from (trifluoroacetoxy)dodecahedrane **135** with hydroxide and methoxide ion in an ion cyclotron resonance mass spectrometer to infer indirectly the existence of **137** in the gas phase [91]. More recently, Prinzbach et al. have developed several new conditions to perform these β-eliminations on the dodecahedrane skeleton [92]. Dodecahedrene **137** of high purity can now be obtained by the reaction of the bromododecahedrane **136** with the strongly basic naked fluoride ion form, the iminophosphorane base "P$_2$F" developed by Schwesinger et al. [93] (Scheme 23) [92].

The fact that dodecahedrene **137** with its highly out of plane bent C=C double bond ($\Psi = 43.5°$ (Fig. 13) [94]) can be isolated, is remarkable on its own, but even

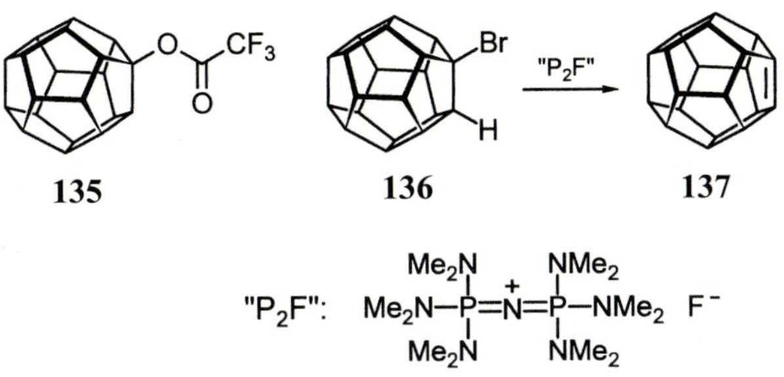

Scheme 23

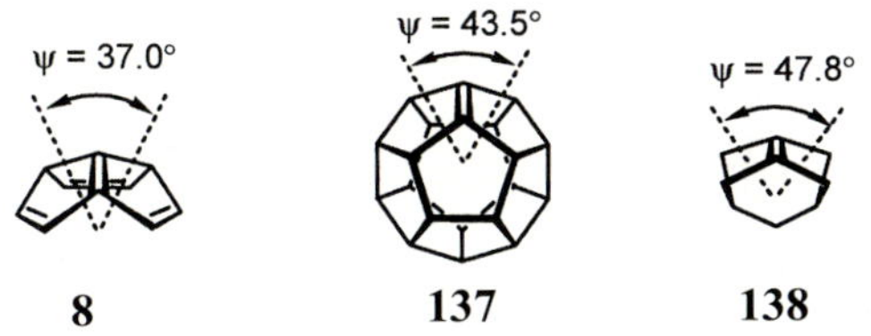

Fig. 13. Highly out-of-plane bent double bonds in comparison

more remarkable is the observation that dodecahedrene **137** dimerizes only under forcing conditions (around 300 °C) [44]. Such a degree of kinetic stabilization was unexpected, but it can be attributed to the efficient steric shielding by the four allylic hydrogen atoms all rigidly held in eclipsed orientations adjacent to the double bond on the surface of **137**. The dihydroacepentalene **8** with a similarly pyramidalized double bond ($\Psi = 37°$) but with only two eclipsed allylic hydrogens protecting it, cannot be isolated or even observed as a monomer at −80 °C (see Sect. 4.3). The tricyclo[3.3.2.0^{3,7}]non-3(7)-ene **138** ($\Psi = 47.8°$) which does have eight allylic hydrogens but not with the same orientation as the four in **137**, could only be generated in an argon matrix at 10 K, and it dimerized upon warming of the matrix to about 40 K [95].

Dodecahedradiene **142** has also been prepared and isolated as a stable compound. The highest yields (50–70%) and best purities (90–95%) of **142** were obtained by flash vacuum pyrolysis of the bislactone **141** (Scheme 24) [92]. The latter is accessible from the dioxopagodanedicarboxylate **139** by a twofold intra-

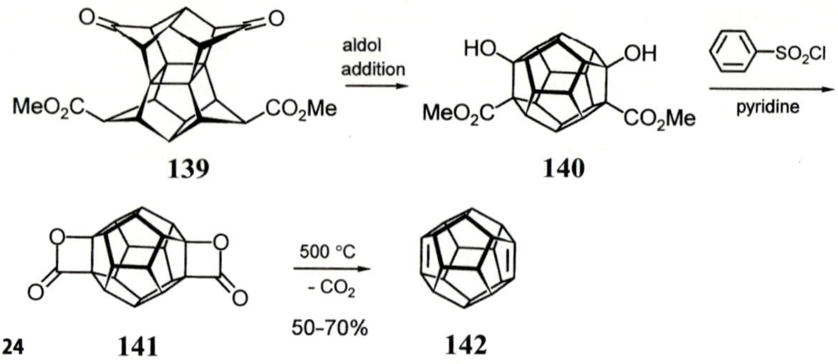

Scheme 24   **139**   aldol addition   **140**   —SO₂Cl / pyridine

141   500 °C / - CO₂   50–70%   **142**

molecular aldol addition to yield **140** and subsequent lactonization [96]. The calculated degree of pyramidalization of the double bonds in **142** ($\Psi = 42.8°$) is slightly smaller than that in dodecahedrene **137** ($\Psi = 43.5°$).

For dodecahedrane derivatives with an even greater degree of unsaturation, several approaches have been reported by Prinzbach et al. [92, 97]. Mass spectrometric evidence for the existence in the vapor phase of dodecahedranes with more than two C=C double bonds has already been presented [44, 92, 97, 98]. The triene **143** (Fig. 14) and the tetraene **144** should have a good chance to exist

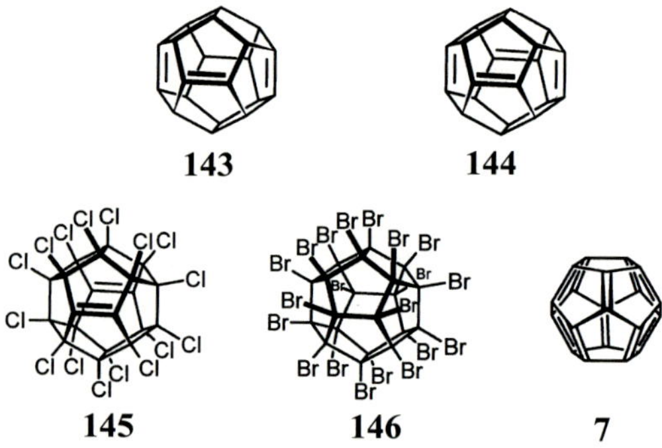

143   **144**

145   **146**   **7**

Fig. 14. Unsaturated dodecahedrane derivatives

at room temperature as all their double bonds are sterically protected by four allylic hydrogen atoms as in dodecahedrene **137** and the diene **142** [92].

Also the fully unsaturated C_{20}-fullerene **7** has been approached by Prinzbach et al. [44, 97]. As suitable precursors the perchlorinated dodecahedradiene $C_{20}Cl_{16}$ **145** and the perbrominated dodecahedrane $C_{20}Br_{20}$ **146** were prepared from **88** under forcing conditions. Due to their low solubility these compounds have so far only been identified by their infrared and mass spectra. Dehalogenation by fragmentation in the mass spectrometer has generated several unsaturated species but the occurrence of C_{20}-fullerene **7** remains to be rigorously proved.

Acknowledgments. The authors acknowledge the tremendous help of Stefan Beußhausen in drawing the figures and schemes and the assistance of Dafydd Owen in literature searching and proof reading. R.H. is grateful to Prof. Steven V. Ley for his moral support and helpful discussions, and to the Deutsche Forschungsgemeinschaft, the Gottlieb Daimler- and Karl Benz-Stiftung as well as the Fonds der Chemischen Industrie for financial support. The authors are indebted to Dr. Burkhard Knieriem, Göttingen, for his careful reading of the final manuscript.

8
References

1. (a) Paquette LA (1979) Topics Curr Chem 79; (b) Paquette LA (1984) Topics Curr Chem 119. Springer, Berlin Heidelberg New York and references cited therein
2. Paquette LA, Doherty AM (1987) Polyquinane chemistry. Springer, Berlin Heidelberg New York and references cited therein
3. Streitwieser A Jr (1961) Molecular orbital theory for organic chemists, Wiley, New York, and references cited therein
4. Haag R, Schröder D, Zywietz T, Jiao H, Schwarz H, Schleyer PvR, de Meijere A (1996) Angew Chem 108:1413; Angew Chem Int Edn Engl 35:1317
5. Kroto HW (1992) Angew Chem 104:113; Angew Chem Int Edn Engl 31:111 and references cited therein
6. Thiec J, Wiemann J (1960) Bull Soc Chim France 1066
7. Sturm E, Hafner K (1964) Angew Chem 76:862; Angew Chem Int Edn Engl 3:749
8. Hollenstein R, von Philipsborn W, Vögeli R, Neuenschwander M (1973) Helv Chim Acta 56:847
9. Scott AP, Agranat I, Biedermann PU, Riggs NV, Radom L (1997) J Org Chem 62:2026 and references cited therein
10. Thiele J, Bahorn H (1906) J Liebigs Ann Chem 348:1
11. Griesbeck AG (1989) J Org Chem 54:4981
12. Burton NA, Quelch GE, Gallo MM, Schaefer HF III (1991) J Am Chem Soc 113:764
13. Wiersum UE, Jenneskens LW (1991) Recl Trav Chim Pays-Bas 110:129
14. Dürr H, Ruge B (1972) Angew. Chem. 84:215; Angew Chem Int Edn Engl 11:225
15. Dürr H, Ruge B, Schmitz H (1973) Angew Chem 85:616; Angew Chem Int Edn Engl 12:577
16. Simmons HE, Fukunaga T (1967) J Am Chem Soc 89:5208
17. Maier G, Pfriem S (1978) Angew Chem 90:551; Angew Chem Int Edn Engl 17:519
18. Irngartinger H, Jahn R, Maier G, Emrich R (1987) Angew Chem 99:356; Angew Chem Int Edn Engl 26:356
19. Maier G, Pfriem S, Schäfer U, Matusch R (1978) Angew Chem 90:552; Angew Chem Int Edn Engl 17:520
20. Emme I, Schreiner PR, de Meijere A (1998) in preparation
21. Schwarz H (1981) Angew Chem 93:1046; Angew Chem Int Edn Engl 20:991 and references cited therein
22. Emme I (1997) Diplomarbeit, Universität Göttingen
23. (a) Gesing ERF, Tane JP, Vollhardt KPC (1980) Angew Chem 92:1057; Angew Chem Int Edn Engl 19:1023; (b) Maier G, Lange HW, Reisenauer HP (1981) Angew Chem 93:1010; Angew Chem Int Edn Engl 20:976
24. Review: Schore NE (1988) Chem Rev 88:1081 and references cited therein
25. Usieli V, Victor R, Sarel S (1976) Tetrahedron Lett 31:2705
26. de Meijere A (1996) Pure Appl Chem 68: 61; Schirmer H, Funke F, de Meijere A (1998) (in preparation)
27. Schirmer H (1996) Diplomarbeit, Universität Göttingen
28. Hafner K, Dönges R, Goedecke E, Kaiser R (1973) Angew Chem 85:362; Angew Chem Int Edn Engl 12:337
29. Bloch R, Marty RA, de Mayo P (1971) J Am Chem Soc 93:3071

30. LeGoff E (1962) J Am Chem Soc 84:3975
31. Hafner K, Süss HU (1973) Angew Chem 85: 626; Angew Chem Int Edn Engl 12:575
32. Armit JW, Robinson R (1922) J Chem Soc 121:827; (1925) 127:1604
33. Baird NC, West RM (1971) J Am Chem Soc 93:3072
34. Bischof P, Gleiter R, Hafner K, Knauer KH, Spanget-Larsen J, Süss HU (1978) Chem Ber 111:932
35. Zywietz TK, Jiao H, Schleyer PvR, de Meijere A (1998) J Org Chem 63: in press and references cited therein
36. Brown RFC, Choi N, Eastwood FW (1995) Aust J Chem 48:185
37. Katz TJ, Rosenberger M (1962) J Am Chem Soc 84:865
38. Stezowski JJ, Hoier H, Wilhelm D, Clark T, Schleyer PvR (1985) J Chem Soc Chem Commun 1263
39. Hafner K (1973) Angew Chem 85:958; Angew Chem Int Edn Engl 12:925
40. Bunel EE, Valle L, Jones NL, Carroll PJ, Barra C, Gonzales M, Munoz N, Visconti G, Aizaman A, Manriwuez JM (1988) J Am Chem Soc 110:6596
41. Jonas K, Gabor B, Mynott R, Angermund K, Heinemann O, Krüger C (1997) Angew Chem 109:1790; Angew Chem Int Edn Engl 36:1712
42. Jonas K, Kolb P, Kollbach G, Gabor B, Mynott R, Angermund K, Heinemann O, Krüger C (1997) Angew Chem 109:1793; Angew Chem Int Edn Engl 36:1714
43. Woodward RB, Fukunaga T, Kelly RC (1964) J Am Chem Soc 86:3162
44. Prinzbach H, Weber K (1994) Angew Chem 106:2329; Angew Chem Int Edn Engl 33:2239 and references cited therein
45. Liebman JF, Paquette LA, Peterson JR, Rogers DW (1986) J Am Chem Soc 108:8267 and references cited therein
46. Winstein S (1959) J Am Chem Soc 81:6524
47. Bischof P, Gleiter R, Heilbronner E (1970) Helv Chim Acta 53:1425
48. de Meijere A, Kaufmann D, Schallner O (1971) Angew Chem 83:404; Angew Chem Int Edn Engl 10:417
49. Kaufmann D, Fick H-H, Schallner O, Spielmann W, Meyer L-U, Gölitz P, de Meijere A (1983) Chem Ber 116:587
50. Holder A (1993) J Comp Chem 14:251 and references cited therein
51. Miller MA, Schulman JM, Disch RL (1988) J Am Chem Soc 110:7681
52. Verevkin SP, Beckhaus H-D, Rüchardt C, Haag R, Kozhushkov SI, Jiao H, Schleyer PvR, Zywietz TK, de Meijere A (1998) J Am Chem Soc (submitted)
53. de Meijere A, Lee C-H, Bengtson B, Pohl E, Sheldrick GM (1998) (in preparation)
54. Mascal M, Hext NM, Shishkin OV (1996) Tetrahedron Lett 37:131
55. Jacobson IT (1967) Acta Chem Scand 21:2235
56. Berger T, Hopf H (unpublished results)
57. Butenschön H, de Meijere A (1985) Chem Ber 118:2757
58. Butenschön H, de Meijere A (1984) Tetrahedron Lett 25:1693
59. Butenschön H, de Meijere A (unpublished results)
60. Butenschön H, de Meijere A (1986) Tetrahedron 42:1721
61. Schröder G, Butenschön H, Boese R, Lendvai T, de Meijere A (1991) Chem Ber 124:2423
62. Lendvai T, Friedl T, Butenschön H, Clark T, de Meijere A (1986) Angew Chem 98:734; Angew Chem Int Edn Engl 25:719
63. Haag R, Fleischer R, Stalke D, de Meijere A (1995) Angew Chem 107:1642; Angew Chem Int Edn Engl 34:1492
64. Haag R, Ohlhorst B, Noltemeyer M, Fleischer R, Stalke D, Schuster A, Kuck D, de Meijere A (1995) J Am Chem Soc 117:10,474
65. Haag R, Schuengel M, Ohlhorst B, Noltemeyer M, Haumann T, Boese R, de Meijere A (1998) Chem Eur J 4: in press
66. Ayalon A, Sygula A, Cheng P-C, Rabinovitz M, Rabideau PW, Scott LT (1994) Science 265:1065
67. Zywietz TK (1996) Diplomarbeit, Universität Göttingen

68. Herberich GE, Kruder C, Englert U (1994) Angew Chem 106:2589; Angew Chem Int Edn Engl 33:2465
69. Haag R, Schüngel M, Poremba P, Edelmann F, de Meijere A (1997) (unpublished results)
70. Streitwieser Jr. A, Kinsley SA (1985) In: Marks TJ, Fragala IL (eds) Fundamental and technological aspects of organo-f-element chemistry. Reidel, Dordrecht p 77; NATO ASI Series C: vol 155
71. Lannoye G, Sambasivarao K, Wehrli S, Cook JM (1988) J Org Chem 53:2327
72. Fu X, Cook JM (1992) Aldrichimica Acta 25: 43 and references cited therein
73. Paquette LA, Wyvratt MJ, Berk HC, Moerck RE (1978) J Am Chem Soc 100:5845
74. Paquette LA, Ternansky RJ, Balogh DW, Kentgen G (1983) J Am Chem Soc 105:5446
75. Lee C-H, Liang S, Haumann T, Boese R, de Meijere A (1993) Angew Chem 105:611; Angew Chem Int Edn Engl 32:559; Lee C-H, Liang S, Kusnetsov MA, de Meijere A (unpublished results)
76. Hirao K-I, Takahashi H, Ohuchi Y, Yonemitsu O (1992) J Chem Res (S):319; (M) 2601
77. Garrat PJ, White JF (1977) J Org Chem 42:1733
78. Olah GA, Buchholz HA, Prakash GKS, Rasul G, Sosnowski JJ, Murray RK Jr, Kusnetsov MA, Liang S, de Meijere A (1996) Angew Chem 108:1623; Angew Chem Int Edn Engl 35:1499
79. Böhm MC, Gleiter R, Schang P (1979) Tetrahedron Lett 2575
80. Brunvoll J, Guidetti-Grept R, Hargiatti I, Keese R (1993) Helv Chim Acta 76:2838
81. Thummen M, Keese R (1997) Synlett:231
82. Kuck D, Bögge H (1986) J Am Chem Soc 108:8107
83. Iyoda M, Otani H, Oda M, Kai Y, Baba Y, Kasai N (1986) J Chem Soc Chem Commun 1794
84. Eaton PE, Srikrishna A, Uggeri F (1984) J Org Chem 49:1728
85. Wu H-J, Wu C-Y (1997) Tetrahedron Lett 38: 2493
86. McKervery MA, Vilbuljan P, Ferguson G, Siew PY (1982) J Chem Soc Chem Commun 912
87. Houk KN, Gandour RW, Strozier RW, Rondan NG, Paquette LA (1979) J Am Chem Soc 101: 6797 and references cited therein
88. Eaton PE, Sidhu RS, Langford GE, Cullison DA, Pietruszewski CL (1981) Tetrahedron 37: 4479
89. Scuseria GE (1996) Science 271:942 and references cited therein
90. von Helden G, Gotts NG, Bowers MT (1993) Nature 363:60
91. Kiplinger JP, Tollens FR, Marshall AG, Kobayashi T, Lagerwall DR, Paquette LA, Bartmess JE (1989) J Am Chem Soc 111:6914
92. Melder J-P, Weber K, Weiler A, Sackers E, Fritz H, Hunkler D, Prinzbach H (1996) Res Chem Intermed 22:667 and references cited therein
93. Schwesinger R, Willaredt J, Schlemper H, Keller M, Schmitt D, Fritz H (1994) Chem Ber 127:2435
94. Haddon RC (1993) Science 261:1545 and references cited therein
95. Borden WT (1996) Synlett 711 and references cited therein
96. Bertau M, Leonhardt J, Weiler A, Weber K, Prinzbach H (1996) Chem Eur J 2:570 and references cited therein
97. Wahl F, Wörth J, Prinzbach H (1993) Angew Chem 105: 1788; Angew Chem Int Edn Engl 32:1722
98. Pinkas R, Melder JP, Weber K, Hunkler D, Prinzbach H (1993) J Am Chem Soc 115:7173

Note added in proof: Compound **99** has recently been approached by a tandem Pauson-Khand reaction; Van Ornum SG, Cook JH (1998) ACS Spring meeting, Dallas.

The Centropolyindanes and Related *Centro*-Fused Polycyclic Organic Compounds

Polycycles between Neopentane C(CH₃)₄ and the Carbon Nucleus C(CC₃)₄

Dietmar Kuck

Fakultät für Chemie, Universität Bielefeld, Universitätsstrasse 25, D-33615 Bielefeld, Germany.
E-mail: kuck@chema.uni-bielefeld.de
Fachbereich Chemie und Chemietechnik, Universität-GH Paderborn, Warburger Straße 100,
D-33098 Paderborn, Germany

Centropolyindanes constitute a complete family of arylaliphatic polycyclic hydrocarbons containing several indane units. Mutual fusion of the five-membered rings leads to three-dimensional, carbon-rich molecular frameworks bearing a central carbon atom, such as benzo-annelated [3.3.3]propellanes, triquinacenes, and [5.5.5.6]- and [5.5.5.5]fenestranes. In this review, the structural concept of centropolyindanes is contrasted to other fused indane hydrocarbons. Besides the syntheses of the parent centropolyindanes and recently described related indane hydrocarbons, the preparation of a large variety of bridgehead and arene substituted centropolyindanes is presented including strained, heterocyclic, and centrohexacyclic derivatives. In appropriate cases, the particular reactivity and some structural features of these unusual, sterically rigid polycyclic compounds are pointed out.

Keywords: Polycyclic hydrocarbons (three-dimensional), indane hydrocarbons, propellanes, triquinacenes, acepentalenes, fenestranes, polyquinanes.

Topics in Current Chemistry, Vol. 196
© Springer Verlag Berlin Heidelberg 1998

1
Introduction

The construction of novel molecular carbon frameworks has been a challenge for organic chemists for some time. Extended structures with planar or distorted-planar fusion of mostly unsaturated six-membered rings have been contrasted to molecular arrangements consisting of saturated rings fused in three dimensions. Within both groups of polycyclic hydrocarbons, organic chemistry has faced several renaissances in the past decades and is probably more vivid than ever. Since Greenberg and Liebman's inspiring book on "Strained organic molecules" was published in 1978 [1], a variety of novel motifs have been developed in polycyclic chemistry, and previously well-known motifs [2] have been varied further, as documented in several recent monographs and reviews [3, 4].

Besides novel planar, graphite-type fragments derived from coronene ([6]-circulene) [5], various quasi-planar carbon frameworks characterized by strong out-of-plane distortions due to severe steric interactions, strained interannular fusion or bridging have been synthesized and their properties studied. Recent examples include the chemistry of [n]circulenes that enclose five- [6–8] or seven-membered rings [9], but also other purely benzenoid hydrocarbons such as perphenylated condensed arenes [10], tetradehydrodianthracenes and the tetradehydrodianthracene dimer [11] which all are characterized by the bent shapes of their polycyclic carbon frameworks. Various fullerene fragments other than corannulene derivatives [12–14], multiply fused biphenylenes [15] as well as helical polyarenes (helicenes) [16] represent further topics of active current research in the field of fused aromatic hydrocarbons. Fullerenes and their chemistry [17, 18] mark an extreme point in this context because they consist of completely closed, bent (and quasi-planar) carbon surfaces forming three-dimensional carbon globes.

In this view, other polycyclic hydrocarbons such as cubane [19, 20] and dodecahedrane [21, 22] and their derivatives are not different. Less known but related examples are bisnoradamantane [23] and trishomocubane [24], consisting of four and, respectively, six cyclopentane rings mutually fused in a "global" manner. In sharp contrast, truly three-dimensional molecular frameworks contain at least one quaternary carbon atom. Adamantane and its homologues [25] represent the by far best known class of polycyclic hydrocarbons containing three-dimensionally fused rings. This leads back to the motif of fusing six-membered rings, this time mostly saturated (and folded) ones making up a part of the diamond network. Of considerable interest are polycyclic hydrocarbons which incorporate both "planar" benzene-type and alicyclic building blocks. Hart's well explored class of iptycenes [26], based on triptycene as the parent unit, represents a major example for combining benzene and cyclohexane rings in three-dimensional structures, others are Stoddart's octahydro[12]cyclazene and [12]collarene [27].

This review comprises the advances in fusing benzene and other arene rings to three-dimensional cores which consist of multiply fused cyclopentane rings. Indane represents the parent building block here, and polycyclic frameworks consisting of several indane units fused in three dimensions have been called the "centropolyindanes" [28, 29]. As will be shown in this chapter, centropolyindanes comprise some other well-known sub-classes of polycyclic hydrocarbons treated in comprehensive reviews, viz. propellanes [30, 31], triquinacenes [32, 33], and fenestranes [34–37] as well as even more exotic polycyclic systems of unusual topology [38].

2
Indane as a Building Block for Three-Dimensional Ring Systems

2.1
Regular Centropolyindanes

Among the various structural types of polyquinanes [32, 39, 40] those bearing more than two five-membered rings fused at a common central atom or neopentane core have been classified as "centropolyquinanes" [41]. Some of the higher members of this highly interesting family are still elusive. Centropoly-*indanes* [28, 29] comprise all ring systems which consist of more than two indane units fused mutually at the edges of their five-membered rings. In Scheme 1 the so-called "regular" centropolyindanes are compiled as a systematic series of benzeno-bridged derivatives of neopentane **1**. Corresponding schemes for centropolyquinanes [41] and for partially benzoannelated centropolyquinanes [42] have also been presented. Following 2,2-dimethylindane (**2**) and the two possible "regular" diindanes, viz. 2,2-spirobiindane (**3**) and the methyl-substituted C_s-diindane **4** (the "*spiro*"- and "*fuso*"-diindanes, cf. Section 3), there are several regular centropolyindanes comprising up to six indane units about the four C–C bonds of the neopentane core. It may be noted that, within the series of compounds **1–11**, the central $C(CH_3)_4$ core loses all of its twelve hydrogen atoms to become, eventually, an all-carbon cluster C_{17}.

The C_{3v}-symmetrical centrotriindane **5** is the mono-*fuso* congener, first synthesized by Thompson in 1966 [43]. The angularly fused centrotriindane **6**, first described in 1984 [44], is the corresponding di-*fuso* isomer, and 10-methyltribenzotriquinacene **7** [44], another $C_{23}H_{18}$ isomer, represents a methyl derivative of tri-*fuso*-centrotriindane. The desmethyl (nor) analogues of **4** and **7** will also be discussed in this chapter.

There are two possible isomers comprising four indane units within this "regular" scheme. These are the tri-*fuso*-centrotetraindane congener **8** [45] and the isomeric tetra-*fuso*-centrotetraindane **9** [46], which has been coined "fenestrindane" owing to its [5.5.5.5]fenestrane framework. Although in **9** all of the

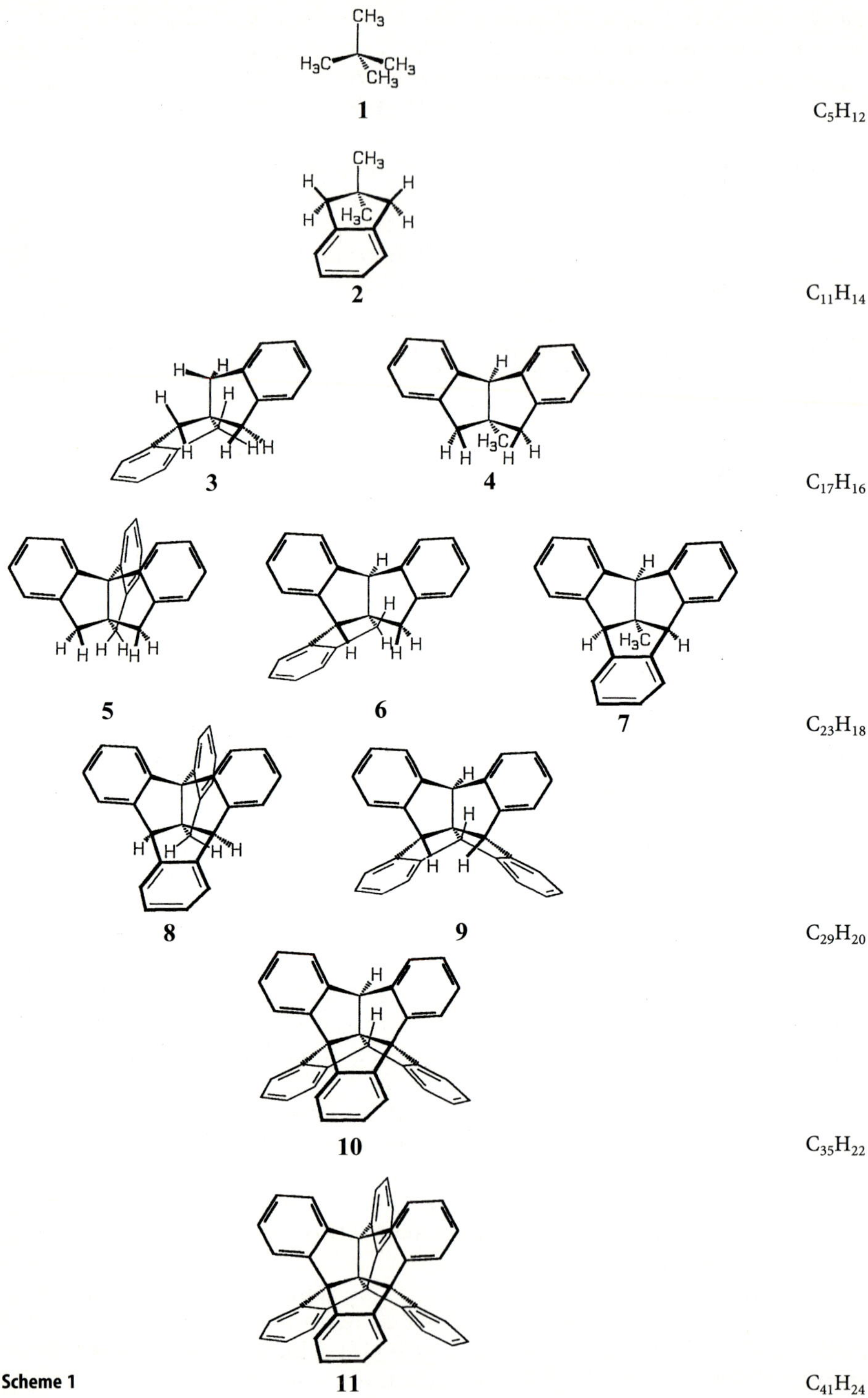

C_5H_{12}

1

$C_{11}H_{14}$

2

$C_{17}H_{16}$

3   **4**

$C_{23}H_{18}$

5   **6**   **7**

$C_{29}H_{20}$

8   **9**

$C_{35}H_{22}$

10

Scheme 1   **11** $C_{41}H_{24}$

four neopentane C–C bonds are used for annelation, two additional, and structurally unique, centropolyindanes exist. These are centropentaindane **10** [47] and centrohexaindane **11** [48], containing five and, respectively, the maximum of six benzene rings fused across the $C(C)_4$ skeleton of neopentane. Centrohexaindane is the only congener without aliphatic C–H bonds and thus the most carbon-rich parent compound ($C_{41}H_{24}$) among the regular centropolyindanes **1**–**11** of the general composition $C_{5+6n}H_{12+2n}$.

2.2
Irregular Centropolyindanes

All of the regular centropolyindanes **5**–**11** have become accessible by various synthetic routes (for recent reviews on selected topics, see [29, 37, 38]), which will be discussed in this chapter. Before doing so, however, some "irregular" centropolyindanes [49] will be mentioned briefly, the majority of which are *not* known to date (Scheme 2). In the following section, some structural relations of the centropolyindanes to well-known di- or polyindanes (hydro)carbon frameworks containing no quaternary atom will be pointed out (Schemes 3 and 4).

Diindane **12** (1,2′-spirobiindane) may be considered the simplest congener of the irregular centropolyindanes. One of the benzene rings is fused with one of the neopentane carbon atoms, thus blocking it against annelation of further

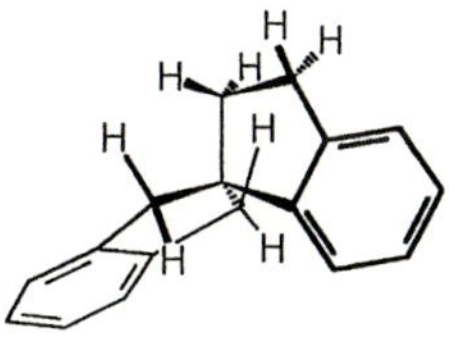

12

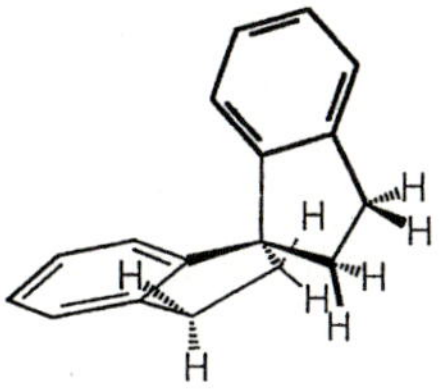

13

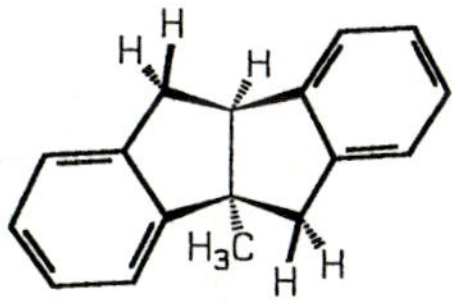

14

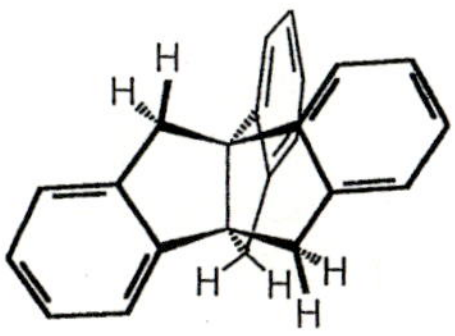

15

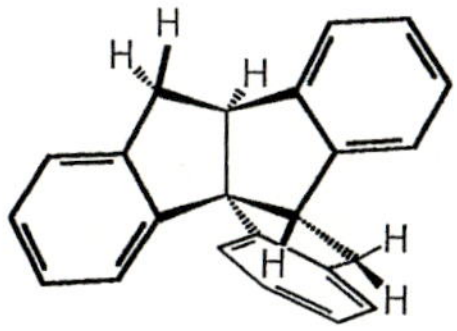

16

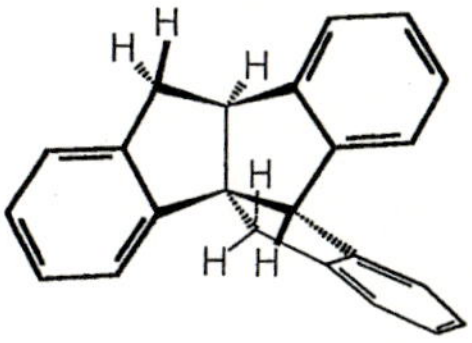

17

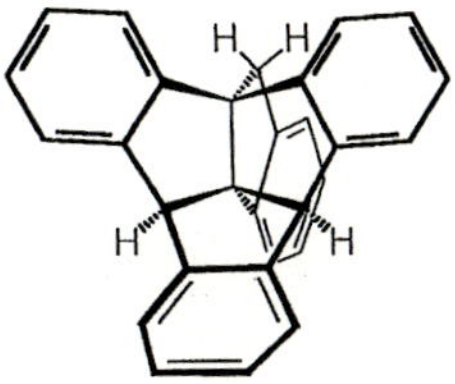

18

Scheme 2

benzene rings, in contrast to **3**. This feature recurs even twice in the 1,1'-spiro-biindanes [50] (cf. **13**) and also in C_2-symmetrical mono-*fuso*-diindanes [51], such as the parent of the methyl derivative **14**. Notably, however, there appears to be no report on the C_s-symmetrical isomer **15** of triptindane (**5**), also bearing two "irregularly" fused neopentane atoms. In contrast, the angular, C_2-symmetrical di-*fuso*-centrotriindane **16** was synthesized in 1980 [52] (see below). In turn, its C_1-symmetrical isomer, di-*fuso* congener **17** has remained unknown. The incorporation of one of the neopentane C atoms into a benzene nucleus causes the fact that tri-*fuso*-centrotetraindane **18** is the highest possible irregular centropolyindane. Apparently, no report exists on it either.

2.3
A Contrast: Polyindanes with Planar or Quasi-Planar Frameworks

As shown above, the unique structural feature of centropolyindanes is the mutual fusion of several cyclopentane rings each of which being annelated to a benzene nucleus. A number of other combinations of these ring elements can be considered as "diindanes", "triindanes" etc. which in most cases lack a saturated tertiary or quaternary carbon atom. As illustrated in Schemes 3 and 4, tetrahydrobenzo[*d,e*]pentalene **19** is one of the very few exceptions. Molecules of this type have a strained, bent molecular shape [1,53]. The tetrahydroindacenes **20** and **21**, acenaphthene (**23**) and fluorene (**24**) are clearly planar, as are the threefold congener "trindane" (**22**) [54] and fluoranthene (**25**) and its congeners [55].

As a dibenzo analogue of **19** and a *peri*-benzoannelated diindane (cf. **4**), fluoradene (**26**) is considerably strained [56], as is pyracyclene (**27**) [57]. Truxene (**29**), the parent "polyindane" of truxenequinone and related well known condensates of 1,3-indanedione, is essentially planar, but has been used as a basis for approaching fullerene fragments by synthetic means [13, 14, 58, 59]. Coran-

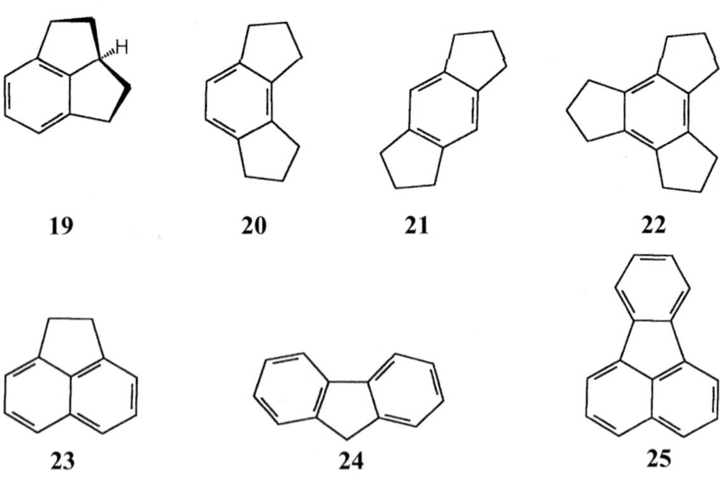

Scheme 3

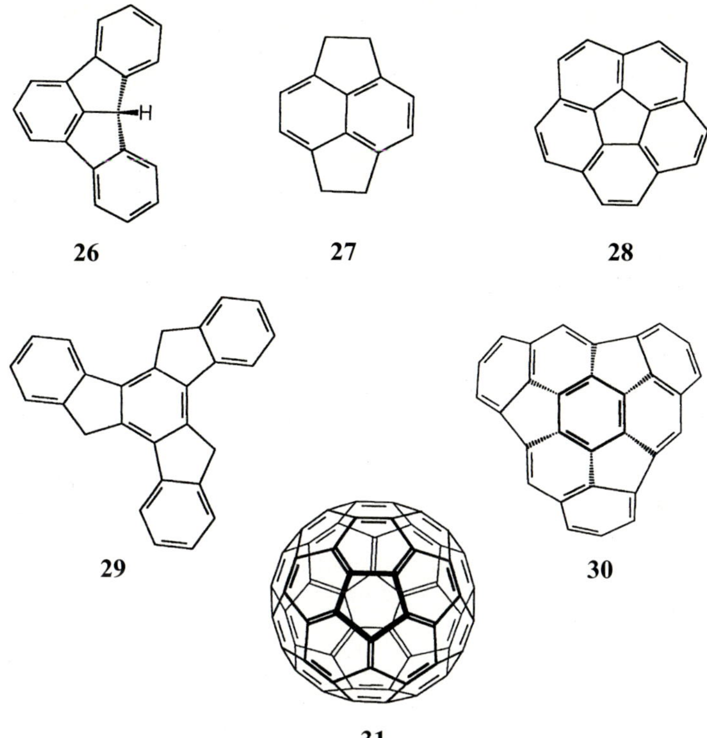

Scheme 4

nulene (**28**) [6, 7] represents the first member of "polyindanes" with definite out-of-plane bending due to the presence of a five-membered ring within a closed assembly of benzene nuclei. Semifullerene **30**, the synthesis of which has been reported recently [14], is a strongly concave hydrocarbon [60]. Finally, C_{60}-fullerene (**31**) [17, 18] may be considered a polyindane having an ideally global, convex molecular surface. In contrast to the centropolyindanes, however, the molecular frameworks of **28**, **30**, and **31** represent quasi-planar combinations of indane units. In this view, buckminsterfullerene **31** forms the extreme case of a hollow cavity, whereas centrohexaindane **11** spans a "massive" centrosymmetrical molecular network.

3
Diindanes: Recent Advances

2,2′-Spirobiindane **3** and C_s-diindane **34** (*cis*-4b, 9, 9a, 10-tetrahydroindeno[1,2-*a*]indene), the parent of **4**, constitute the diindane building blocks of all higher centropolyindanes. The synthesis and chemistry of **3** has been well established [61], and only few papers appeared in recent years [62]. Some attempts to use derivatives of **3** to construct centropolyindanes have all failed [29, 37]. In contrast, the derivatives of mono-*fuso*-diindane **4** or its parent compound **34** have

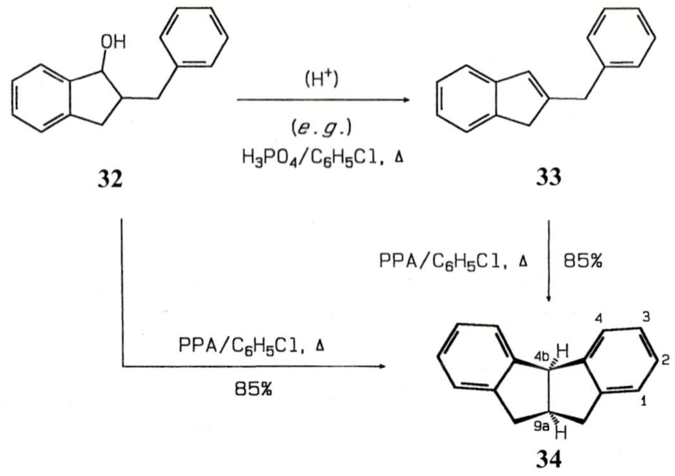

Scheme 5

proven much more useful. Several syntheses have been published for **34** [63, 64] but recent observations led to a new and very efficient access to this hydrocarbon (Scheme 5) [65].

2-Benzyl-1-indanol **32**, readily available in a few steps from 1-indanone, has been converted into 2-benzylindene **33** in various ways [66–68]. Treatment of **32** with orthophosphoric acid in chlorobenzene, for example, gives **33** as the sole product. However, use of polyphosphoric acid in the same solvent effects clean cyclodehydration of both **32** and **33** giving the mono-*fuso*-diindane **34** in high yield [65]. Several "*centro*"-alkylated derivatives of **34**, including the methyl

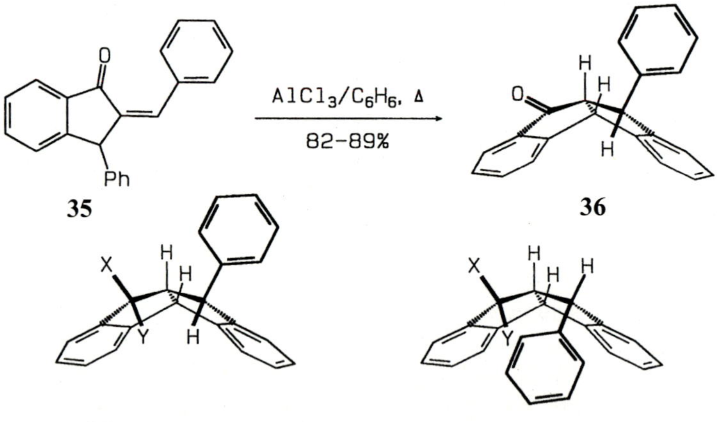

37 X = H, Y = OH
38 X = OH, Y = H
39 X = Y = H

40 XY = O
41 X = H, Y = OH
42 X = Y = H

Scheme 6

derivative **4** [61], were prepared by cyclodehydration of appropriate tricarbonylchromium-complexed 2-benzyl-1-indanols [68].

Diindanone **36**, which is easily obtained from cinnamic acid [64], was found to be another versatile starting point (Scheme 6). The previously unknown *anti*-stereochemistry of this ketone was derived from the mass spectrometric fragmentation of the related alcohols **37** and **38** and confirmed by synthesis [69] (see below). In a sequence of conversions, the *syn*-diindanone **40** and the related alcohol **41** were prepared and characterized. On this basis, the two epimeric phenyldiindanes **39** and **42** were also synthesized [69].

The recent synthesis of diindane **45** [70] is remarkable because of its particular 4b, 9a-diaryl substitution pattern (Scheme 7). One-pot two-fold benzylation of desoxybenzoin **43** under forced conditions afforded formation of the quaternary centre in **44**, which then underwent "bicyclodehydration" (for this term see [29]) in polyphosphoric acid to generate the clamped 1,1,1,2-tetraarylethane **45**. X-ray structural analysis revealed a remarkably long bond distance (161.8 pm) for the central C–C bond of the diindane unit. It has been traced to the efficient through-bond coupling of the π orbitals of the co-facially *syn*-oriented bridgehead aryl groups via the diindane junction [70, 71]. Similar effects were observed in some other rigid centropolyindanes (see below).

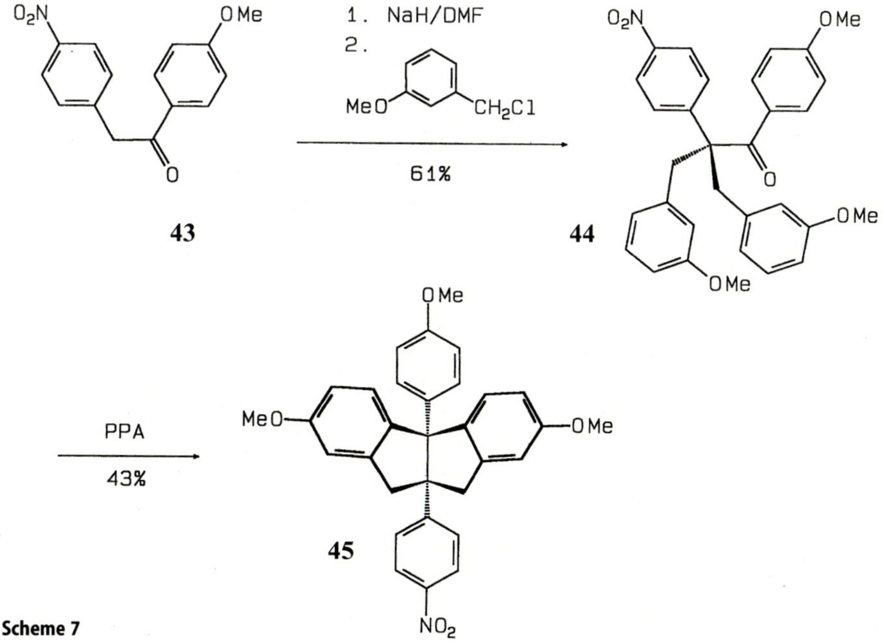

Scheme 7

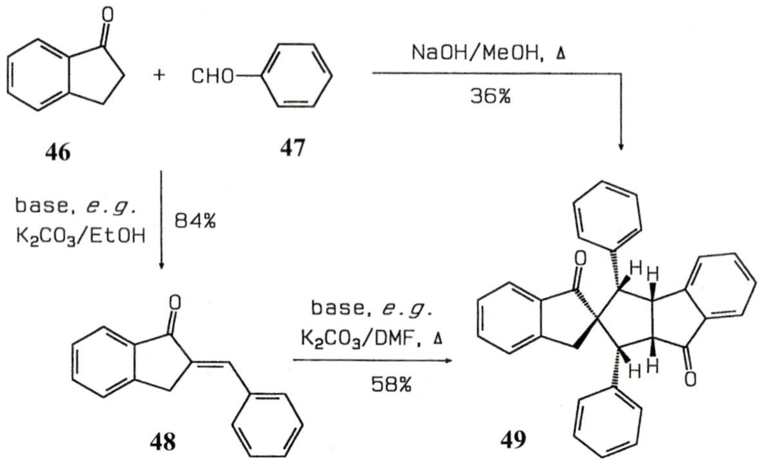

Scheme 8

The condensation of 1-indanones and benzaldehydes giving 2-benzylidene-1-indanones is a classic approach to C_s-diindanes. However, the condensation of **46** and **47** and their derivatives under basic conditions often leads to dimers of **48**, the structure of which has been a matter of long debate [72]. Recent studies [73, 74] confirmed the constitution and configuration of the major product formed either directly from **46** and **47** [74] or by directed dimerization of **48** [73] as the dibenzotriquinanedione **49** (Scheme 8). This dimer represents an "extended" diindane bearing a central *spiro-* and *fuso-*annelated *cis-*diphenylcyclopentane unit with the two phenyl groups oriented *syn* to the *fuso-*diquinane moiety and *cis* to the *spiro-*indanone carbonyl group.

4
Centrotriindanes and Related Systems

4.1
Mono-*fuso*-centrotriindanes: Triptindanes and Related Benzoannelated Propellanes

Among the large class of propellanes, benzoannelated derivatives have remained rare for a long time. In 1966, Thompson published the first synthesis of C_{3v}-symmetrical tribenzo[3.3.3]propellanes, the parent hydrocarbon of which was called "triptindan" **5** [43] (Scheme 9). (According to the revised spelling of indan*e* compounds, **5** has been addressed now as "triptindane".) Starting from 2,2-dibenzylated 1-indanone **50**, at least one *meta*-methoxy group is required to electronically activate the first ring closure. Thus, mixtures of methoxytriptindanes such as **51** and **52** were obtained from which the substituent was removed in a three-step procedure involving hydrogenolysis of the corresponding phenyltetrazolyl ether(s). In contrast, 2,2-dibenzylindanone **53** was found to be unreactive even in polyphosphoric acid at 160 °C.

Thompson's strategy was applied in our laboratory to synthesize a number of symmetrically substituted triptindanes, which could be used for the construc-

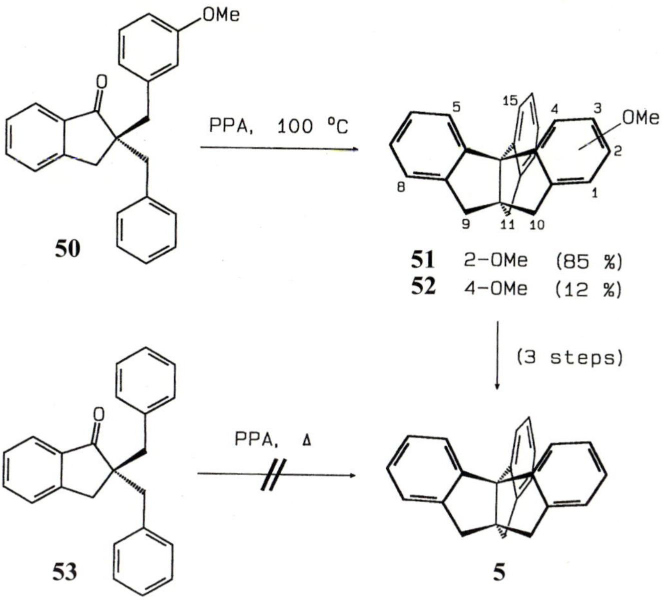

Scheme 9

tion of bridged propellanes of theoretical interest [75]. 2,2-Bis(3,5-dimethoxy-benzyl)- and 2,2-bis(3,5-dimethylbenzyl)-1-indanones bearing another pair of appropriate substituents at positions 5 and 7, as well as analogues with "mixed" substituent patterns, were prepared and subjected to cyclodehydration [76] (Scheme 10). With indanones bearing at least one electronically activated methoxybenzyl group (cf. **54, 56** and **58**), the relatively mild ion-exchange resin Amberlyst 15 (A-15) in refluxing toluene effects bicyclization in high yield. With 3,5-dimethylbenzyl groups (cf. **60** and **62**), polyphosphoric acid was required to generate the centrotriindane framework. In this series, hexamethyltript-indane **63** represents a most remarkable propellane because of the strong steric interaction of the three methyl groups pointing to each other. As a particular challenge, we have envisaged the incorporation of a cyclopropane unit between the "cavity" positions C(4), C(5) and C(15) of **63** or related trisubstituted triptin-danes [77]. This three-fold ring closure would produce an electronically interesting bullvalene derivative but has not yet been achieved.

Triptindanes such as **5, 55** and **63** exist in a dynamic equilibrium of two C_3-symmetrical rotamers [77]. Steric hindrance caused by the substituents in positions 4, 5 and 15 (cf. Scheme 10) increases steadily in the series **55 < 57 < 61 < 63**, i.e. with increasing steric bulk of the substituents pointing into the cavity (or three-dimensional "bay") of the triptindane framework. A number of other triptindanes bearing electron-releasing substituents in the aromatic rings have been prepared as potential precursors for the synthesis of the corresponding three-fold *para*-benzoquinones [74, 77].

It appears that detailed investigations of proximity effects within the set of three "bay" substituents would reveal a wealth of interesting phenomena with

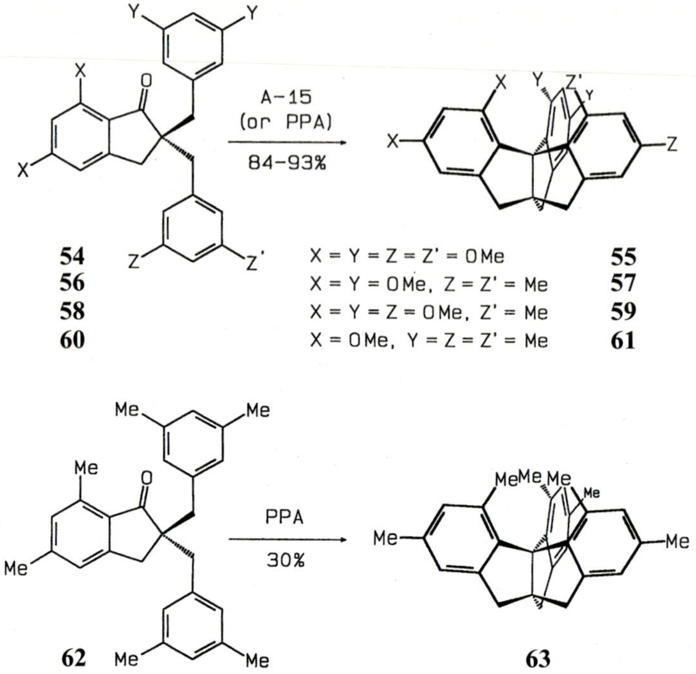

Scheme 10

both physical–organic and synthetical aspects. Six-fold ether cleavage of **55** led to the tris(resorcinol) **64** [76] (Scheme 11). Application of the tetrazolyl ether method produces the tris(phenol) **65**, which upon hydrogenolysis yielded **66** as a unique triptindane bearing three polar cavity substituents. This polyindane is prone to, for example, conversion into unusual heterocycles by capping or complexing it with heteroatomic centres.

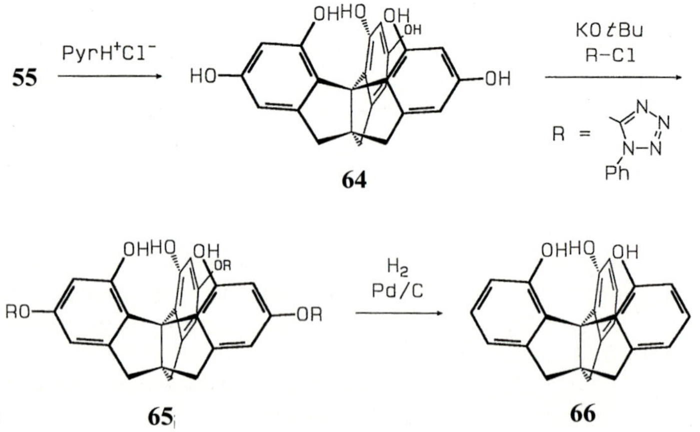

Scheme 11

An unexpected proximity effect has been observed upon bicyclization of the indanone **67** bearing the methoxy groups in *para* position [74] (Scheme 12). Instead of the corresponding 1,4,5,8,12,15-hexamethoxy isomer of **55**, phenol **68** was isolated in good yield, indicating the unusually facile cleavage of one of the methoxy groups by assistance of the two adjacent ones present in the overcrowded bay.

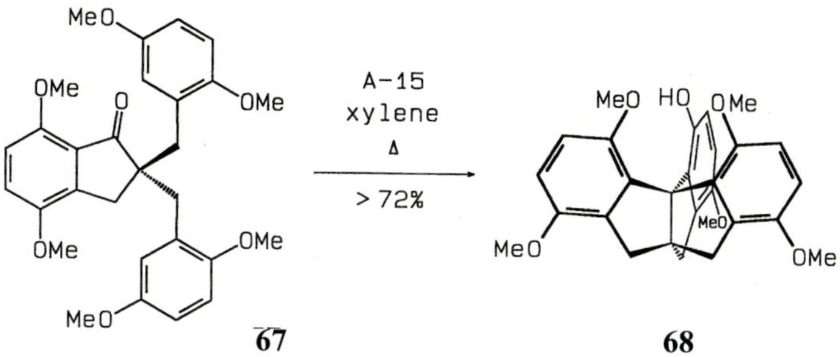

Scheme 12

An important breakthrough towards a broader utilization of the triptindane framework was achieved by the finding that 1,3-indanediones bearing two benzyl groups undergo the desired bicyclization reaction more readily than the corresponding 1-indanones [78] (Scheme 13). Most strikingly, the unsubstituted diketone **69** was converted into 9-triptindanone **70** in excellent yield. Notably, electronic activation appears to originate here from the electrophilic partner, i. e. from the presence of the second carbonyl group at the indanone ring. Being an interesting building block in itself, **70** can also be easily oxidized to triptindanetrione **72** via dibromoketone **71** [78]. This α,α',α''-tricarbonylmethane compound is a highly versatile building block for further synthesis of complex polycyclic frameworks, as will be shown below.

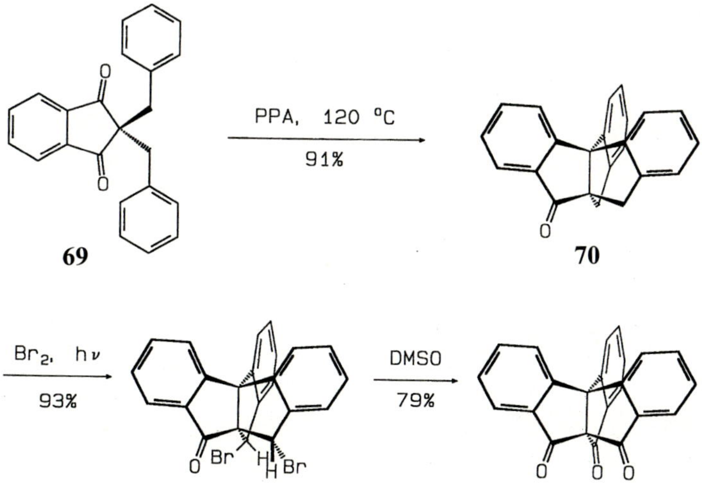

Scheme 13

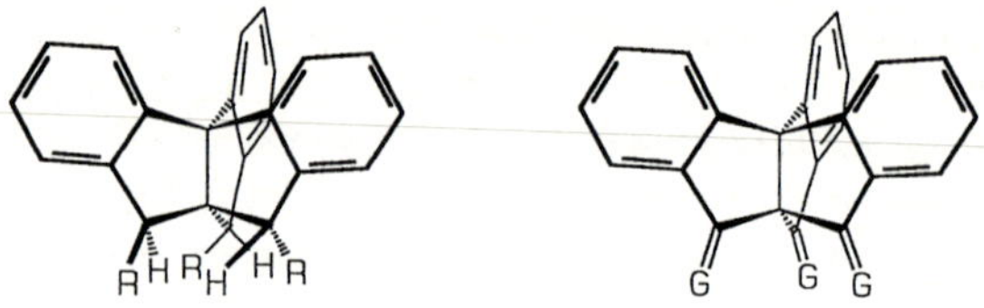

73 R = OH
74 R = Br
75 R = N$_3$
76 R = NH$_2$

77 G = CH$_2$
78 G = CHPh
79 G = CH$_2$CH$_2$
80 G = CH$_2$O

Scheme 14

One set of interesting triptindane derivatives is obtained by re-establishing the *sp^3*-hydridized centres at the alicyclic portion of 5 (i.e. at C-9, C-10 and C-11), and another one by modifying the doubly bonded substituents (Scheme 14). Reduction of the achiral trione **72** or treatment with metal-organic synthons gives *C$_3$*-symmetrical, and thus chiral, triols such as **73** [79]. Further *C$_3$*-symmetrical analogues such as the tribromide **74** and the triazide **75** have been prepared. The corresponding triamine **76** appears to be unusually sensitive towards oxidation [79]. Triptindane derivatives of this type might show interesting chiral recognition properties if suitable tentacular groups were attached to the chiral centrotriindane basis.

Conversion of triptindanetrione **72** to three-fold styrenes such as **77** and stilbenes such as **78** occurs readily under dehydrating conditions [78, 79]. Cyclopropanation of **77** gave the corresponding three-fold spirane **79** whereas the preparation of the corresponding tris(epoxide) **80** by epoxidation of **77** or by methenylation of **72** proved unsuccessful so far [74, 78]. Treatment of **77** with dioxiranes is under current investigation [80].

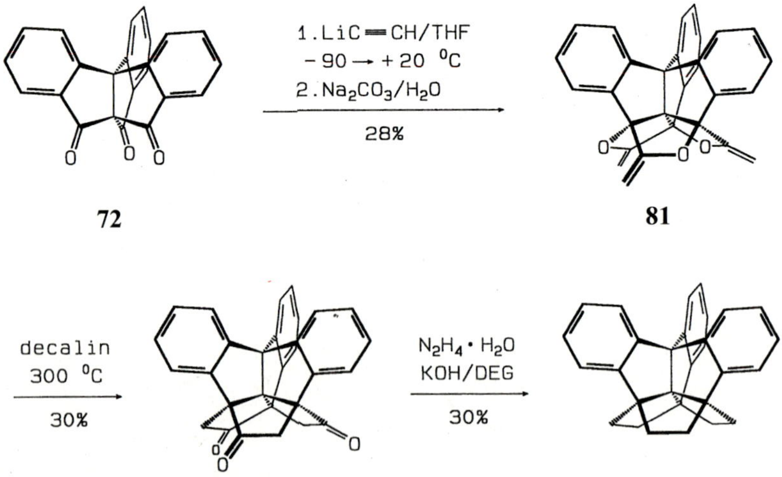

72

81

82

83

Scheme 15

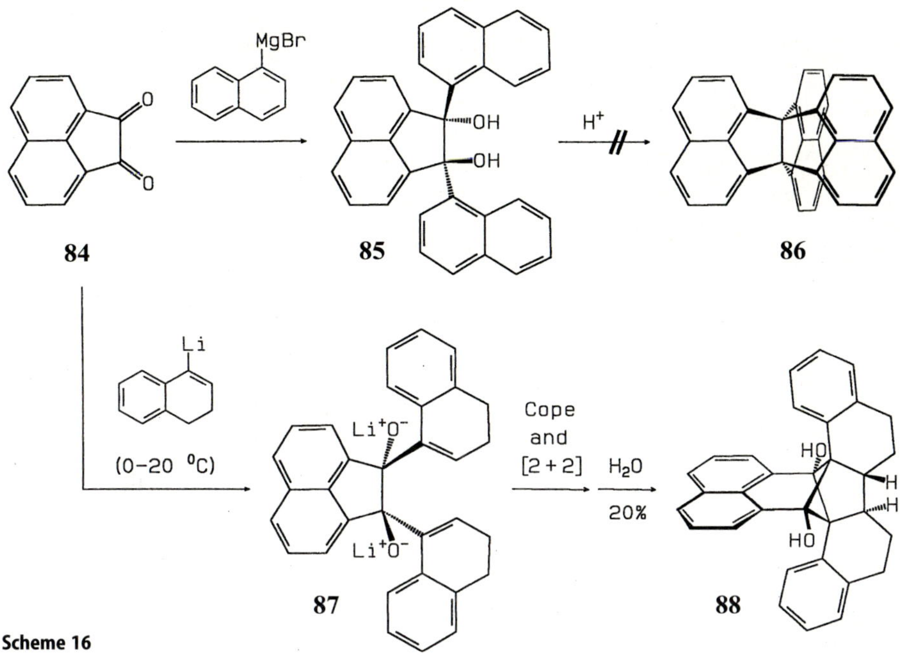

Scheme 16

The synthesis of **79, 80** and related spiranes has been of interest because of their potential conversion to centrohexacyclic derivatives bearing three additional bridges across the neopentane carbon atoms C-9, C-10 and C-11, in analogy to some unusual non-benzoannelated analogues such as the Simmons–Paquette molecule [37, 81, 82]. In fact, triptindanetrione **72** has been used in several cases to construct three additional five-membered rings (Schemes 15 and 56). A surprisingly efficient three-fold bridging with aliphatic diatomic units has been achieved by treatment of **72** with an excess of lithium acetylides. In this way, centrohexacyclic triptindanes **81–83** and related compounds have become accessible, as shown in Scheme 15 [83].

An interesting synthetic target related to triptindane is the tris(naphtho)[3.3.3]propellane **86** (Schemes 16 and 17). Alder et al. [84] reported on first attempts to construct this "two-fold" triptindane bearing a highly strained C–C bond. Cyclodehydration of acenaphthenediol **85**, prepared previously from acenaphthene quinone (**84**) [85], or the corresponding pinacolone formed as an intermediate in the acidic medium, did not occur in the desired way (Scheme 16). Attempts to utilize the corresponding bis(dihydro) derivative (i.e. the diol derived from **87**) resulted in the formation of the C_2-symmetrical polycycle **88**, possibly by di-oxy Cope rearrangement followed by a "criss-cross" [2 + 2] cycloaddition [84].

Recently, an elegant synthesis of **86** was developed by Dyker et al. [86, 87] (Scheme 17). Pd^0-catalyzed coupling of acenaphthylene (**89**) and 1,8-diiodonaphthalene provides an efficient access to acenaphth[1,2-*a*]acenaphthylene **90**

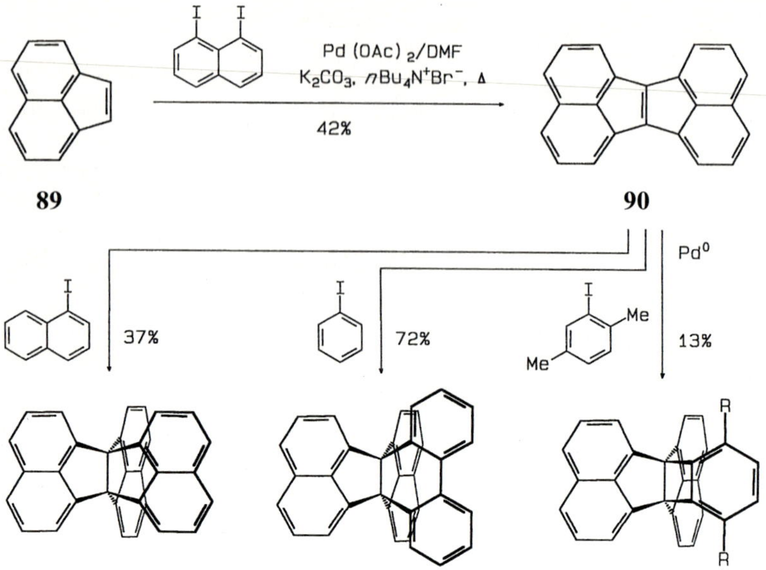

Scheme 17

[86], a strained olefin which readily undergoes further directed annelation reactions either by Diels–Alder [88] or Heck-type reactions [86,87]. Thus, propellane **86** was obtained by Pd^0-catalyzed condensation with 1-iodonaphthalene in good yield. Annelation of an entire phenanthrene unit was carried out with particular efficiency using iodobenzene to give **91**. In contrast, use of the sterically congested 2-iodo-1,4-dimethylbenzene enabled the incorporation of a single arene nucleus to form the benzocyclobutene derivative **92** in low yield. Finally, the non-methylated parent propellane **93** was obtained in good yield (57%) by performing a [2 + 2] cycloaddition of **90** with benzyne. X-ray structural analysis of **86** and **91**, representing fully clamped hexaarylethane derivatives, showed again an extreme elongation of the central C–C bonds [162.1 (±0.1) pm (**86**) and 161.1 (±0.1) pm (**91**)] [86, 87]. A slight torsion about the propellane bond was found for **86**, whereas **91** is considerably distorted along the propellane axis. Despite of its strained character the central C–C bond of **86** was found to be reluctant towards oxidation, in line with its perpendicular orientation to the adjacent arene orbitals [89].

Triptindanone **70** and triptindanetrione **72** provide an independent access to 1,8-naphtho-annelated [3.3.3]propellanes [79, 90] (Scheme 18). Monobenzylidenetriptindane **94**, readily obtained from **70**, was subjected to photocyclodehydrogenation giving **95**, a monophenanthro analogue of **5**, in excellent yield. Unfortunately, the analogous three-fold conversion of tris(benzylidene)triptindane **78** failed under these conditions. However, three-fold cyclodehydrogenation of **78** to **96** was achieved using palladium on charcoal at 310 °C furnishing

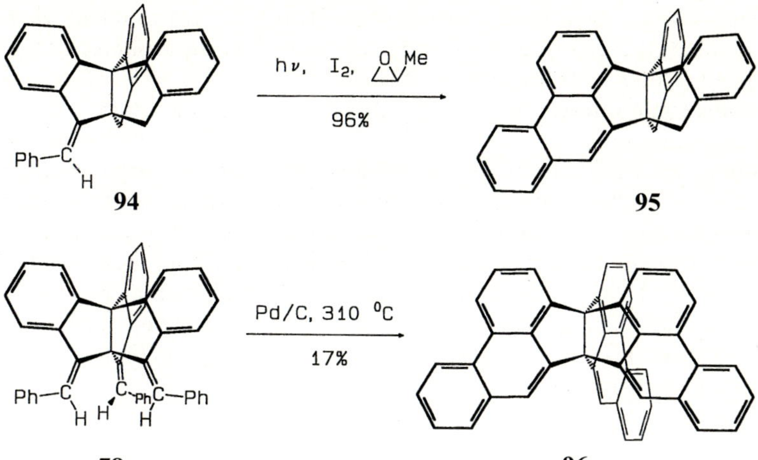

Scheme 18

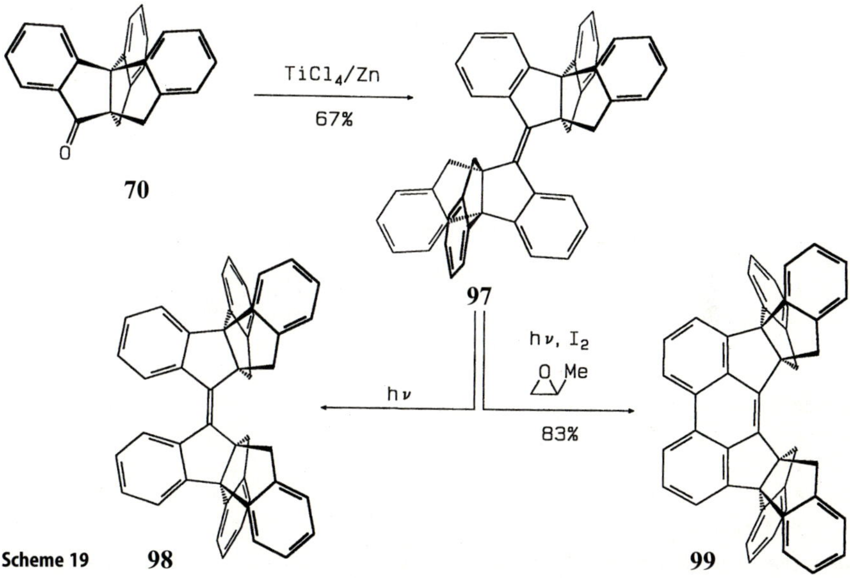

Scheme 19

this C_{3v}-symmetrical tribenzo derivative of **86**, or hexabenzo derivative of tript-indane **5**, in low yield.

McMurry coupling of **70** under carefully controlled conditions yields 9,9′-di(triptindanylidene) **97** as the *trans* isomer [79, 91] (Scheme 19). Photoiso-merization of this stilbene derivative leads to the relatively labile *cis* isomer **98** which, under oxidizing conditions, undergoes cyclodehydrogenation to give the

bis(triptindane) **99** in good yield. In this "two-fold" congener of **95**, the rare
di(cyclopenta)[*j, k; l, m*]phenanthrene unit, representing a fullerene fragment, is
annelated to two centrodiindane entities.

Finally, some interesting heterocyclic benzopropellanes have been prepared
from triptindanone **70** and triptindanetrione **72** [79, 92] (Scheme 20). Using
polyphosphoric acid, the *cis*-oxime **100** was converted stereospecifically into the
propellane-type benzamide **101**, whereas the *trans* isomer yields the corresponding propellane anilide (not shown) along with **101**. Under the same conditions,
all-*cis*-triptindanetrioxime **102** furnishes the three-fold benzamide **103** in
moderate yield. This remarkable heterocyclic propellane may be considered as
an orthoamide of tris(2-carboxyphenyl)acetic acid. As expected, X-ray analysis
of **103** [92] reveals a high degree of torsion about the propellane axis due to the
heavily interacting benzo nuclei.

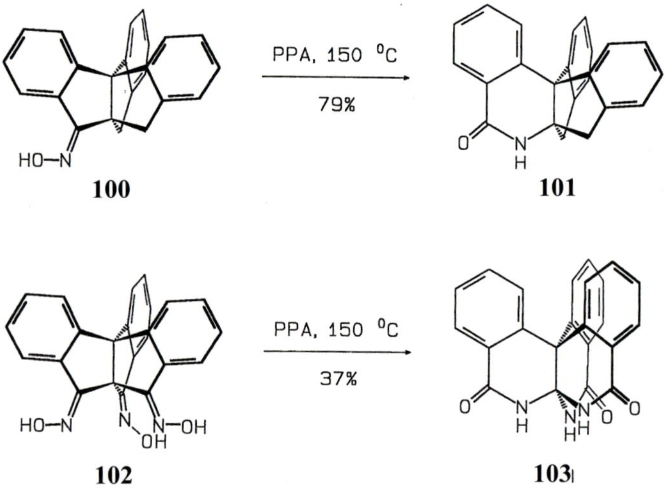

Scheme 20

4.2
Di-*fuso*-centrotriindanes and Their Derivatives

The angular (di-*fuso*-)centrotriindane **6** is readily synthesized from 2,2-diben-
zyl-1,3-indandione **69** via the corresponding indanediol **104** by using standard
cyclodehydration conditions (Scheme 21) [44, 93]. Similar to related cases (see
below), this C_2-symmetrical, and thus chiral, triindane is isolated in excellent
yield. The isomeric di-*fuso*-type, irregular centrotriindane **16** was synthesized in
low yield by Ten Hoeve and Wynberg by bicyclodehydration of 1,3-dibenzyl-2-
indanone (**105**) or, much more efficiently, via the corresponding morpholinium
bromide [52].

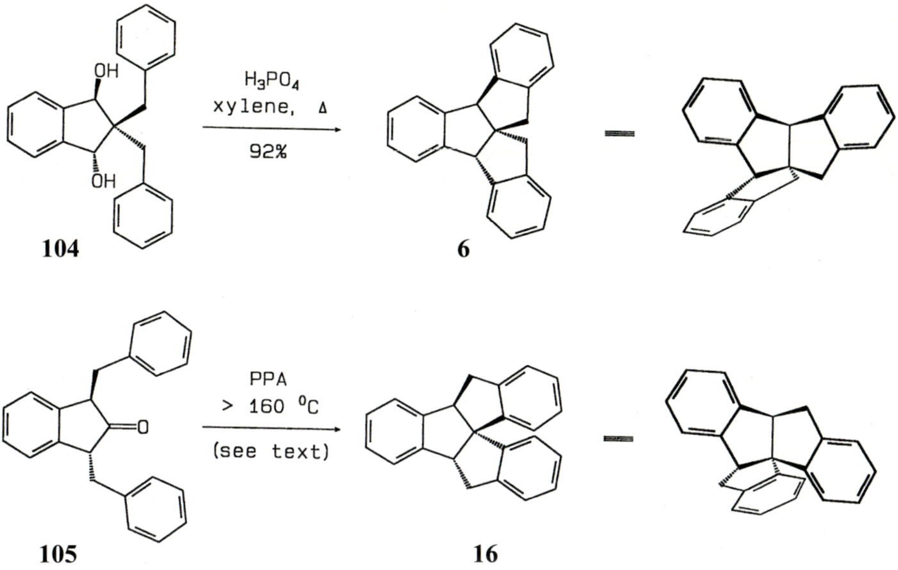

Scheme 21

Derivatives of the C_2-symmetrical mono-*fuso*-diindane (*cis*–4b, 5, 9b, 10-tetrahydroindeno[2, 1-*a*]indene, cf. **14**), were used to prepare some irregular centrotriindanes [49] (Scheme 22). Diketone **106**, readily available in a few steps [51,94], had to be benzylated in two separate synthetic steps and was then reduced to the corresponding diindanediols **107** and **108**. Interestingly, treatment of the stereoisomeric diols **107** under various dehydration conditions (e. g. H_3PO_4, xylenes at 140 °C) did not generate triindene **109** but led to polymerization. Under these conditions, the doubly benzylated diindanediols **108** gave the Wagner-Meerwein rearranged product **111** instead of the desired tetraindane **112**. Obviously, the product formed by [1,2]-benzyl shift, triindene **110**, is again too much strained to survive under the drastic reaction conditions. Similar effects have been observed in the synthesis of tribenzotriquinacene [65, 69] (cf. Scheme 26).

Notably, use of chlorobenzene as a solvent instead of xylenes allowed us to synthesize tetraindane **112** in moderate yield [49] (Scheme 22). This hydrocarbon contains two C_s– and one C_2-diindane units fused in a two-fold angular way (as well as two 1,2′-spirobiindanes **12**) and it appears to be more strained than expected. According to our schematics for centropolyindanes, **112** has to be classified as a (two-fold) irregular centrotriindane. In this view, both **111** and **112** represent the first derivatives of the irregular triindane **17**.

A variety of regular di-*fuso*-centrodiindanes related to **6** have become accessible by two-fold cyclodehydration. Starting from 4,7-diphenyl-substituted 2,2-dibenzyl-1,3-indanediones or various spiro[cyclohexane-1,2′-indane]diones (see below), the corresponding triindanes such as **113** [95] and **115** [93, 96], respectively, were synthesized in high yields (Scheme 23). In a related case, 2,2-dibenzyl-

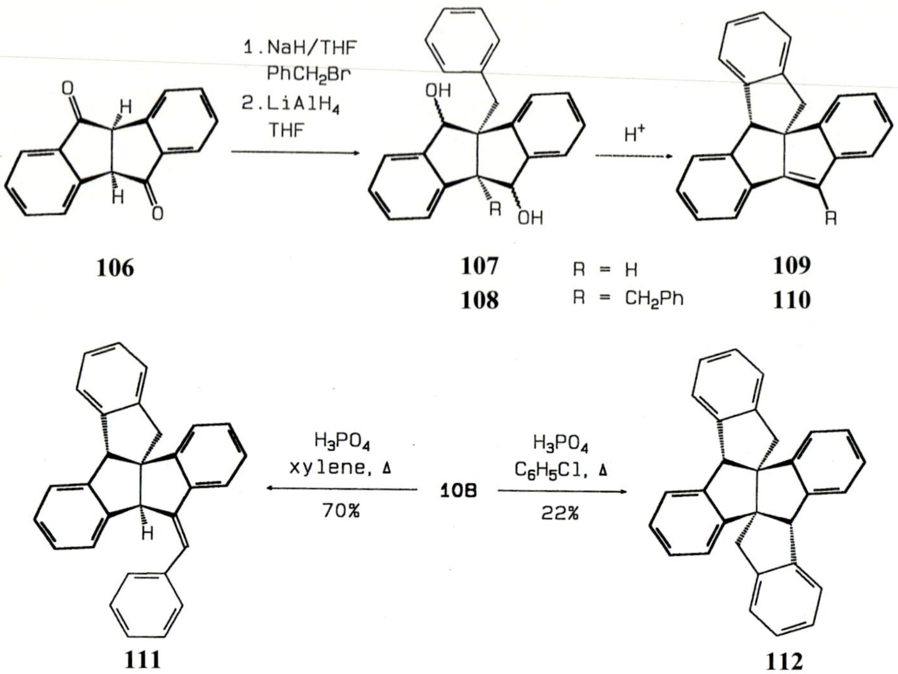

1.NaH/THF
  PhCH₂Br
2.LiAlH₄
  THF

OH

OH

H⁺

106

107   R = H   **109**
108   R = CH₂Ph   **110**

H₃PO₄
xylene, Δ

70%

108

H₃PO₄
C₆H₅Cl, Δ

22%

111

112

Scheme 22

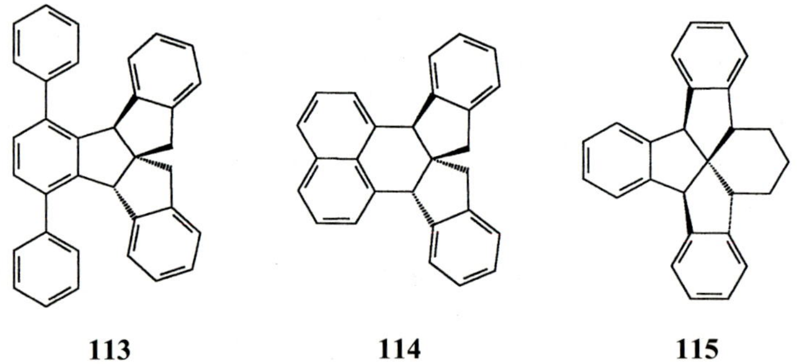

113   **114**   **115**

Scheme 23

dihydrophenalene-1,3-dione furnished tetrahydrodiindenophenalene **114** in moderate yield [97]. The ready availability of tribenzo[5.5.5.6]fenestranes with the framework of **115** has been a clue to the development of centropolyindane chemistry and also contributed considerably to the chemistry of fenestranes [37, 93].

In contrast, the chemistry of the angular triindane **6** itself has not been developed very far. This is in part due to the difficulties to functionalize this centrotriindane in a controlled way. Dibromide **116** (Scheme 24) was found to be

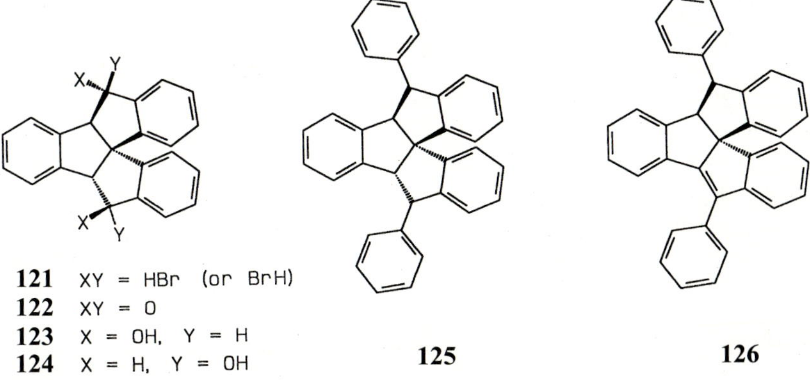

116 X=Y=Br   **118** Z=H, Z'=OH   **120**
117 X=OH, Y=H   **119** Z=Z'=O

Scheme 24

extremely sensitive and isolated in only 10% yield [98]. The monoalcohol **117** was obtained by oxyfunctionalization of **6** using dimethyldioxirane and methyl(trifluoromethyl)dioxirane along with the secondary alcohol **118** and the triindanone **119** [99]. Notably, oxygen insertion by dioxiranes into the benzylic and benzhydrylic C H bonds occurs with similar ease and **117–119** were all isolated in moderate yields. As has been observed frequently for centropolyindanes [100] and centropolyquinanes [101], "exhaustive" bromination of all the peripheral bridgehead positions is much easier than partial bromination. In accordance with this experience, four-fold bromination of **6** can be achieved to give the tetrabromide **120** as a mixture of stereoisomers in >88% yield under carefully controlled conditions [47, 102] (see below).

The irregular centrotriindane **16** was functionalized by bromination to the stereoisomeric dibromides **121** [52] (Scheme 25). Subsequent hydrolysis or oxidation to the diketone **122** followed by reduction gave triindanediols **123** or **124** respectively, with defined stereochemistry. Two-fold Friedel–Crafts reaction of **123** in benzene yielded a C_2-symmetrical triindane, **125**, whereas the dibromides afforded mainly the diphenyltriindene **126**. This compound represents one of

121 XY = HBr (or BrH)
122 XY = O
123 X = OH, Y = H
124 X = H, Y = OH   **125**   **126**

Scheme 25

the rare *difuso*-centrotriindanes containing a strained bridgehead double bond (see above).

4.3
Tri-*fuso*-centrotriindanes: Tribenzotriquinacenes, Tribenzodihydroacepentalenes and Related Polycycles

The synthesis and chemistry of tribenzotriquinacenes has been developed in several ways since its beginning in 1984 [44]. In most cases, synthesis is performed by two-fold cyclodehydration of various 2-benzhydryl-substituted 1,3-indanediols using suitable Brønsted acids in aromatic solvents. This holds for both the parent compound **128** and several *centro*-substituted derivatives such as **7**. Unsubstituted tribenzotriquinacene **128** was also prepared in an independent multistep sequence.

Two-fold cyclodehydration of 2-benzhydryl-1,3-indanediol **127** under optimized conditions affords only low yields of the parent compound **128** [65] (Scheme 26). However, this centrotriindane can be isolated completely from the reaction mixture owing to its extremely low solubility. A major portion of the starting material reacts to the singly cyclized diindene **130** bearing the double bond at the diquinane junction. This isomer appears to be the most stable one, it was also obtained by heating diindanols **37** and **41** in dipolar medium. The instability of **129**, the isomer expected to be formed from both **127** and from **37** and **41**, is reminiscent of the behaviour of the elusive triindenes **109** and **110** (Scheme 22).

In view of the very direct but anyway inefficient access to tribenzotriquinacene **128**, a stepwise synthesis was developed [69, 103] (Scheme 27), based on diindanone **36** [64] (cf. Scheme 6). Introduction of an additional double bond

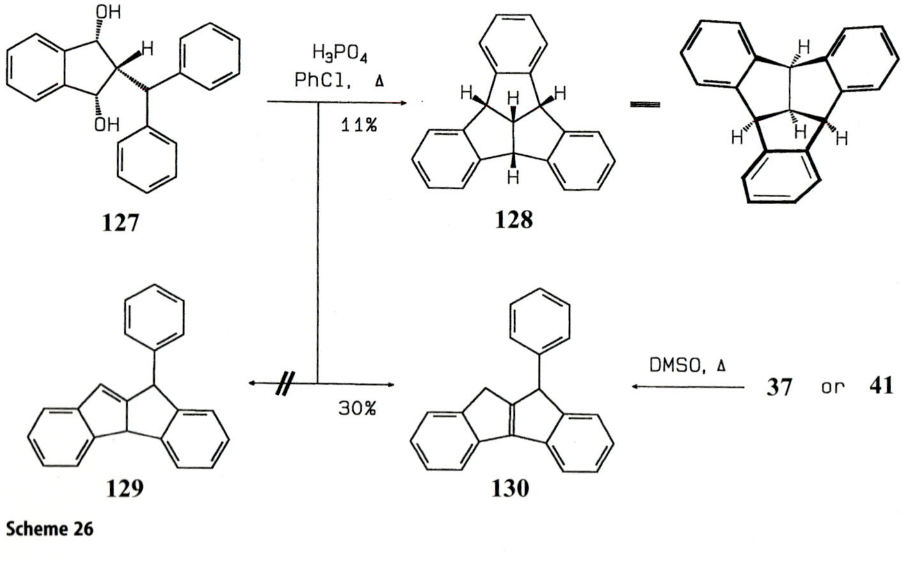

Scheme 26

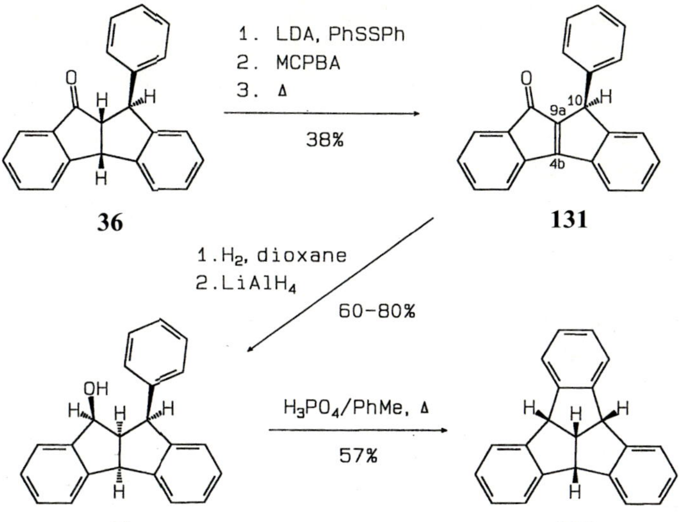

1. LDA, PhSSPh
2. MCPBA
3. Δ

38%

36

9a 10
4b

131

1.H$_2$, dioxane
2.LiAlH$_4$

60–80%

OH

41

H$_3$PO$_4$/PhMe, Δ

57%

128

Scheme 27

into the diquinane core was achieved by conversion of **36** into the phenylsulfinyl derivative followed by thermal *cis*-elimination. This "dehydrogenation" sequence yields the $\Delta^{4b,9a}$-diindenone **131** together with the $\Delta^{9a,10}$-isomer (not shown in Scheme 27). Re-hydrogenation of these diindenones furnished the C(10)-epimer of **36**, which was reduced to the *syn*-alcohol **41**. Finally, single cyclodehydration of **41** gave the target triindane **128** in good yield. In total, the stepwise synthesis of **128** is superior (total yields are 15–19% from **36**) to that involving two-fold cyclodehydration (11% from **127**), but much more cumbersome [69].

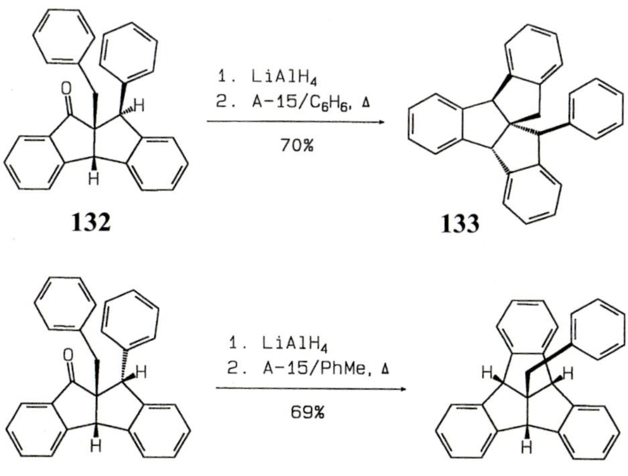

1. LiAlH$_4$
2. A–15/C$_6$H$_6$, Δ

70%

132

133

1. LiAlH$_4$
2. A–15/PhMe, Δ

69%

Scheme 28   **134**

135

A related, stepwise cyclization procedure was developed for the synthesis of the *centro*-benzyl-substituted tribenzotriquinacene **135** (Scheme 28). Bridgehead benzylation of **36** and its C(10)-epimer gave diindanones **132** [69] and **134** [45] respectively. Two-step conversion of the former compound by reduction and cyclodehydration with Amberlyst 15 gave the di-*fuso*-triindane **133** [69] bearing an *anti*-phenyl substituent, whereas a similar sequence starting from **134** furnished the tri-*fuso*-triindane **135** [45]. The synthesis of **135** clearly demonstrates the importance of the pre-orientation of the phenyl group to be attacked during the cyclodehydration step. The epimer of **133** is not formed because of entropically favored cyclization involving the *syn*-phenyl group rather than the pendant benzyl substituent. The ease of this cyclization has also been used for the synthesis of a next-higher congener, tri-*fuso*-centrotetraindane **8** [45] (cf. Scheme 44).

In contrast to the synthesis of **128**, the preparation of tribenzotriquinacenes is relatively efficient when 2,2-disubstituted 1,3-indanediols are subjected to two-fold cyclodehydration [65] (Scheme 29). Somewhat varied conditions have been used, depending on the starting material. Best yields (>30%) were obtained for the *centro*-methyl derivative **7**, whereas the yield of the benzyl-substituted analogue **135** is relatively low. The moderate yields of the tribenzotriquinacenes shown in Scheme 29 have to be attributed to several unfavorable

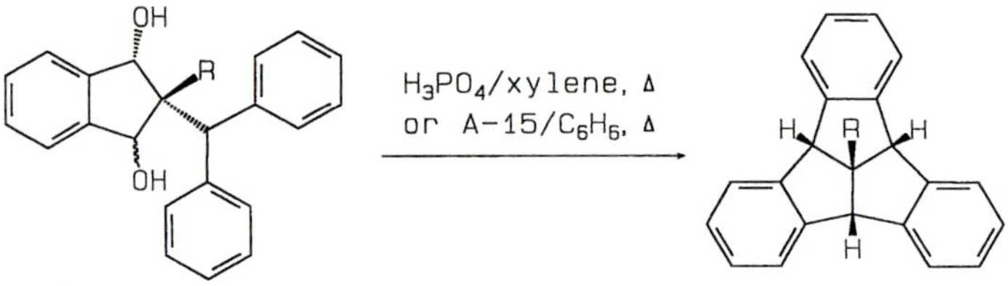

136	R = Me	**7**	(33%)
137	R = Et	**138**	(27%)
139	R = CH₂Ph	**135**	(18%)
140	R = CHPh₂	**141**	(23%)

Scheme 29

factors and experimental limits of this approach have been found also [65]. Nevertheless, their syntheses is particular straightforward. Also remarkable in this context is the formation of tribenzotriquinacene **141** bearing a second, pendant bulky benzhydryl group at the central bridgehead position. In contrast, the 10-allyl analogue was obtained in very low yield and the 10-phenyltribenzotriquinacene remained elusive [65].

The chemistry of tribenzotriquinacenes has been developed into a rich field within a short period of time, demanding for even further exploitation. Most of the work described here concerns the reactions of multiply bridgehead-substituted derivatives of the parent centrotriindanes, **128** and **7**.

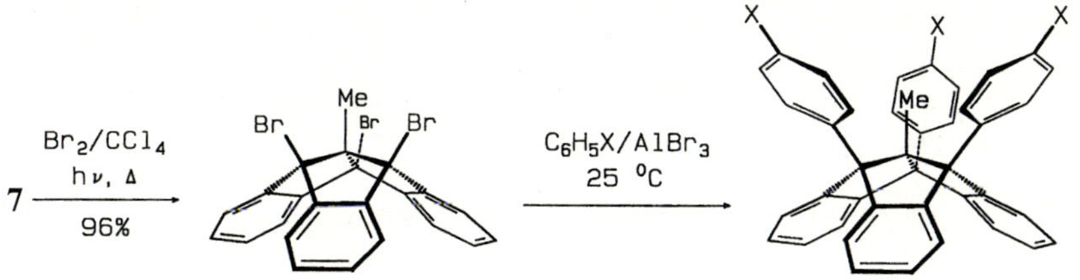

Br$_2$/CCl$_4$
hν, Δ
7 ────────→
96%

142

C$_6$H$_5$X/AlBr$_3$
25 °C
────────→

143   X = H   (40%)
144   X = OMe   (91%)

AlMe$_3$
(Y=Br, R'=Me)
EtMgBr
(Y=Br, R'=Et)
────────→

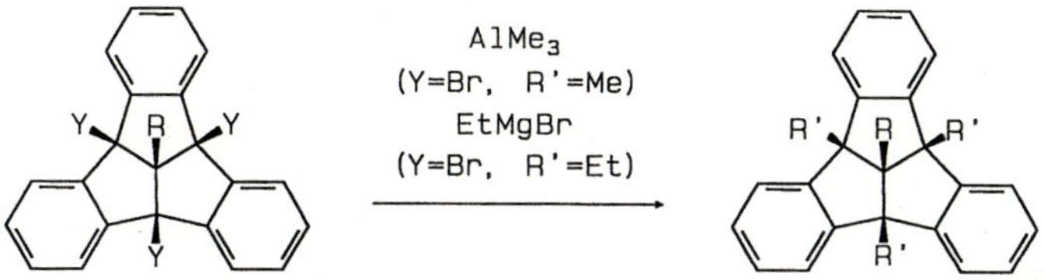

145   R = H,   Y = Br
146   R = Et,   Y = Br
147   R = Me,   Y = CN
148   R = Me,   Y = CH$_2$CHCH$_2$

149   R = H,   R' = Me
150   R = R' = Me
151   R = Et,   R' = Me
152   R = Me,   R' = Et

Scheme 30

Bromination at all three benzhydrylic bridgehead positions of **7** is almost quantitative to give **142** [104] (Scheme 30), a triquinacene bearing the four bridgehead substituents in fully eclipsed orientation. Similar to **142**, the tribromides **145** and **146** were obtained in high yields. Starting from these intermediates, various four- and three-fold carbon substituted tribenzotriquinacenes (Scheme 30) as well as heterofunctionalized analogues (see below) have become available.

Friedel–Crafts reaction of **142** in benzene [104] and anisole [100] has been achieved to give the three-fold triarylmethanes **143** and **144** respectively, bearing a sterically highly shielded methyl group which may be functionalized by thermolytic [104] or photolytic activation through the adjacent aryl groups. The steric shielding by, and the orientation of, the three aryl groups is reflected by the high-field resonance of the methyl protons at $\delta = -0.3$ ($\Delta\delta = -2.1$ ppm relative to **7**).

Reaction of **142** with trimethylsilyl cyanide and allyltrimethylsilane under SnCl$_4$ catalysis yields the tris(cyanide) **147** and the tris(allyl)tribenzotriquinacene **148** [98]. Substitution of the bromine atoms takes place very efficiently by treatment of **142**, **145** and **146** with trimethylaluminum, to give trimethyltribenzotriquinacene **149** and the tetraalkyl-substituted analogues **150** and **151**, respectively, in excellent yields (>90%) [100]. Whereas reaction of **142** with triethylaluminum gives rise to complete reduction to **7**, use of ethylmagnesium bromide affords **152** in low (20%) yield. A variety of further carbon-bonded sub-

153 X = Cl	**156** X = F	**159** X = OMe
154 X = N$_3$	**157** X = Cl	**160** X = (OCH$_2$CH$_2$)$_2$OMe
155 X = NH$_2$	**158** X = N$_3$	**161** X = OOtBu

Scheme 31

stituents such as 4-phenoxyphenyl [100] and α-furyl groups [98] have also been attached to the tribenzotriquinacene backbone.

Exchange of the bromine atoms in **142** and **145** for other halides or hetero-atomic groups is similarly facile [100]. Among those listed in Scheme 31, the tri-azides **154** and **158** are interesting because of their further conversion. Reduction of **154** with lithium aluminium hydride furnishes the C_{3v}-symmetrical triamine **155** [105] (cf. Scheme 33 for the C_s-symmetrical isomer). The homologue **158** was converted into unusual heterocycles by thermolysis [100] (cf. Scheme 43). Three-fold bridgehead ethers such as **159** and **160** were obtained by alcoholysis of **142**, and Ag^I ion assisted reaction of **142** with *tert*-butyl hydroperoxide affor-ded the tris(peroxide) **161** [100].

Another set of tribenzotriquinacenes functionalized at the peripheral bridgehead positions is shown in Scheme 32. The triamines **162** and **163** were obtained in excellent yield by aminolysis of the methyltribenzotriquinacene **142** with ammonia and dimethylamine, respectively, in benzene solution [100]. Solvolysis of **142** in the neat alkyl (R = Me, Et, Pr, nBu) and benzyl mercaptane gave the three-fold thioethers **164** and **165** [106]. Oxidation of these compounds with *meta*-chloroperbenzoic acid (MCPBA) led to complete decomposition [98], whereas use of dimethyldioxirane gave high yields of the tris(sulfones) **166** and **167** as thermally stable compounds [106]. ^{1}H- and ^{13}C-NMR spectrometry as well as X-ray structural analysis revealed the dynamic behaviour of the sulfones,

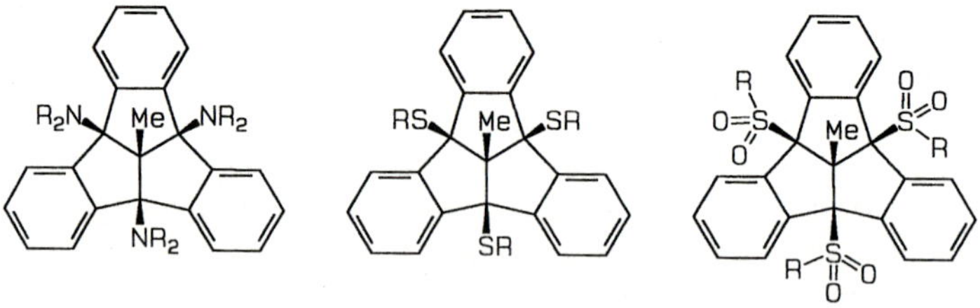

162 R = H	**164** R = n-alkyl	**166**
163 R = Me	**165** R = CH$_2$Ph	**167**

Scheme 32

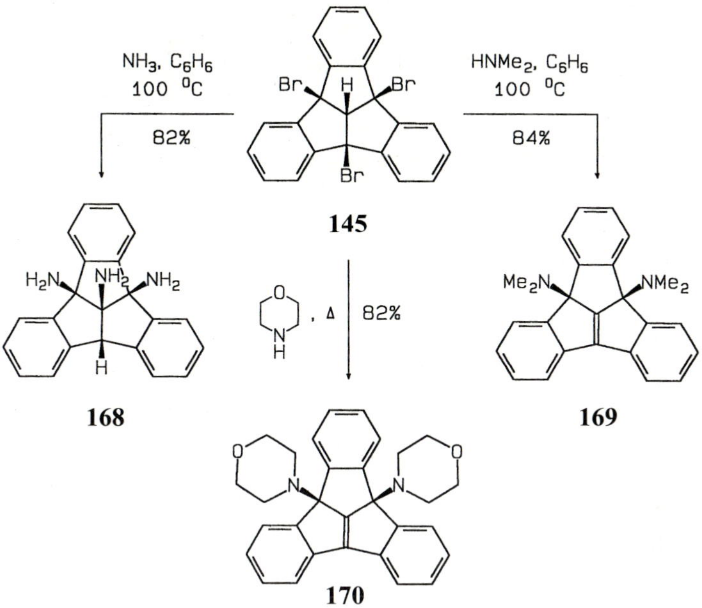

Scheme 33

which preferably adopt a chiral (i.e. C_3– rather than C_{3v}-symmetrical) ground-state conformation [106], as indicated for one of the enantiomers in Scheme 32. The chemistry of related trifunctionalized tribenzotriquinacenes is under current investigation.

The reactions of the tribromotribenzotriquinacene **145** with ammonia and secondary amines are most remarkable (Scheme 33). In the former case, the 1,4,10-triamino derivative **168** is formed instead of the expected 1,4,7-tri-amino isomer **155** [107]. In contrast, treatment of **145** with dimethyl amine in benzene and with morpholine generates the tribenzodihydroacepentalenes **169** [107] and **170** [108] respectively, as stable compounds containing a highly strained C–C double bond [109]. This behaviour is in line with the corresponding reactions of 1,4,7-tribromotriquinacene [110]. With regard to the tendency of **145** to form dihydroacepentalene derivatives such as **169** and **170**, the unusual course of the ammonolysis to **168** has to be attributed to a chelate-assisted orientation of an ammonium ion prior to addition to the relatively unshielded and reactive C–C double bond of the intermediate diamino-substituted tribenzodihydroacepentalene.

The strain in **169** and related tribenzodihydroacepentalenes is reflected by a strong out-of-plane distortion of the central double bond. The bis(morpholino) derivative **170** [100, 108, 109] represents an extreme case in that the bulky bridgehead substituents optimize steric shielding and maximize the degree of pyramidalization (see below). The less shielded bis(dimethylamino) analogue **169** has been used as a basis for various addition reactions leading to further bridgehead-substituted tribenzotriquinacenes [100, 107] (Scheme 34). Reaction

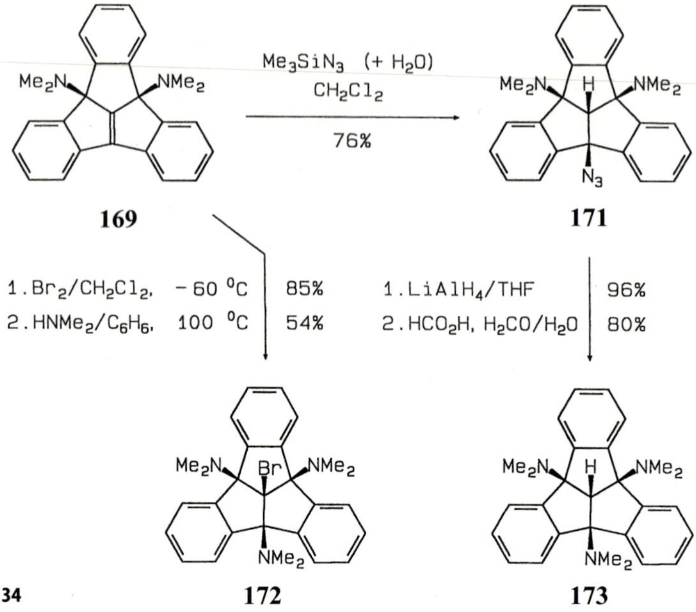

Scheme 34

with trimethylsilyl azide in moist dichloromethane yielded azide **171**. Thus, hydrazoic acid, in contrast to ammonia, adds in the conventional way, as do other monobasic acids in related cases (cf. Scheme 38). The constitution of **171** was proved by reduction and subsequent methylation to give the C_{3v}-symmetrical tris(dimethylamino)tribenzotriquinacene **173** [107]. Another interesting conversion of **169** concerns addition of bromine to the central double bond giving the corresponding 1,10-dibromo-4,7-bis(dimethylamino)tribenzotriquinacene. Subsequent aminolysis with dimethyl amine yielded the fully bridgehead-heterofunctionalized tribenzotriquinacene **172** [107]. So far, attempts to substitute the central bromine atom for a hydroxy or an additional amino group have remained unsuccessful. Unusual chemical and physical properties may be expected from such bent molecules "coated" with polar substituents at the convex surface.

Owing to its strained double bond but despite the steric shielding by the dimethylamino groups, tribenzodihydroacepentalene **169** readily undergoes cycloaddition with 1,3-dipolar reagents [107] (Scheme 35). Addition of phenyl azide leads to the 1,10-triazole **176**, a triquinacene bearing, in fact, four bridgehead nitrogen substituents. By treatment with diazomethane and diazopropane, **169** is converted into the related pyrazoles **174** and **175**, which both undergo photolytic desazotation with concomitant two-fold desamination. Thus, the first cyclopropa-annelated triquinacenes, **177** and **178**, were obtained, albeit in relatively low yields [107]. Force-field (MM+) calculations suggest the presence of two almost inverted bridgehead carbon atoms at C(1) and C(10). Also interesting is the finding that photolysis of the diazopropane adduct **175** forms the ring-opened isomer of **178**, 10-(propen-2-yl)tribenzotriquinacene **179**, in considerable yields. This compound represents one of the extremely rare tri-

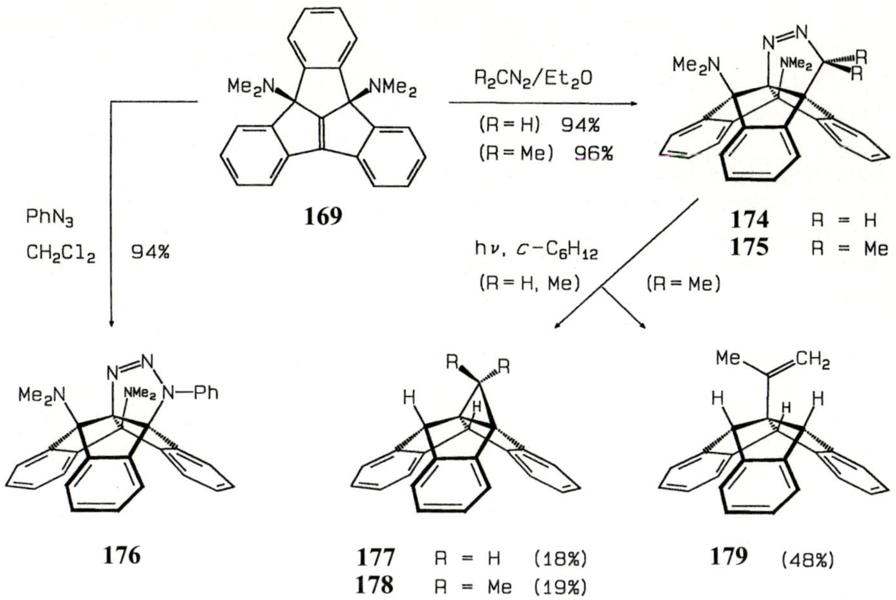

Scheme 35

quinacenes bearing an α-unsaturated *centro* substituent [33, 107]. These compounds are interesting because they promise an access to derivatives of the yet unknown triquinacene 10-carboxylic acid.

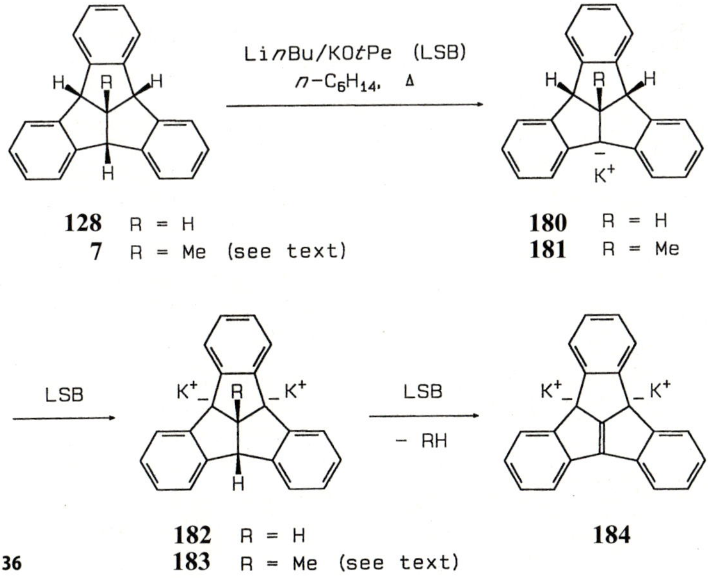

Scheme 36

Extended studies have been undertaken to convert tribenzotriquinacenes into tribenzoacepentalene **212** (cf. Scheme 42) and its derivatives [103, 108, 109, 111, 112]. Making use of the generally stabilizing effect of benzoannelation on reactive intermediates and strained products, deprotonation of **128, 7** and other related *centro*-substituted tribenzotriquinacenes (cf. Scheme 29) with Lochmann–Schlosser bases (LSB) were investigated. As expected, deprotonation of **128** with *n*-butyllithium/potassium *t*-pentoxide takes place more readily than with triquinacene [103] and leads to the benzhydrylic mono- and dianions as the potassium salts **180** and **182** (Scheme 36). Similar single and double deprotonation takes place with the *centro*-alkyl derivatives such as **7** to give, for example, potassium salts **181** and **183**. Most remarkable is the subsequent formation of dipotassium tribenzodihydroacepentalenediide **184** by (formal) elimination of dihydrogen or the corresponding alkane. The relative ease of the elimination process has been monitored by ^{1}H NMR spectrometry, reflecting the relative stabilities of the corresponding anions ($H^- \gg PhCH_2^- > Me^- > Et^-$) expelled from the *centro* position [108, 109].

Starting with 10-methyltribenzotriquinacene **7**, the most readily available congener among the tri-*fuso*-centrotriindanes, controlled deprotonation leads to a number of interesting bridgehead derivatives [108] (Scheme 37). The mono-anion salt **181** can be quenched by electrophiles such as alkyl sulfates and alkyl or silyl halides to give 1,10-disubstituted tribenzotriquinacenes **185** – **188**. Since partial functionalization (e.g. bromination) of the benzhydrylic bridgeheads of **7** is difficult [100], the deprotonation/alkylation strategy represents a valuable alternative. Further substitution of the neopentane core of **7** is also possible by subsequent single or two-fold deprotonation, as shown by the facile conversion of 1, 10-dimethyltribenzotriquinacene **185** into the 1,4,7,10-tetramethyl derivative **150** (Scheme 37, cf. Scheme 30).

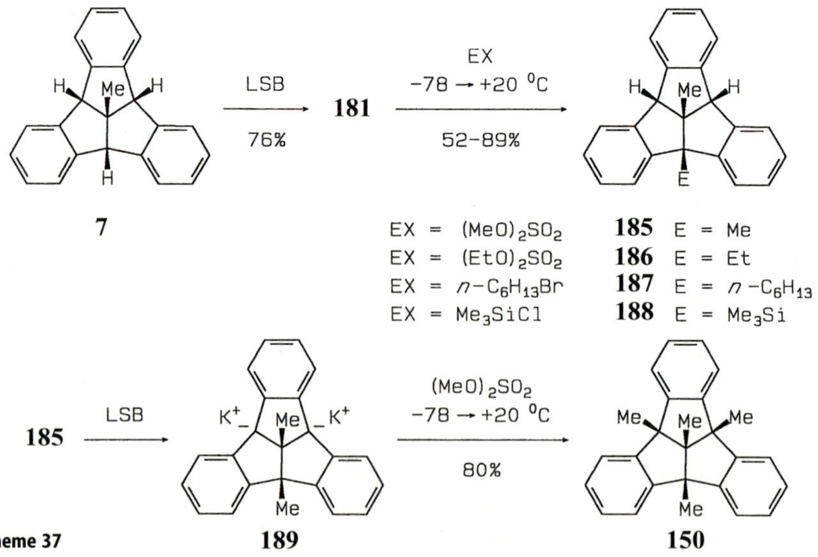

$$EX = (MeO)_2SO_2 \quad \mathbf{185} \quad E = Me$$
$$EX = (EtO)_2SO_2 \quad \mathbf{186} \quad E = Et$$
$$EX = n\text{-}C_6H_{13}Br \quad \mathbf{187} \quad E = n\text{-}C_6H_{13}$$
$$EX = Me_3SiCl \quad \mathbf{188} \quad E = Me_3Si$$

Scheme 37

The facile formation of dipotassium tribenzoacepentalenediide **184** has opened efficient and versatile syntheses of further 1,4-disubstituted tribenzo-1,4-dihydroacepentalenes such as **190**–**193** [108, 109] (Scheme 38). These analogues of **169** and **170** are not accessible via tribromide **145** (cf. Scheme 33). Use of the hexane-soluble Lochmann–Schlosser base Li(2-ethylhexyl)/KO*t*Bu allows one to synthesize the bis(trimethylsilyl) compound **190** in almost quantitative yield. The functionalized tribenzodihydroacepentalenes **190**–**193** are relatively unstable at room temperature, but addition of protic acids can be performed leading to further "mixed" 1,4,7-trisubstituted tribenzotriquinacenes such as **194** and **195** (Scheme 38, cf. Scheme 34). Most remarkable is the conversion of the labile bis(stannyl) derivative **193** (obtained in 42% yield) into the corresponding dilithium salt **196**, which co-crystallizes with six molecules of dimethoxyethane (DME). X-ray crystal structure analysis revealed that **196** forms solvent-separated ion triplets consisting of ribbon-like layers of tribenzoacepentalenediide ions alternating with layers of [Li$^+$ 3 DME] chelates [108]. In contrast, the analogous non-benzoannelated salt forms contact-ion triplets [33].

As mentioned above, the reactivity of 1,4-tribenzodihydroacepentalenes **169**, **170** and **190**–**193** is due to the presence of the strained central double bond. X-ray structure analysis of **170**, **190** and **191** showed the extremely strong pyramidalization at C(7) and C(10). Out-of-plane distorsion is most pronounced at the central carbon atoms of **170** and **191**. In these cases, the double bonds stick out of the planes C(1)-C(10)-C(4) by 47.2 ($\pm$0.3)° and 45.8 ($\pm$0.5)° [108, 109].

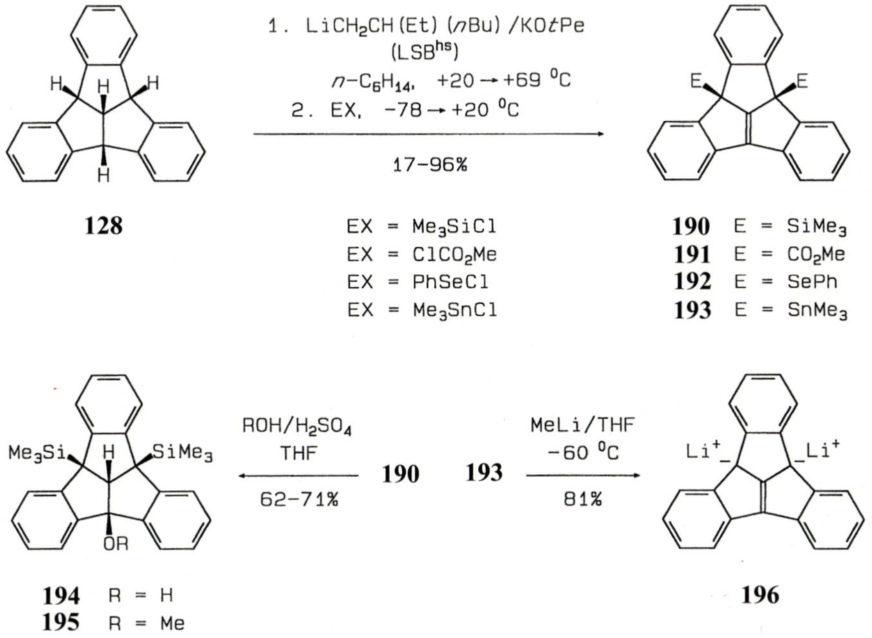

128

1. LiCH$_2$CH(Et)(*n*Bu)/KO*t*Pe (LSB^{hs})
   n-C$_6$H$_{14}$, +20 → +69 °C
2. EX, −78 → +20 °C

17–96%

EX = Me$_3$SiCl   **190**   E = SiMe$_3$
EX = ClCO$_2$Me   **191**   E = CO$_2$Me
EX = PhSeCl   **192**   E = SePh
EX = Me$_3$SnCl   **193**   E = SnMe$_3$

ROH/H$_2$SO$_4$, THF   **190**   **193**   MeLi/THF −60 °C

62–71%   81%

194   R = H
195   R = Me

196

Scheme 38

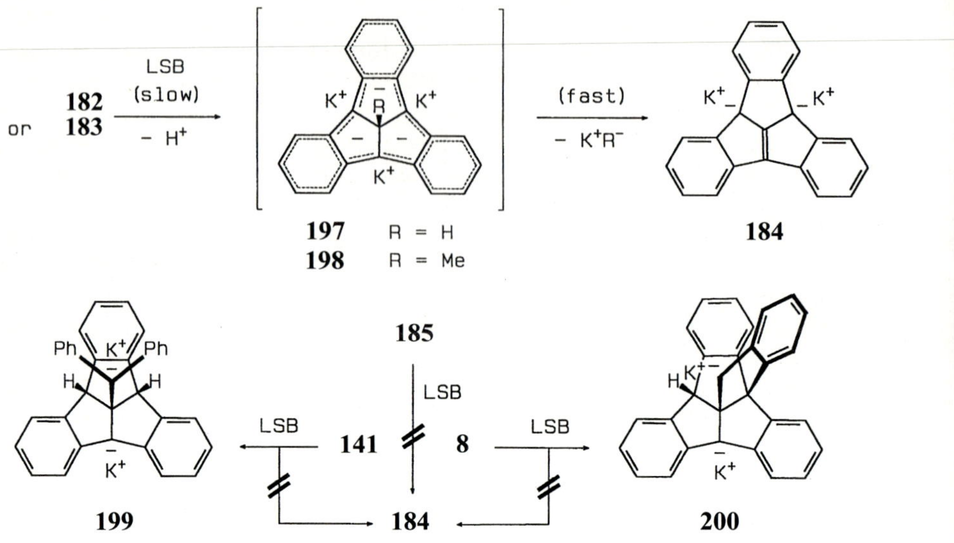

Scheme 39

A mechanism accounting for the intriguing conversion of tribenzotriquinacenes to the dipotassium salt **184** by deprotonation/elimination has been
suggested [108,111]. The different tendencies of the tribenzotriquinacenes **128**,
7, **138** and **135** to expel the *central* atom or group account for a slow, rate-deter-
mining deprotonation of dianion salts such as **182** and **183** to generate the respective tripotassium trianions, e. g. **197** and **198** (Scheme 39). In the final step,
the trianions expel a hydride or alkide ion to produce **184**. Further evidence for
this mechanism is provided by the fact that the *centro*-benzhydryl derivative **141**
(cf. Scheme 29) does not undergo elimination. Rather, the dianion **199** is formed,
preventing cleavage of the otherwise relatively labile C–C bond. Moreover, tribenzotriquinacenes bearing alkyl substituents at at least one of the peripheral
bridgehead positions, such as 1,10-dimethyltribenzotriquinacene **185** and also
tri-*fuso*-centrotetraindane **8**, do not form the dipotassium salt **184** [108].

Further experiments have revealed a first insight into the chemistry of unsubstituted tribenzodihydroacepentalene **201** [108, 109] (Scheme 40). Careful hydrolysis of the dipotassium salt **184** at low temperatures gives solutions of **201**,
which can be trapped by adding dienes such as 1,3-diphenylisobenzofuran to
yield adduct **202**. Without the trapping reagent **201** dimerizes by [2 + 2] cycloaddition across the central double bond to form **204** in almost quantitative yield.
As demonstrated by X-ray structure analysis, **204** is the product of a head-to-
head addition suggesting the intermediacy of the relatively stable two-fold
benzhydrylic diradical **203**. Bis(tribenzotriquinacene) **204** represents another
derivative of C_s-diindane **34** (Scheme 5) bearing a long C–C bond. In this case,
the C(1)–C(2) bond distance was found to be 160.2 ($\pm$0.6) pm, suggesting a
through-bond coupling of the diindane π orbitals via the σ^* orbital of the
C(1)–C(2) bond. In contrast to **45** (Scheme 7), a lateral bond is involved in **204**.
Selective cleavage of the C(1)–C(2) bond and trapping of the diradical **203** under

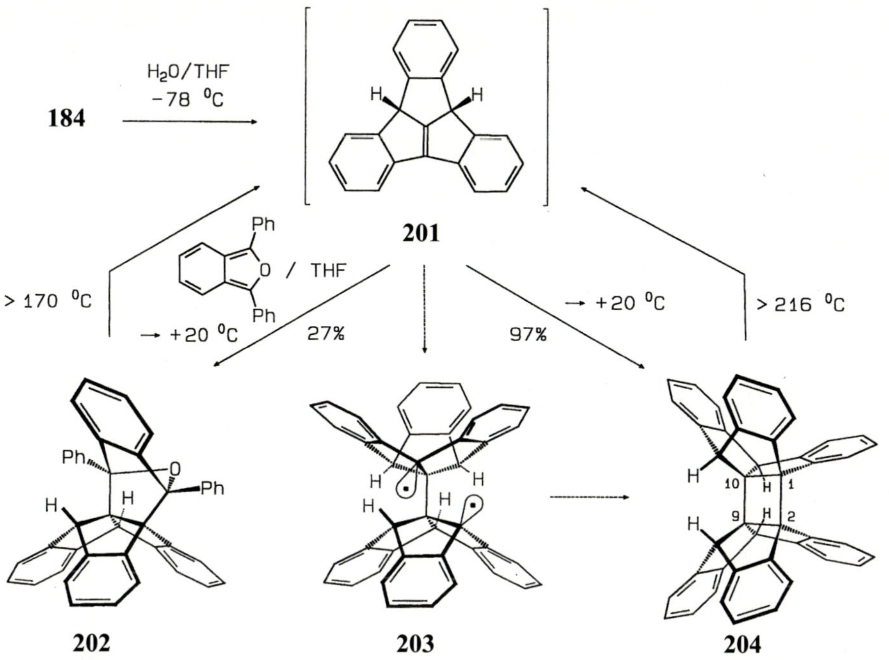

Scheme 40

reducing conditions should lead to the hypothetical 10,10′-dimer of tribenzotri-quinacene-10-yl. So far, all attempts have remained unsuccessful [108,112].

The cycloaddition products **202** and **204** may be used as a source of **201** [108]. Thermal cycloreversion of **202** and **204** starts at >170 °C and >216 °C, respectively. When dimer **204** is mixed with an excess of anthracene or tetracyclone and heated to 220 °C, the new cycloadducts **205** and **206**, respectively, are formed in good yields (Scheme 41). The anthracene adduct is again remarkable since it represents another 1,1,2,2-tetraarylethane derivative with a rigid molecular framework. X-ray structure analysis showed a bond distance of 158.7 (± 0.5) pm for the C(1)–C(2) bond, which is again oriented laterally to a *fuso*-diindane (and a 9,10-dihydroanthracene) unit.

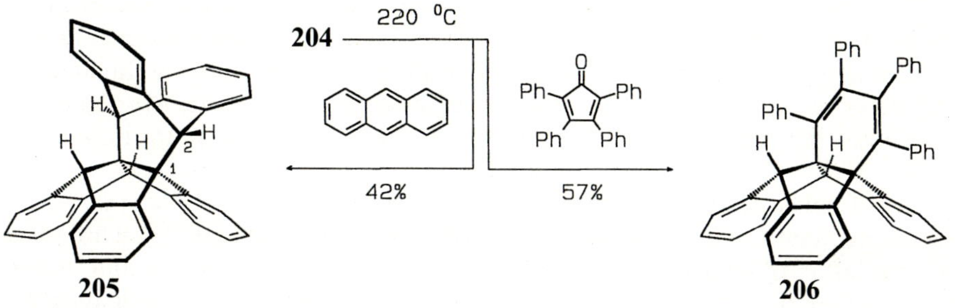

Scheme 41

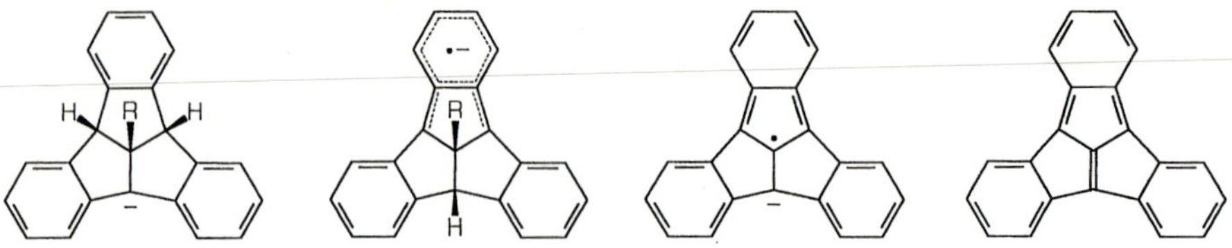

207　　R = H　　**209**　　　　**211**　　　　**212**
208　　R = Me　　**210**

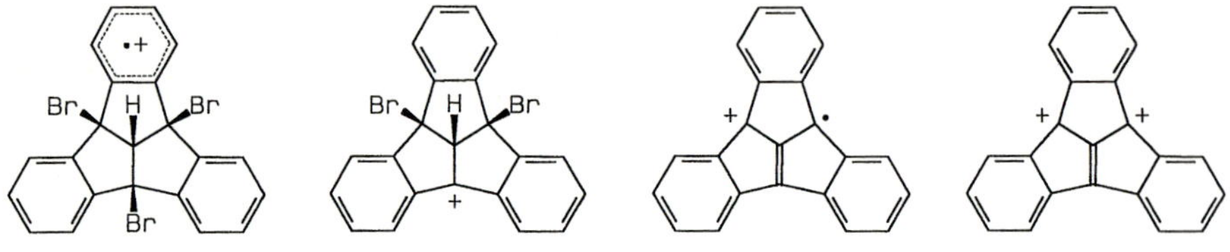

213　　　　　　**214**　　　　　　**215**　　　　　　**216**

Scheme 42

The readiness of tribenzotriquinacenes to undergo dehydrogenation or elimination of bridgehead atoms or groups has led to efforts to generate the fully dehydrogenated parent, tribenzoacepentale (**212**). However, all attempts have remained unsuccessful [108]. Nevertheless, the reactivity of the related tribenzotriquinacene derived anions and cations parallels in part the behaviour in solution and may give hints to further investigation towards the generation of **212** (Scheme 42). Gas-phase deprotonation of **128** and **7** by negative chemical ionization (NCI) using F^- or OH^- as reactant ions gives anions **207** and **208** [103]. Under forcing conditions (i.e. relatively low reagent gas pressures), single dehydrogenation products $[M-2\,H]^{\cdot-}$ (**209** and **210**) were formed along with the dehydrogenation/elimination product **211**, i.e. tribenzoacepentalene radical anion $[128-4\,H]^{\cdot-} \equiv [7-2\,H-CH_4]^{\cdot-}$. These processes were found to occur to a much greater extent with the tribenzotriquinacenes than with non-benzoannelated triquinacene [103].

The gas-phase fragmentation of *cationic* tribenzotriquinacene derivatives is also remarkable. For example, the electron-impact (EI) mass spectrum of tribromotribenzotriquinacene **145** is void of the peak corresponding to radical cation **213** due to extremely facile loss of a bromine atom to give the benzhydrylic cation **214**. However, loss of the residual bromines is very pronounced, and both the radical cation **215** and the dication **216** give rise to dominating peaks in the EI mass spectrum [100].

As a final facet of the chemistry of tribenzotriquinacenes, another interesting access to heterocyclic congeners is presented here [100] (Scheme 43). Three-fold desazotation of triazide **158** was effected by heating the compound in decaline. Quinonaphthiridine **217** and Pyridinodiazepine **218** were generated in a ca. 2:3 ratio and isolated in moderate yields as red and yellow products $[\lambda_{\max}(\mathbf{217})=$

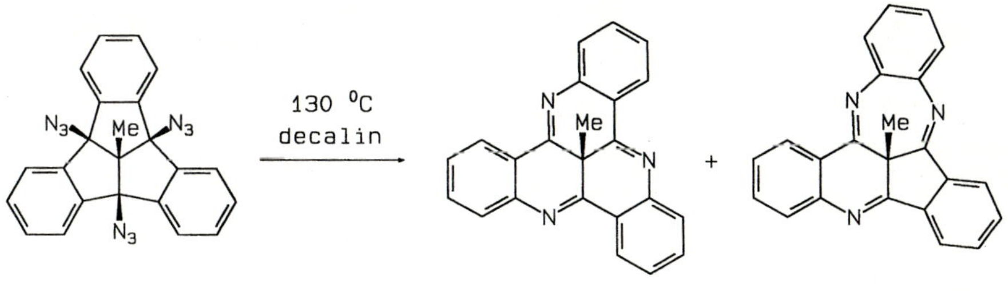

158   **217** (18%)   **218** (30%)

Scheme 43

365 nm, $\lambda_{max}(218) = 330$ nm]. Thus, azido-substituted centropolyindanes such as **158** could open new aspects for the synthesis of complex condensed azaarenes [113].

5
Centrotetraindanes

As shown at the outset (Scheme 1), there are only two ways to fuse four indane units at a neopentane core within the schematics of the regular centropoly-indanes. Both tri-*fuso*-centrotetraindane **8** and tetra-*fuso*-centrotetraindane **9** have been synthesized. In contrast, the *irregular* centrotetraindane **18** (Scheme 2), being the highest possible centropolyindane of this family, is unknown. The C_s-symmetric structure of tri-*fuso*-centrotetraindane **8** comprises all three kinds of regular centrotriindanes, i.e. **5**, **6** and **7** (or **128**), whereas the tetra-*fuso* isomer **9**, being a [5.5.5.5]fenestrane derivative, has a higher formal symmetry (D_{2d}) and containes exclusively di-*fuso*-centrotriindanes subunits (**6**). Two synthetic routes have been developed for **8**, which both are finally based on the mono-*fuso*-diindanes. In contrast, the synthesis of **9** represents an independent approach, starting from a *spiro*-annelated 1,3-indanedione.

5.1
Tri-fuso-centrotrindane

The first synthesis of **8** [45] (Scheme 44) consists of a two-fold cyclization (bicy-clodehydration) of 9a-benzyl-10-phenyldiindanone **134**, obtained by benzyla-tion of diindanone **40** (cf. Schemes 6 and 27), with polyphosphoric acid. This step is analogous to the syntheses of triptindanes, discussed in Section 4.1, and related conversions such as that of **44** to **45**. However, the bicyclodehydration of **134** is particularly remarkable since, in this case, there is no need of an electro-nically activating substituent (e.g. *m*-OCH$_3$). In contrast to 2,2-dibenzyl-1-inda-none **53**, which does not cyclize even in PPA [43] (Scheme 9), the formation of **8** is achieved at relatively mild temperatures. Again, the origin of this behaviour can be attributed to the favourable pre-orientation of the *syn*-phenyl group at C(10). This effect has already been encountered in the synthesis of 10-benzyltri-

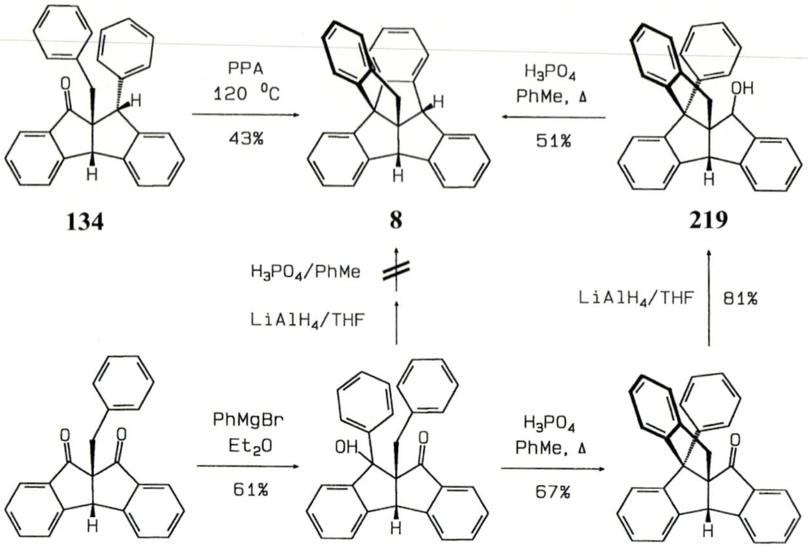

Scheme 44

benzotriquinacene **135** from **134** (Scheme 28) and is also important in one of the syntheses of centrohexaindane **11** [104] (cf. Scheme 56). Overall, the first synthesis of **8** requires six steps from **36** or nine steps from 1-indanone.

The alternative synthesis of **8** [45] is only slightly shorter (Scheme 44). C_s-diindane-9,10-dione, readily available in three steps from 1-indanone [64], was used as the starting material. Bridgehead benzylation of this diketone gives **220**, to which phenylmagnesium bromide adds without retro-aldol cleavage to yield ketol **221**. Whereas reduction of the carbonyl group of **221** was successful giving the corresponding diol, subsequent two-fold cyclodehydration to **8** failed. However, a three-step sequence starting with cyclodehydration of **221** to **222**, reduction to **219** followed by another cyclodehydration giving **8** was successful.

The chemistry of centrotetraindane **8** has not been explored in detail. It may be noted, however, that the rigidity of the tribenzotriquinacene backbone renders the framework of **8** also conformationally rigid, giving rise to a single, C_s-symmetric conformer, as confirmed by X-ray structure analysis [114]. Bridgehead-functionalized derivatives of **8** should provide a wealth of further carbon-rich polycyclic compounds.

5.2
Tetra-*fuso*-centrotetraindane (Fenestrindane) and Related Benzoannelated Fenestranes

The chemistry of tetra-*fuso*-centrotetraindane **9**, which has been coined "fenestrindane" [46] owing to its fenestrane core, and related compounds has been explored to a much greater extent [93, 98, 99]. A recent review has focused on benzoannelated fenestranes [37]. Therefore, we will restrict the discussion on

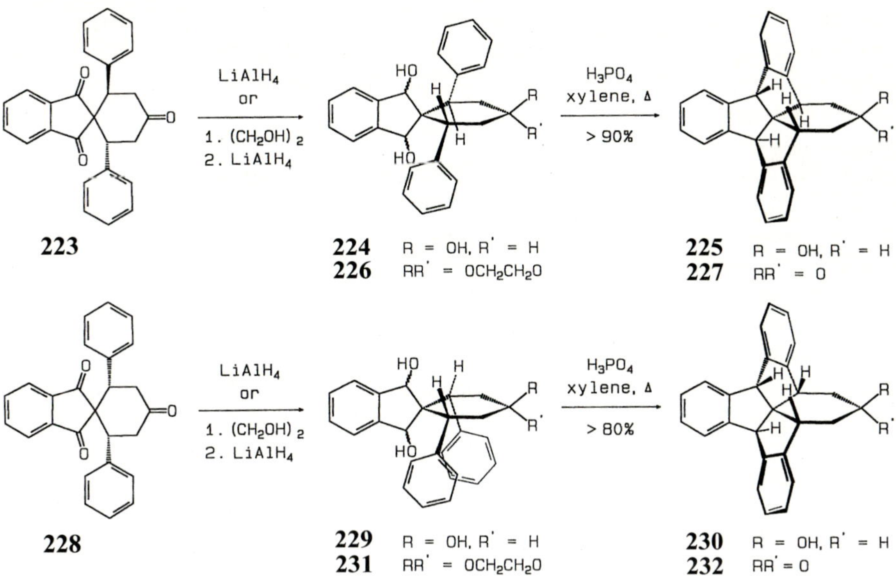

Scheme 45

the main synthetic aspects, some recently found stereochemical extensions of this area.

The starting point for the synthesis of fenestrindane **9** has been the facile access to spiro[cyclohexane-1,2′-indane]trione **223** [115, 116] (Scheme 45). The *trans*-diphenyl compound **223** and the *cis*-diphenyl isomer **228** are both obtained by two-fold Michael addition of 1,3-indanedione to dibenzylideneacetone. The *trans* isomer **223** is the product of kinetic control and the stereoorientation of the two phenyl groups in this compound is ideal to allow two-fold cyclodehydration of the corresponding *spiro*-condensed 2,2-dibenzyl-1,3-indandiols **224** and **226** [93] (cf. **104** → **6**, Scheme 21). Both the "triol route" via **224**, leading to tribenzo[5.5.5.6]fenestranol **225**, and the "acetal route" via **226**, leading directly to tribenzo[5.5.5.6]fenestranone **227**, afford high yields. Notably, the *trans* configuration of **223** translates nicely into the all-*cis*-fenestrane framework of **225** and **227**. Subsequent elaboration of the six-membered ring leads, in a few further steps, to all-*cis*-fenestrindane **9** (see below).

Surprisingly, the *cis*-diphenyl stereoisomers of **224** and **226** also undergo the two-fold cyclodehydration process. Spirotriketone **228**, being the product of the thermodynamically controlled Michael addition, provides, by conversion into the triol **229** or acetal-diol **231**, a similarly efficient access to the corresponding *cis,cis,cis,trans*-[5.5.5.6]fenestranes **230** and **232** [96] (Scheme 45). These recently obtained results were completely unexpected in view of the increased strain of fenestrane frameworks containing at least one *trans*-fused pair of rings [35, 117]. Force-field (MM+) and semiempirical MO (AM1) calculations suggest that **230** and **232** are less stable than the epimers **225** and **227**, respectively, by ca. 12 kcal mol^{-1} (50 kJ mol^{-1}) [37].

The second stage of the synthesis of all-*cis*-fenestrindane **9** begins by contraction of the cyclohexanone ring of **227** by two-fold bromination followed by Favorskii rearrangement [93] (Scheme 46). Subsequent decarboxylation of the acrylic acid furnishes the all-*cis*-tribenzo[5.5.5.5]fenestrene **233** in moderate overall yield. This compound as well as the precursors **225** and **227** have been converted into a number of related fenestranes including all-*cis*-tribenzo-[5.5.5.5]- and all-*cis*-tribenzo[5.5.5.6]fenestrane [93]. Finally, benzoannelation of **233** using tetrachlorothiophene-*S,S*-dioxide and subsequent reductive aromatization furnished fenestrindane **9**. The overall yield of the eight-step synthesis of **9** from *spiro*-triketone **223** is ca. 15%.

The development of benzoannelated fenestranes has opened several interesting perspectives. Some of them concern structural problems of fenestrane chemistry [34–37, 40, 117], mostly related to the planar-tetracoordinate carbon problem [118–120]. In this context, the generation of strained *cis,cis,cis,trans*-[5.5.5.5]fenestranes such as "*epi*-fenestrindane" (**234**) and "fenestrindene" (**235**, Scheme 47) are particularly challenging. According to computational results (MM+ and AM1), **234** should be strained by some additional 36 kcal mol^{-1} (150 kJ mol^{-1}) relative to **9** [37]. There are various potential approaches to **235**, which would represent an interesting benchmark species in the exciting field of the yet elusive fully bridgehead unsaturated fenestranes [40, 117, 118].

In an approach to fenestrane frameworks embedded in a larger polycondensed aromatic periphery, we have synthesized several naphtho-annela-

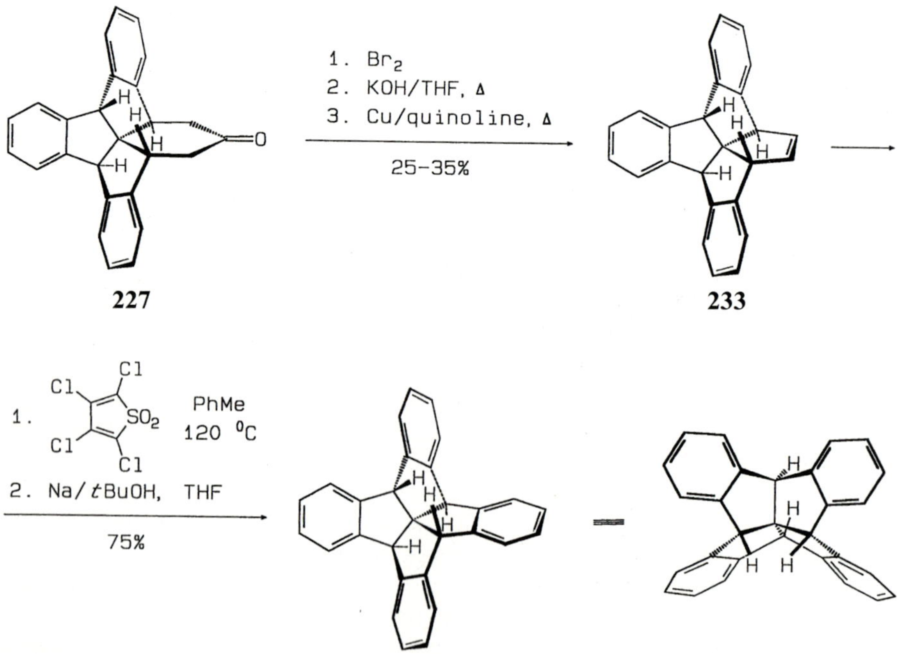

Scheme 46

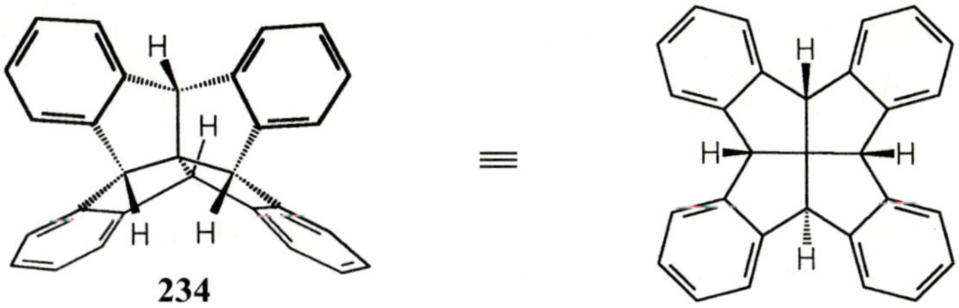

234

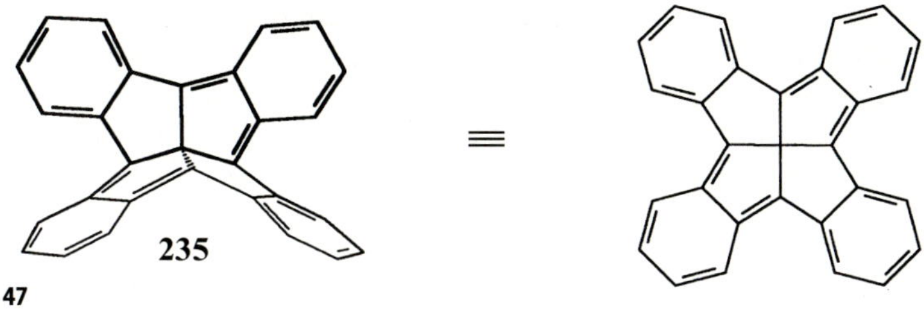

≡

235

≡

Scheme 47

ted all-*cis*-[5.5.5.6]fenestranones such as **236** and **237** [97] and all-*cis*-[5.5.5.6]fenestranones bearing two or four phenyl substituents at *ortho* positions relative to the fenestrane core, e.g. **238–240** [95] (Scheme 48). Isomers **236** and **237** exhibit markedly different magnetic deshielding effects in the ¹H-NMR spectra. The spectra of **238–240** show strong high-field shifts of the *ortho* protons vis-à-vis the outer benzene rings of the *para*-terphenyl units. The *cis,cis,cis,trans* isomer of **240** was also synthesized with surprising ease [95].

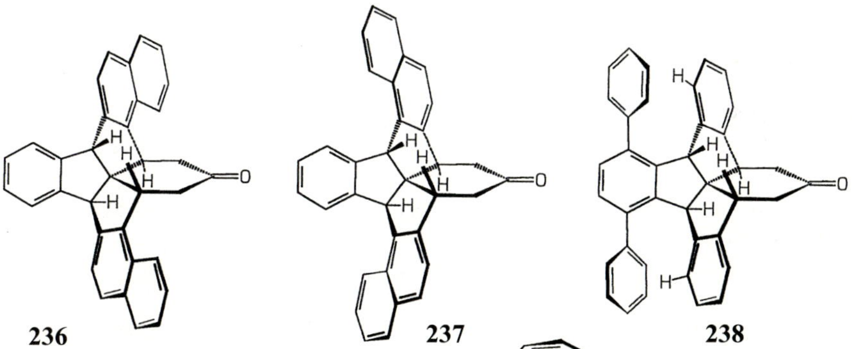

236

237

238

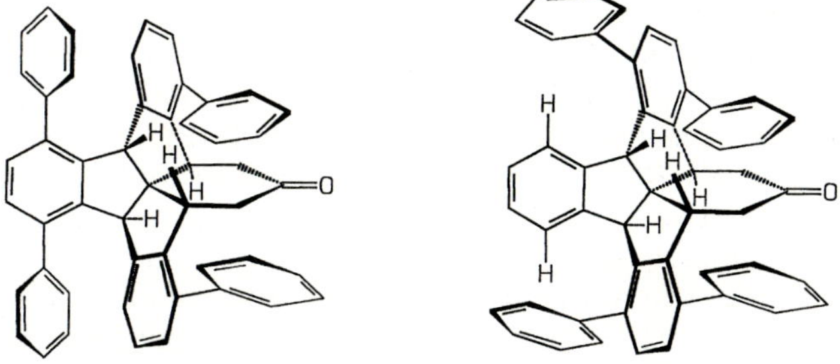

Scheme 48   **239**   **240**

Thieno-annelated analogues (Scheme 49) of [5.5.5.6]fenestranones **227** and **232** were synthesized and revealed interesting differences to the homocyclic compounds [96]. For example, cyclodehydration of the di(2-thienyl) analogue of the *cis*-diphenylspirodiol acetal **231** furnished the all-*cis*-benzodithieno[*b*] [5.5.5.6]fenestranone **241**, instead of the expected *cis,cis,cis,trans* isomer. Closer inspection revealed that epimerization occurs *during* the second cyclization step in this case. In contrast, cyclodehydration of the di(3-thienyl) analogue of **231** led stereospecifically to *cis,cis,cis,trans*-fenestranone **243**. Starting with the related *trans*-di(3-thienyl) substituted spirodiol acetal, i.e. the analogue of **226**, all-*cis*-benzodithieno[*b*][5.5.5.6]fenestranone **242**, a constitutional isomer of **241**, was obtained as the major product.

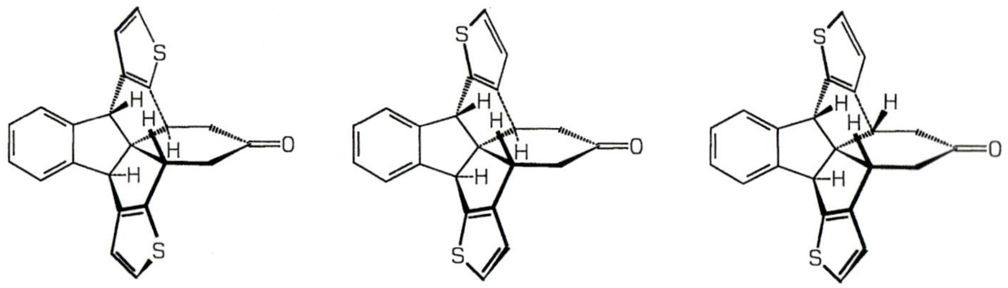

241   **242**   **243**

Scheme 49

5.3
Bridgehead-Substituted Fenestrindanes

All-*cis*-fenestrindane **9** is a stable, high-melting hydrocarbon of moderate solubility in many organic solvents. Whereas NMR spectrometry reflects the high molecular symmetry (apparent D_{2d}) of the fenestrane framework, X-ray structural analysis revealed the lower symmetry (S_4) of **9** in the solid state [37, 46]. Four-fold bridgehead-substituted derivatives of **9** are easily accessible (see below) owing to the benzhydrylic character of the bridgehead C–H bonds. Depending on the nature of the substituents, the S_4-symmetric distortion of the fenestrindane skeleton is enhanced and the degenerate equilibrium observed in solution becomes apparently static [121].

Complete bridgehead substitution of **9** is most readily achieved by bromination (Scheme 50). Tetrabromide **244** [104] is a stable compound at ambient temperature and has been converted into the tetrahalides **245** and **246** as well as into the tetraalcohol **247** [100]. Direct oxyfunctionalization of **9** using dioxiranes, in particular methyl(trifluoromethyl)dioxirane, directly furnished tetraol **247** in 55% yield [99]. Notably, single and double hydroxylation using this reagent was also achieved [122] whereas controlled partial bromination of **9** is difficult [98]. Tetraacetate **248** [99] and tetrakis(trifluoroacetate) **249** [100] were also prepared.

Tetrabromide **244** served as the base to prepare a number of other bridge-head-heterofunctionalized fenestrindanes, such as the tetraazide **250** [121], from which the tetraamine **251** [98, 123] has become accessible. In many cases, the S_N1-type conversions of **244** parallel those of tribromotribenzotriquin-acenes such as **142**, as do the syntheses of **244** and **142** from the parent centro-polyindanes (cf. Schemes 30 and 50). Thus, tetraazide **250** was obtained in 94% yield by $SnCl_4$ catalysis using trimethylsilyl azide. However, solvolysis reactions of **244** were found to be unsuccessful in some instances. Aminolysis of **244** with ammonia and dimethylamine were found to proceed incompletely [100]. There-fore, tetraamine **251** had to be prepared by reduction of **250** using lithium alumi-num hydride. As a further example, four-fold thiolysis of **244** failed, in contrast to three-fold thiolysis of **142** giving **164** (e.g. R = Me, Scheme 32. Tetrakis-(methylthio)fenestrindane **252** was again obtained by performing an Sn^{IV}-cataly-zed reaction of **244** with Me_3SiSMe [98, 121]. Accordingly, treatment of **244** with trimethylsilyl cyanide furnished tetracyanofenestrindane **253** in 56% yield [98, 121] (cf. **147**, Scheme 30). As a further parallel, four-fold bridgehead me-thylation of **244** was achieved using trimethylaluminum in *n*-heptane (cf. **150**, Scheme 30), giving **254** in 73% yield [100, 121].

Fenestrindanes **244–254** represent the first four-fold bridgehead substituted fenestranes. X-ray structure analyses of two fully bridgehead-substituted fene-stranes, viz. **244** and **254**, has revealed for the first time experimental data on the extent of flattening and S_4-symmetric distortion of the all-*cis*-[5.5.5.5]fenestra-tetraene core by bridgehead substitution [37, 100, 124]. The advantage offered by the stabilizing effects of benzoannelation on the reactive and synthetic fene-strindane intermediates, as compared to the corresponding species derived from "simple" fenestranes or fenestratetraenes [35, 40], is also obvious from further

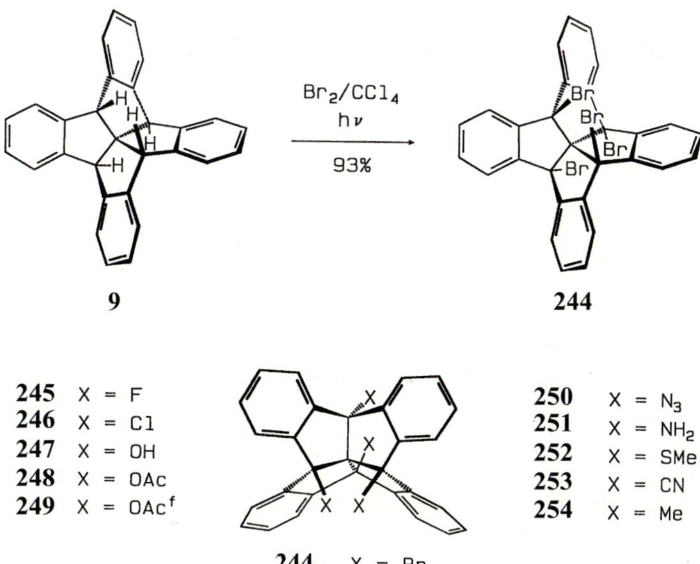

Scheme 50

synthetic conversions. It has allowed us to use fenestranes as a base for the synthesis of unprecedented centrohexacyclic derivatives of **9**, both as heterocyclic and homocyclic congeners. Thus, bis(endoperoxide) **255** was prepared from **244** by Ag^I ion-assisted condensation with *tert*-butyl hydroperoxide in moderate yield (28%) [100, 121]. The well crystallizing compound explodes upon heating to >200 °C. Two-fold bridging of the fenestrane framework in **256** by disulfide units was achieved by heating the tetrabromide **244** in elemental sulfur or by SnCl$_4$-catalyzed reaction with disilthiane, Me$_3$Si–S–SiMe$_3$ [98, 121]. Similarly, bridging of **9** by two four-membered units was possible either by solvolysis of **244** in ethanedithiol or by Lewis-acid catalyzed reaction with Me$_3$Si–SCH$_2$CH$_2$–SiMe$_3$, to generate the [5.7.5.7.5.5.]centrohexacyclane **258** (Scheme 51) [123]. The interesting bis(aminal) **257**, a [5.6.5.6.5.5.]centrohexacyclane bearing two six-membered rings, was prepared by two-fold C$_1$-bridging of tetraaminofenestrindane **251** [123].

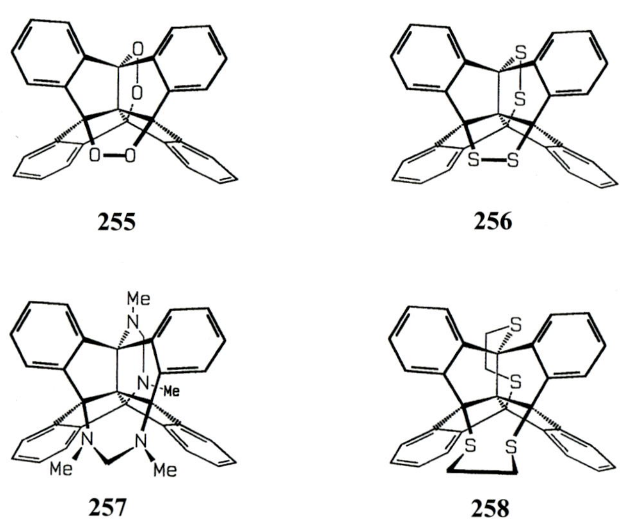

<table>
<tr><td></td><td>255</td><td></td><td>256</td></tr>
<tr><td>**Scheme 51**</td><td>257</td><td></td><td>258</td></tr>
</table>

6
Centropentaindane

The *aufbau* strategy by which two further rings are incorporated into a fenestrane after four-fold bridgehead substitution proved also successful in the homocyclic series. It provided the first access to the highest members of the centropolyindane family (cf. Scheme 1), centropentaindane **10** and centrohexaindane **11**, as well as to some related hydrocarbons. Tetrabromotriindanes **120** and **259**, generated as crude mixtures of stereoisomers and both being by far less stable than tetrabromotetraindane **244**, were condensed with two molecules of benzene under Lewis-acid catalysis with aluminum tribromide (Scheme 52). Thus, centropentaindane **10** was obtained in excellent yield (88% overall yield starting from triindane **6**) [47, 102]. A similar conversion via tribenzo-[5.5.5.5]fenestrane **259** furnished pentabenzohexaquinane **260**, the highest member of the family of partially benzoannelated centrohexaquinanes [42, 123].

Centropentaindane **10** is a high-melting compound bearing two residual bridgehead C–H bonds. Similar to tri-*fuso*-tetraindane **8**, centropentaindane has a rigid molecular structure. The presence of two (mutually fused) tribenzotriquinacene units in **10** gives rise to an almost perfect C_{2v}-symmetrical conformation. From another viewpoint, the benzo bridge across the fenestrindane backbone of **10** fixes its conformation in an almost perfect D_{2d}-symmetrical skeleton, as confirmed by X-ray structure analysis [102].

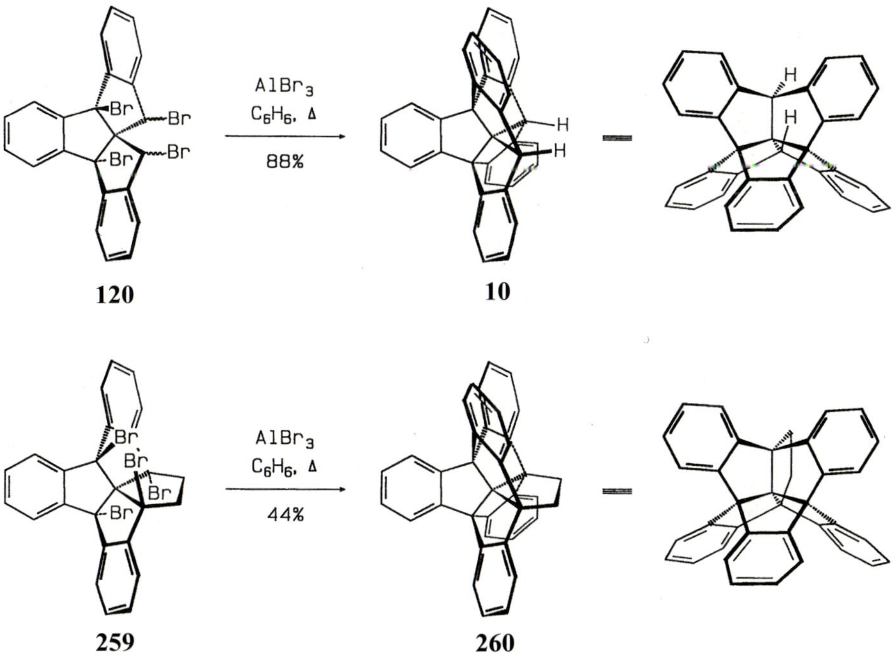

Scheme 52

An independent synthesis of centropentaindane **10** was developed by starting from 10-benzhydryltribenzotriquinacene **141** [102] (Scheme 53). Two-fold cyclization was achieved by catalytic dehydrogenation using palladium on charcoal at 500 °C. Radical induced cleavage of the pendant neopentane C–C bond occurs as a side reaction, generating diphenylmethane and dihydroacepentalene **201**, which undergoes rapid hydrogenation to tribenzotriquinacene **128**. This fragmentation may be regarded as the radical variant of the base induced (anionic) elimination of central substituents from tribenzotriquinacenes, a reaction found to be blocked especially in the case of **141** (Scheme 39).

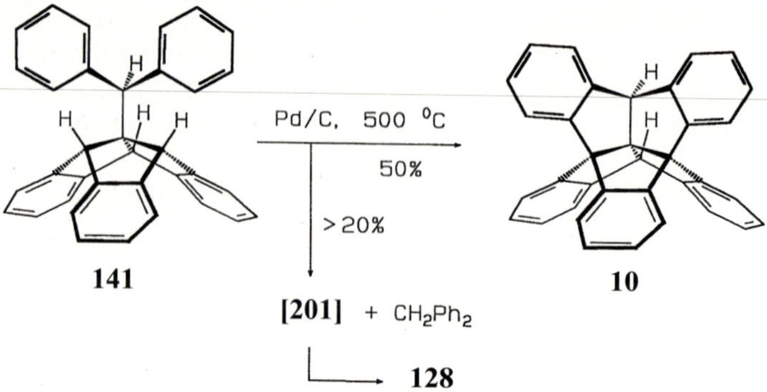

141 10

Scheme 53

Centropentaindane **10** can be easily functionalized by bromination of one of the bridgehead positions [47, 102] (Scheme 54). Hydrolysis of monobromide **261** leads to centropentaindanol **262** and Friedel–Crafts condensation with benzene or anisole affords the corresponding aryl derivatives **263** and **264**, respectively. The latter compounds may be viewed as *seco* derivatives of centrohexaindane **11** or as *syn*-arylated C_s-diindanes (cf. **42**, Scheme 6) being capped by a triptindane. The pendant aryl group is fixed vis-à-vis to the remaining single bridgehead methine group. Functionalization of that C–H bond should be extremely difficult due to steric reasons.

In spite of the extreme rigidity of the centropentaindane framework, the two perfectly 1,3-*syn* oriented bridgehead hydrogens can be substituted both using

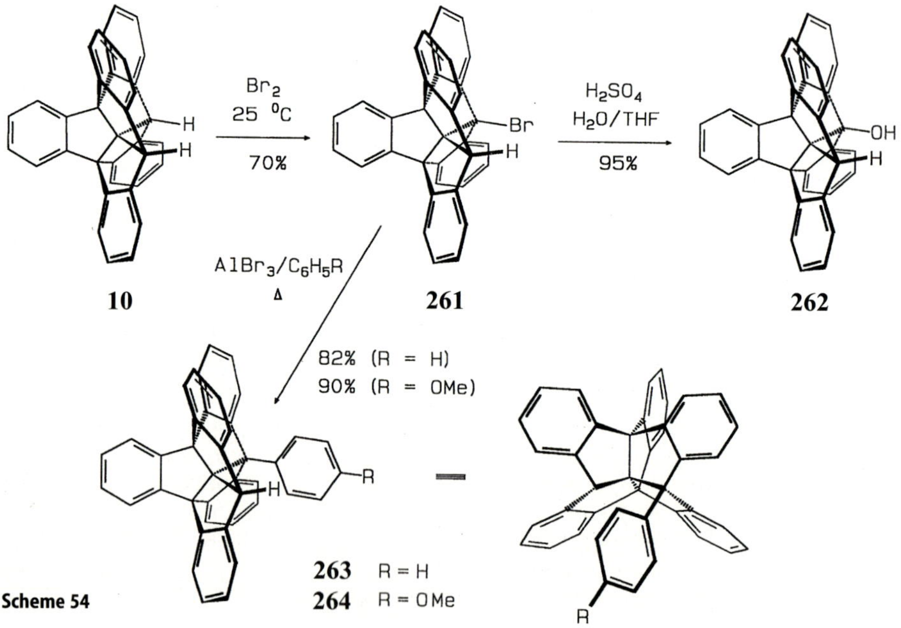

10 261 262

263 R = H
264 R = OMe

Scheme 54

harsher conditions as compared to monobromination. In fact, dibromocentro-pentaindane **265** (Scheme 55) is remarkably labile due to the extreme steric interaction of the two bulky bromine atoms, as corroborated by force-field calculations [102]. However, isolation of the compound under inert atmosphere was successful. Under conditions similar to those used for substitution of tetrabromofenestrindane **244**, the dibromo analogue **265** was converted to dimethylcentropentaindane **266**, endoperoxide **267** and endodisulfide **269** in 50–60% yield. Diol **268** was obtained by reduction of **267**, whereas controlled hydrolysis of **265** failed.

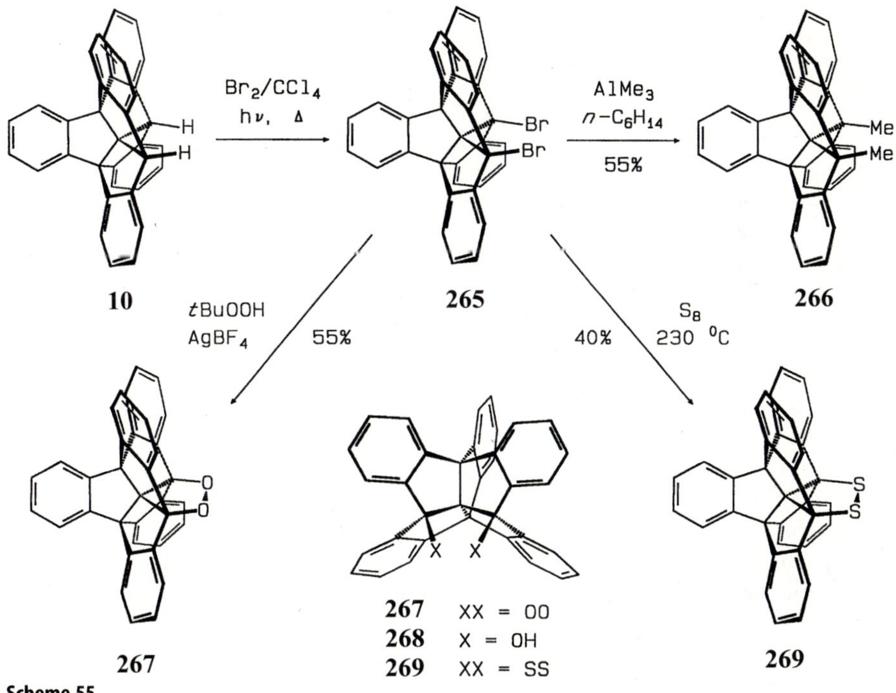

Scheme 55

7
Centrohexaindane

The synthesis of the highest member of the regular centropolyindanes, centro-hexaindane **11**, was first achieved when tetrabromofenestrindane **244** had become available [48]. Meanwhile three independent syntheses of **11** have been developed based on triptindanetrione **72**, tetrabromofenestrindane **244** and dibromocentropentaindane **265** [38, 104] (Scheme 56). Three-fold addition of phenyllithium to **72** yields the corresponding triol which can be converted to **11** by three-fold cyclodehydration [83]. The surprising efficiency of this annelation sequence, by which a tribenzotriquinacene is fused to the given triptindane

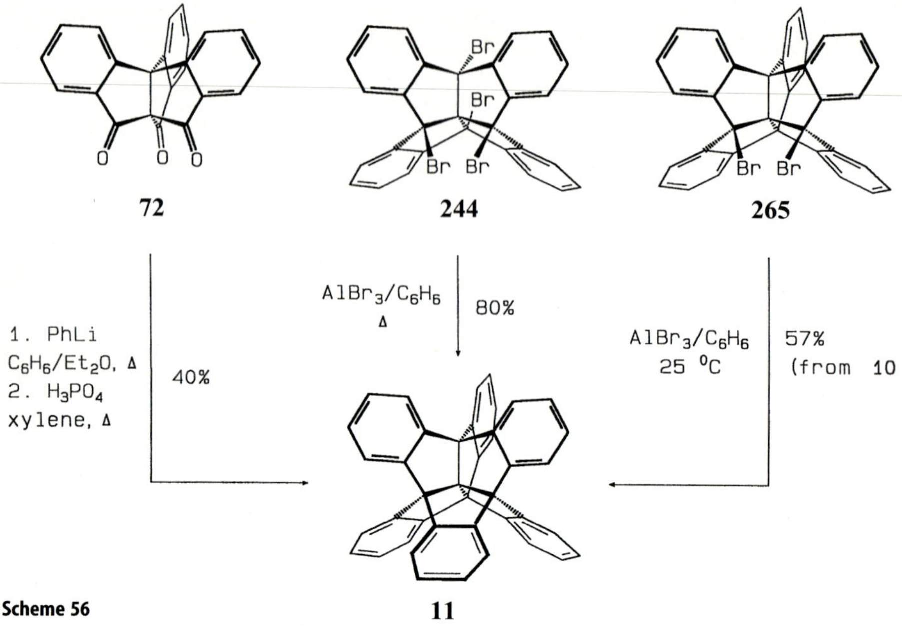

72   **244**   **265**

1. PhLi
C$_6$H$_6$/Et$_2$O, Δ

2. H$_3$PO$_4$
xylene, Δ   40%

AlBr$_3$/C$_6$H$_6$
Δ   80%

AlBr$_3$/C$_6$H$_6$
25 °C   57%
(from **10**

Scheme 56   **11**

backbone in two steps, highlights the favourable cyclodehydration of *syn*-phenyl substituted diindanols such as **41** (Scheme 27) and **219** (Scheme 44, cf. Scheme 28) and bicyclodehydration of diindanone **134** (Scheme 44). The overall yield of centrohexaindane **11** achieved by this so-called "propellane route", a six-step sequence starting from 1,3-indanedione, is 25% [104].

The so-called "fenestrane route" to centrohexaindane **11** is completed by two-fold condensation of tetrabromofenestrindane **244** with benzene under Lewis-acid catalysis. The high yield of this conversion may be traced to the intrinsic stability of the fenestrane framework bearing four benzhydrylic bridgeheads. The fenestrindane route comprises eleven steps from 1,3-indanedione and affords a 10–12% overall yield [104].

The third route has been called the "broken fenestrane route" since it involves two "broken fenestranes", viz. di-*fuso*-triindanes **5** and **120** (Scheme 52), instead of the respective "intact" fenestranes **9** and **244**, as crucial intermediates. In this case, the final step consists of a two-fold Friedel–Crafts reaction of dibromocentropentaindane **265** incorporating a single solvent molecule. In total, the broken fenestrane route comprises seven steps from 1,3-indanedione and furnishes centrohexaindane **11** in 40% overall yield [102,105]. Pursuing these three routes, and maybe a further one [104] to **11**, should make available a variety of centrohexaindanes with different substitent patterns at the molecular periphery.

8
Miscellaneous Derivatives of Centropolyindanes

As a novel family of three-dimensional arylaliphatic polycyclanes with particularly stable molecular frameworks, the centropolyindanes offer a wide range of synthetic extensions. Some of those studied recently have not been discussed in the previous sections but will be presented here in a brief potpourri (Scheme 57).

Tricarbonylchromium complexes of several diindanes and triindanes have been prepared and their structural details have been determined by X-ray structure analysis. In addition to the $Cr(CO)_3$ complexes of 2,2'-spirobiindane [125] and its derivatives, which were used as diastereomeric synthetic intermediates [126], the mono- and bis-complexes of C_s-diindane 34 [127] and its C_2-symmetric isomer (cf. 14 and 106) [128] have been studied. Previously, the stereoisomeric *mono*-$Cr(CO)_3$ complexes of 4 as well as some analogues were synthesized [68]. Also, several $Cr(CO)_3$ complexes of alkylated C_2-dihydrodiindenes have been reported [129]. Triptindane 5 has been converted into a mono-, two bis- and the tris-complex 270 (Scheme 57) [130], and 10-methyltribenzotriquinacene 7 furnished a total of six different $Cr(CO)_3$ complexes. Among these the two possible stereoisomeric tris-complexes were characterized [131]. Recently, cationic tricarbonylmanganese complexes of several centropolyindanes have been prepared and their structures characterized [132].

Most of the parent centropolyindanes have been subjected to partial reduction of the benzo rings [133, 134]. Birch reduction of fenestrindane 9, for example, leads to the [5.5.5.5]fenestratetraene 272 bearing four annelated 1,4-

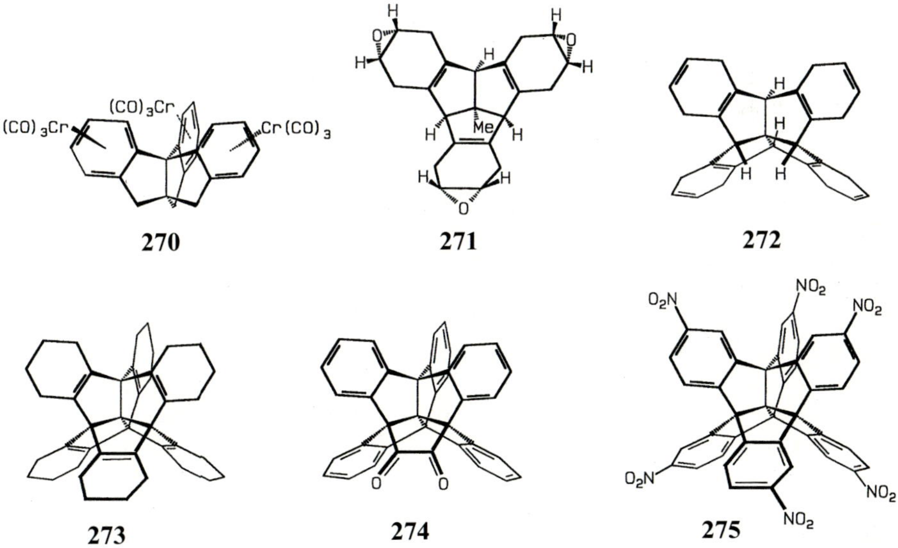

270 **271** **272**

273 **274** **275**

Scheme 57

cyclohexadiene rings. Among other congeners, triptindane **5** and 10-methyltribenzotriquinacene **7** give the corresponding hexaenes. All these polyenes could be epoxidized regio- and stereoselectively [134]. For example, the tris(epoxide) **271** bearing the oxyfunctionalization at the *outer* periphery has been synthesized [135]. In addition, multiple Benkeser reduction has been achieved with several centropolyindanes. For example, centrohexaindane **11** was converted into the first centrohexaquinacene derivative **273** [133].

Partial oxidative degradation of the benzo nuclei of several centropolyindanes has also been performed, albeit with low efficiency [42]. For example, ring cleavage by oxidation with ruthenium(VIII) or by ozonolysis of centrohexaindane **11** leads to diketone **274** in low yield. X-ray structure analysis of the deeply red crystals corroborates the almost perfect *cis*-coplanar orientation of the α,β-dicarbonyl chromophore clamped into the rigid molecular backbone.

Aromatic substitution of centropolyindanes are being performed mainly with tribenzotriquinacenes **7** and **150** and with centrohexaindane **11** [136]. Under carefully controlled conditions, selective monofunctionalization of each benzo nucleus has been achieved, such as six-fold nitration of **11** giving mixtures of constitutional isomers (e.g. **275**) [136]. Also, three-fold *ortho* disubstition at the six equivalent peripheral positions of tribenzotriquinacenes has been performed successfully, making use of the steric hindrance of the six inner positions. Current investigations concentrate on the selective aromatic substitution of various centropolyindanes. In the case of centrohexaindane **11**, these experiments should open ways to, for example, twelve-fold functionalized derivatives of T_d symmetry (**276**, Fig. 1). In turn, these compounds could be used as

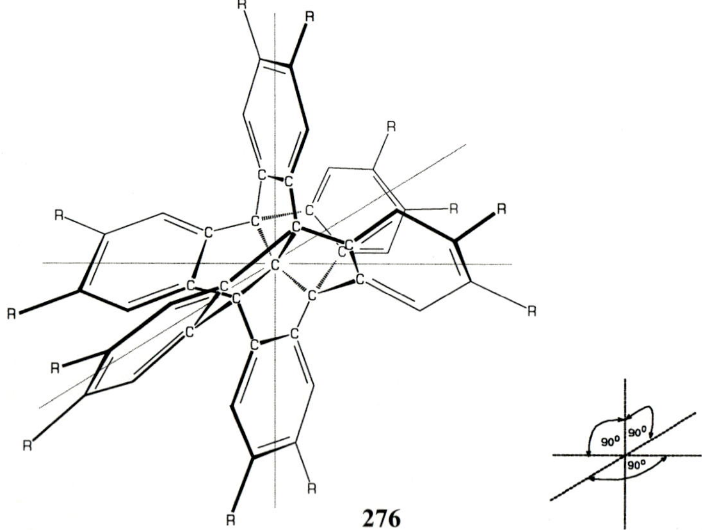

Fig. 1. Representation of hypothetical derivatives **276** of centrohexaindane **11** bearing twelve "peripheral" substituents R. The three axes indicated are embedded in each of the 2,2'-spirobiindane units and cross at right angles in the centre of the highlightened centrohexacyclic C_{17} nucleus

building blocks containing a carbon-rich molecular centre (C_{17}) for the construction of up to six extended aromatic chromophores and tentacular groups extended at right angles in the three-space.

9
Conclusion

Centropolyindanes differ from other polycyclic indane compounds in the rigid fusion of up to six indane units along the C–C bonds of their five-membered rings. This unique structural feature has led to the development of truly three-dimensional, carbon-rich polycyclic frameworks. All "regular" centropolyindanes are synthetically available, some of them by several and/or even surprisingly short routes. Owing to the facile substitution or functionalization of the bridgehead positions, the parent centropolyindanes can be converted to a large variety of related multiply ring-fused structures. Centropolyindanes bearing substituents at the inner or outer arene positions have also been synthesized by starting from suitable precursors. Selective introduction of substituents at the outer (peripheral) positions of the aromatic rings, possibly combined with appropriate incorporation of additional building blocks fixed at the bridgehead positions, will open several new fields in this area of polycyclic hydrocarbons.

10
References

1. Greenberg A, Liebman JL (1978) Strained Organic Molecules. Academic Press, New York
2. Clar E (1964) Polycyclic Hydrocarbons, vol 1 and 2. Academic Press, London, and Springer, Berlin Göttingen Heidelberg
3. Vögtle F (1989) Reizvolle Moleküle der Organischen Chemie. Teubner, Stuttgart
4. Dodziuk H (1995) Modern Conformational Analysis. VCH Publishers, New York; Osawa E, Yonemitsu O (eds 1992) Carbocyclic Cage Compounds: Chemistry and Applications. VCH Publishers, New York; Cram DJ, Cram JM (1994) Container Molecules and Their Guests. Royal Society of Chemistry, Cambridge; Ho T-L (1995) Symmetry: A Basis for Synthetic Design. Wiley, New York; Ho T-L (1992) Tandem Organic Reactions. Wiley, New York; Olah GA (ed 1990) Cage Hydrocarbons. Wiley, New York; Eaton PE (ed 1986) Synthesis of Non-Natural Products: Challenge and Reward. Tetrahedron Symposia-in-Print no 26. Pergamon Press, Oxford
5. Müller M, Mauermann-Düll H, Wagner M, Enkelmann V, Müllen K (1995) Angew Chem 107:1751 (Angew Chem, Int Ed Engl 34:1583); Stabel A, Herwig P, Müllen K, Rabe JP (1995) ibid 107:1768 (34:1609); Iyer VS, Wehmeier M, Brand JD, Keegstra MA, Müllen K (1997) ibid 109:1676 (36:1604); Müller M, Iyer VS, Kübel C, Enkelmann V, Müllen K (1997) ibid 109:1679 (36:1607)
6. Scott LT (1996) Pure Appl Chem 68:291; Scott LT, Hashemi MM, Meyer DT, Warren HB (1991) J Am Chem Soc 113:7082; Scott LT, Hashemi MM, Brather MS (1992) ibid 114:1920; Scott LT, Brather MS, Hagen S (1996) ibid 118:8743; Kiyobayashi T, Nagano Y, Sakiyama M, Yamamoto K, Cheng PC, Scott LT (1995) J Am Chem Soc 117:3270
7. Rabideau PW, Sygula A (1995) Polynuclear Aromatic Hydrocarbons with Curved Surfaces: Hydrocarbons Possessing Carbon Frameworks Related to Buckminsterfullerene. In: Thummel RP (ed) Advances in Theoretically Interesting Molecules, vol 3. JAI Press,

Greenwich CT, p 1; Rabideau PW, Sygula A (1996) Acc Chem Res 29: 235; Abdourazak AH, Sygula A, Rabideau PW (1993) J Am Chem Soc 115:3010; Rabideau PW, Abdourazak AH, Folsom HE, Marcinow Z, Sygula A, Sygula R (1994) ibid 116:7891; Sygula A, Abdourazak AH, Rabideau PW (1996) ibid 118:2271; Sygula A, Folsom HE, Sygula R, Abdourazak AH, Marcinow Z, Fronczek FR, Rabideau PW (1994) J Chem Soc, Chem Commun 2571; Clayton MD, Marcinow Z, Rabideau PW (1996) J Org Chem 61:6052

8. Borchardt A, Fuchicello A, Kilway VK, Baldridge KK, Siegel JS (1992) J Am Chem Soc 114: 1921

9. Yamamoto K, Sonobe H, Matsubara H, Sato M, Okamoto S, Kitaura K (1996) Angew Chem 108:69 (Angew Chem, Int Ed Engl 35:69); Yamamoto K, Saitho Y, Iwaki D, Ooka T (1991) 103:1202 (30:1173); Yamamoto K, Harada T, Okamoto Y, Chikamatsu H, Nakazaki M, Kai Y, Nakao T, Tanaka M, Harada S, Kasai N (1988) J Am Chem Soc 110:3578

10. Qiao X, Padula MA, Ho DM, Vogelaar NJ, Schutt CE, Pascal RA Jr (1996) J Am Chem Soc 118:741; Pascal RA Jr, McMillan WD, Van Engen D, Eason RG (1987) ibid 109:4660

11. Kammermeier S, Jones PG, Herges R (1996) Angew Chem 108:2834 (Angew Chem, Int Ed Engl 35:2669)

12. Sarobe M, Flink S, Jenneskens LW, van Poecke BLA, Zwikker JW (1995) J Chem Soc, Chem Commun 2415

13. Scott LT, Necula A (1996) J Org Chem 61:386; Hagen S, Brather MS, Erickson MS, Zimmermann G, Scott LT (1997) Angew Chem 109:407 (Angew Chem, Int Ed Engl 36:406)

14. Abdourazak AH, Marcinow Z, Sygula A, Sygula R, Rabideau PW (1995) J Am Chem Soc 117:6410

15. Vollhardt KPC, Mohler DL (1996) The Phenylenes: Synthesis, Properties, and Reactivity. In: Halton B (ed) Advances in Strain in Organic Chemistry, vol 5. JAI Press, Greenwich CT, p 121; Vollhardt KPC (1993) Pure Appl Chem 65:153; Mohler DL, Vollhardt KPC, Wolff S (1995) Angew Chem 107:601 (Angew Chem, Int Ed Engl 34:563); Boese R, Matzger AJ, Mohler DL, Vollhardt KPC (1995) ibid 107:1630 (34:1478); Cierckes R, Vollhardt KPC (1986) J Am Chem Soc 108:3150

16. Willmore ND, Liu L, Katz TJ (1992) Angew Chem 104:1081 (Angew Chem, Int Ed Engl 31: 1093); Laali KK, Houser JJ (1994) J Chem Soc, Perkin Trans 2:1303; Staab HA, Diehm M, Krieger C (1994) Tetrahedron Lett 35:8357; Laarhoven WH, Prinsen WJC (1984) Top Curr Chem 125: 63; Martin RH (1974) Angew Chem 86:727 (Angew Chem, Int Ed Engl 13:649)

17. Kroto HW (1992) Angew Chem 104:113 (Angew Chem, Int Ed Engl 31:111); Kroto HW, Heath JR, O'Brien SC, Curl RF, Smalley RE (1985) Nature 318:162; Kroto HW, Allaf AW, Balm SP (1991) Chem Rev 91:1213; Krätschmer W, Lamb LD, Fostiropoulos K, Huffman DR (1990) Nature 347:354; Wudl F (1992) Acc Chem Res 25:157; Hawkins JM (1992) ibid 25:150; Diederich F, Whetten RL (1992) ibid 25:119; Curl RF, Smalley RE (1988) Science 242:1017; Hirsch A (1994) The Chemistry of the Fullerenes. Thieme, Stuttgart

18. Curl RF (1997) Angew Chem 109:1636 (Angew Chem, Int Ed Engl 36:1566); Kroto HW (1997) ibid 109:1648 (36:1578); Smalley RE (1997) ibid 109:1666 (36:1594)

19. Eaton PE (1992) Angew Chem 104:1447 (Angew Chem, Int Ed Engl 31:1421)

20. Higuchi H, Ueda I (1992) Recent Developments in the Chemistry of Cubane. In: Osawa E, Yonemitsu O (eds) Carbocyclic Cage Compounds: Chemistry and Applications. VCH Publishers, New York

21. Paquette LA (1989) Chem Rev 89:1051; Paquette LA (1990) The [n]Peristylane Polyhedrane Connection. In Olah GA (ed) Cage Hydrocarbons. Wiley, New York, p 313

22. Prinzbach H, Fessner WD (1987) Novel Organic Polycycles an Adventure in Molecular Architecture. In: Chizhov O (ed) Organic Synthesis: Modern Trends. Blackwell, Oxford, p 23; Fessner WD, Prinzbach H (1990) The Pagodane Route to Dodecahedrane. In Olah GA (ed) Cage Hydrocarbons. Wiley, New York, p 353

23. Nakazaki M, Naemura K, Harada H, Narukaki H (1982) J Org Chem 47:3470

24. Fessner WD, Prinzbach H (1986) Tetrahedron 42:1797; Nakazaki M, Naemura K, Arashiba M (1978) J Org Chem 43:689; Underwood GR, Ramamoorthy B (1970) Tetrahedron Lett 4125

25. McKervey MA (1980) Tetrahedron 36:971; McKervey MA (1974) Chem Soc Rev 3:479; Schleyer PvR (1990) My Thirty Years in Hydrocarbon Cages: From Adamantane to Dode-

cahedrane. In: Olah GA (ed) Cage Hydrocarbons. Wiley, New York, p 1; McKervey MA, Rooney JJ (1990) Catalytic Routes to Adamantanes and Its Homologues. Ibid. Wiley, New York, p 39

26. Hart H, Bashir-Hashemi A, Luo J, Meador MA (1986) Tetrahedron 42:1641
27. Ashton PR, Isaacs NS, Kohnke FH, Slawin AMZ, Spencer CM, Stoddart JF, Williams DJ (1988) Angew Chem 100:981 (Angew Chem, Int Ed Engl 27:966)
28. Kuck D (1990) Centropolyindan(e)s: Benzoannelated Polyquinanes with a Central Carbon Atom. In: Hargittai I (ed) Quasicrystals, Networks, and Molecules of Fivefold Symmetry. VCH Publishers, New York, p 289
29. Kuck D (1996) Synlett 949
30. Ginsburg D (1975) Propellanes. Structure and Reactions. Verlag Chemie, Weinheim
31. Tobe Y (1992) Propellanes. In: Osawa E, Yonemitsu O (eds) Carbocyclic Cage Compounds: Chemistry and Applications. VCH Publishers, New York
32. Paquette LA (1979) Top Curr Chem 79:41; Paquette LA (1984) ibid 119:1; Paquette LA, Doherty AM (1987) Polyquinane Chemistry, Synthesis and Reactions. Springer, Berlin Heidelberg New York Tokyo; for a more recent review on certain aspects of oligoquinane chemistry see: Haag R, de Meijere A (1998) Top Curr Chem 196:137–165
33. Most recent papers on non-benzoannelated triquinacenes and related systems: Haag R, Fleischer R, Stahlke D, de Meijere A (1995) Angew Chem 107:1642 (Angew Chem, Int Ed Engl 34:1492); Zuber R, Carlens G, Haag R, de Meijere A (1996) Synlett 542
34. Venepalli BR, Agosta WC (1987) Chem Rev 87:399; Agosta WC (1992) Inverted and Planar Carbon. In: Patai S, Rappoport Z (eds) The Chemistry of Alkanes and Cycloalkanes. Wiley, New York, p 927
35. Keese R (1987) Planarization of Tetracoordinate Carbon. Synthesis and Structure of [5.5.5.5]Fenestranes. In: Chizhov O (ed) Organic Synthesis: Modern Trends. Blackwell, Oxford, p 43; Luef W, Keese R (1993) Planarizing Distortions in Carbon Compounds. In Halton B (ed) Advances in Strain in Organic Chemistry, vol 3. JAI Press, Greenwich CT, p 229
36. Krohn K (1991) Fenestranes a Look at "Structural Pathologies". In Mulzer J, Altenbach HJ, Braun M, Krohn K, Reissig HU (eds) Organic Synthesis Highlights. VCH Publishers, Weinheim, p 371
37. Kuck D (1998) Benzoannelated Fenestranes. In: Thummel RP (ed) Advances in Theoretically Interesting Molecules, vol 4. JAI Press, Greenwich CT, p 81–155 (in press)
38. Kuck D (1997) Liebigs Ann Rec 1043
39. Trost BM (1982) Chem Soc Rev 11:141
40. Gupta AK, Fu X, Snyder JP, Cook JM (1991) Tetrahedron 47:3665
41. Gund P, Gund TM (1981) J Am Chem Soc 103:4458
42. Gestmann D, Pritzkow H, Kuck D (1996) Liebigs Ann 1349
43. Thompson HW (1968) J Org Chem 33:621; Thompson HW (1966) Tetrahedron Lett 6489
44. Kuck D (1984) Angew Chem 96:515 (Angew Chem, Int Ed Engl 23:508)
45. Kuck D, Seifert M (1992) Chem Ber 125:1461
46. Kuck D, Bögge H (1986) J Am Chem Soc 108:8107
47. Kuck D, Schuster A, Gestmann D (1994) J Chem Soc, Chem Commun 609
48. Kuck D, Schuster A (1988) Angew Chem 100:1222 (Angew Chem, Int Ed Engl 27:1192)
49. Kuck D, Eckrich R, Tellenbröker J (1994) J Org Chem 59:2511
50. Brewster JH, Prudence RT (1973) J Am Chem Soc 95:1217; Hill RK, Cullison DA (1973) J Am Chem Soc 95:1229
51. Mittal RSD, Sethi SC, Dev S (1973) Tetrahedron 29:1321; Ogura F, Nakao A, Nakagawa M (1980) Bull Chem Soc Jpn 53:291
52. Ten Hoeve W, Wynberg H (1980) J Org Chem 45:2930; Ten Hoeve W (1979) Proefschrift (Doctoral thesis), Rijksuniversiteit te Groningen
53. Rapoport H, Pasky JZ (1956) J Am Chem Soc 78:3788; Lindner HJ, Eilbracht P (1973) Chem Ber 106:2268
54. Mayer R (1956) Chem Ber 89:1443; Katz TJ, Slusarek W (1980) J Am Chem Soc 102:1058

55. Cho BP, Kim M, Harvey RG (1993) J Org Chem 58:5788; Cho BP, Harvey RG (1987) ibid 52: 5668; Cho BP, Harvey RG (1987) ibid 52:5679; Cho BP, Harvey RG (1987) Tetrahedron Lett 28:861
56. Bladauski D, Broser W, Hecht HJ, Rewicki D, Dietrich H (1979) Chem Ber 112:1380; Dietrich H, Bladauski D, Grosse M, Roth K, Rewicki D (1975) Chem Ber 108:1807
57. Simmons GL, Lingafelter EC (1961) Acta Cryst 14:872; Anderson AG Jr, Wade RH (1952) J Am Chem Soc 74:2274
58. Liebermann L, Bergami O (1889) Ber Dtsch Chem Ges 24:782
59. Fabre C, Rassat A (1989) C R Acad Sci Paris, Sér II 308:1223; Kelle T (1996) Dissertation, Universität Bielefeld; Loguercio D Jr (1988) PhD thesis, University of California at Los Angeles
60. Sygula A, Rabideau PW (1994) J Chem Soc, Chem Commun 1497; Sygula A, Rabideau PW (1994) ibid 2271; McKee ML, Herndon WC (1987) J Mol Struct (Theochem) 153:75
61. Neudeck HK, Schlögl K (1977) Chem Ber 110:2624; Dynesen E (1975) Acta Scand B 29:77; Falk H, Fröstl W, Schlögl K (1974) Monatsh Chem 105:574; Falk H, Fröstl W, Hofer O, Schlögl K (1974) Monatsh Chem 105:598
62. Lemmen P, Ugi I (1987) Chem Scr 27:297; Hofer O, Neudeck H (1987) Monatsh Chem 118: 1163; Kabuto K, Yasuhara F, Yamaguchi S (1984) Tetrahedron Lett 25:5801; Lemmen P (1983) Liebigs Ann Chem 668; Lemmen P (1982) Chem Ber 115:1902
63. Smits JMM, Noordijk JH, Beurskens PT, Laarhoven WH; Lijten FAT (1986) J Cryst Spectrosc Res 16:23; Laarhoven WH, Lijten FAT, Smits JMM (1985) J Org Chem 50:3208; Cristol JS, Jarvis BB (1967) J Am Chem Soc 89:401
64. Baker W, McOmie JFW, Parfitt SD, Watkins DAM (1957) J Chem Soc 4026
65. Kuck D, Lindenthal T, Schuster A (1992) Chem Ber 125:1449
66. Kuck D (1984) Z Naturforsch B 39:369; Campbell N, Davison PS, Heller HG (1963) J Chem Soc 996
67. Barnes RA, Beitchman BD (1954) J Am Chem Soc 76:5430
68. Lamarche JM, Laude B (1982) Bull Soc Chim France 97
69. Kuck D, Neumann E, Schuster A (1994) Chem Ber 127:151
70. Anstead GM, Srinivasan R, Peterson CS, Wilson SR, Katzenellenbogen JA (1991) J Am Chem Soc 113:1378
71. Allinger NL, Chen K, Katzenellenbogen JA, Wilson SR, Anstead GM (1996) J Comput Chem 17:747
72. Wendelin W, Schermanz K, Breitmaier E (1988) Monatsh Chem 119:355; Witschard G, Griffin CE (1964) J Org Chem 29:2335; Hassner A, Cromwell NH (1958) J Am Chem Soc 80:893; Chatterjea JN, Prasad K (1957) J Ind Chem Soc 34:375; Schmidt B (1900) J Prakt Chem 62:545; Kipping FS (1894) J Chem Soc 65:480
73. Houlihan WJ, Shapiro MJ, Chin JA (1997) J Org Chem 62:1529
74. Paisdor B (1989) Dissertation, Universität Bielefeld
75. Lipkowitz K, Larter RM, Boyd DB (1980) J Am Chem Soc 102:85
76. Kuck D, Paisdor B, Grützmacher HF (1987) Chem Ber 120:589
77. Paisdor B, Grützmacher HF, Kuck D (1988) Chem Ber 121:1307
78. Paisdor B, Kuck D (1991) J Org Chem 56:4753
79. Hackfort T (1998) Dissertation, Universität Bielefeld
80. Kuck D, Großer M (1997) unpublished results
81. Simmons HE III (1980) PhD thesis, Harvard University; Simmons HE III, Maggio JE (1981) Tetrahedron Lett 22:287; Maggio JE, Simmons HE III, Kouba JK (1981) J Am Chem Soc 103:1579; Benner SA, Maggio JE, Simmons HE III (1981) J Am Chem Soc 103:1581
82. Paquette LA, Vazeux M (1981) Tetrahedron Lett 22:291; Paquette LA, Williams RV, Vazeux M, Browne AR (1984) J Org Chem 49:2194
83. Kuck D, Paisdor B, Gestmann D (1994) Angew Chem 106:1326 (Angew Chem, Int Ed Engl 33:1251)
84. Alder RW, Colclough D, Grams F, Orpen AG (1990) Tetrahedron 46:7933
85. Moriconi EJ, O'Connor WF, Kuhn LP, Keneally EA, Wallenberger FT (1959) J Am Chem Soc 81:6472

86. Dyker G, Körning J, Nerenz F, Siemsen P, Sostmann S, Wiegand A, Jones PG, Bubenitschek P (1996) Pure Appl Chem 68:323; Dyker G (1991) Tetrahedron Lett 32:7241; Dyker G (1993) J Org Chem 58:234

87. Dyker G, Körning J, Bubenitschek P, Jones PG (1997) Liebigs Ann Recueil 203; Dyker G, Körning J, Jones PG, Bubenitschek P (1993) Angew Chem 105:1805 (Angew Chem, Int Ed Engl 32:1733)

88. Dyker G, Körning J, Jones PG, Bubenitschek P (1996) Tetrahedron 52:14777

89. Crockett R, Gescheidt G, Dyker G, Körning J (1996) Liebigs Ann 1533

90. Hackfort T, Kuck D (1997) manuscript in preparation

91. Hackfort T, Exner C, Kuck D (1997) manuscript in preparation

92. Kuck D, Hackfort H, Pritzkow H (1997) manuscript in preparation

93. Kuck D (1994) Chem Ber 127:409

94. Chuen CC, Fenton SW (1958) 23:1538; Blood CT, Linstead RP (1952) J Chem Soc 2263

95. Seifert M (1991) Dissertation, Universität Bielefeld; Seifert M, Barth D, Kuck D (to be published)

96. Kuck D, Bredenkötter B (1997) manuscript in preparation

97. Seifert M, Kuck D (1996) Tetrahedron 52:13167

98. Krause RA (1992) Dissertation, Universität Bielefeld; Kuck D, Krause RA (to be published)

99. Kuck D, Schuster A, Fusco C, Fiorentino M, Curci R (1994) J Am Chem Soc 116:2375

100. Schuster A (1991) Dissertation, Universität Bielefeld; Kuck D, Schuster A (to be published)

101. Butenschön H, de Meijere A (1985) Chem Ber 118:2757

102. Kuck D, Schuster A, Gestmann D, Posteher F, Pritzkow H (1996) Chem Eur J 2:58

103. Kuck D, Schuster A, Ohlhorst B, Sinnwell V, de Meijere A (1989) Angew Chem 101:626 (Angew Chem, Int Ed Engl 28:595)

104. Kuck D, Schuster A, Paisdor B, Gestmann D (1995) J Chem Soc, Perkin Trans 1:721

105. Kuck D, Exner C (1997) unpublished results

106. Kuck D, Bruder A, Krause RA, Pritzkow H (1997) manuscript in preparation

107. Schuster A, Kuck D (1991) Angew Chem 103:1717 (Angew Chem, Int Ed Engl 30:1699)

108. Haag R, Ohlhorst B, Noltemeyer M, Fleischer R, Stahlke D, Schuster A, Kuck D, de Meijere A (1995) J Am Chem Soc 117:10474

109. Haag R, Ohlhorst B, Noltemeyer M, Schuster A, Kuck D, de Meijere A (1993) J Chem Soc, Chem Commun 727

110. Butenschön H, de Meijere A (1985) Helv Chim Acta 68:1658; Butenschön H, de Meijere A (1984) Tetrahedron Lett 25:1693

111. Haag R, Kuck D, Fu XY, Cook JM, de Meijere A (1994) Synlett 340

112. Haag R (1995) Dissertation, Universität Göttingen

113. Tucker SA, Acree WE Jr, Upton C (1993) Appl Spectrosc 47:201; Bloomfield DG, Upton C, Vipond HJ (1986) J Chem Soc, Faraday Trans 1:857; Upton C (1986) J Chem Soc, Perkin Trans 1:1225; Partridge MW, Vipond HJ (1962) J Chem Soc 632

114. Kuck D, Pritzkow H (1994) unpublished results

115. Shternberga IYa, Freimanis YaF (1968) J Org Chem USSR 4:1044; Popelis YuYu, Pestunovich VA, Shternberga IYa, Freimanis Ya F (1972) ibid 8:1907; Shternberga IYa, Freimanis YaF (1972) Latv PSR Zinat Akad Vestis, Kim Ser 207

116. Ten Hoeve W, Wynberg H (1979) J Org Chem 44:1508

117. Luef W, Keese R (1987) Helv Chim Acta 70:543

118. Hoffmann R (1971) Pure Appl Chem 28:181; Hoffmann R, Alder RW, Wilcox CF (1970) J Am Chem Soc 92:4992

119. McGrath MP, Radom L (1993) J Am Chem Soc 115:3320; Lyons JE, Rasmussen DR, McGrath MP, Nobes RH, Radom L (1994) Angew Chem 106:1722 (Angew Chem, Int Ed Engl 33:1667)

120. Röttger D, Erker G (1997) Angew Chem 109:841 (Angew Chem, Int Ed Engl 36:812)

121. Kuck D, Schuster A, Krause RA (1991) J Org Chem 56:3472

122. Fusco C, Fiorentino M, Dinoi A, Curci R, Krause RA, Kuck D (1996) J Org Chem 61:8681

123. Kuck D, Krause D, Gestmann D, Posteher F, Schuster A (1998) Tetrahedron 54 (in press)

124. Pohl S, Saak W, Schuster A, Kuck D (1991) unpublished results
125. Langer E, Lehner H (1973) Tetrahedron 29 : 375
126. Meyer A, Neudeck H, Schlögl K (1977) Chem Ber 110:1403; Meyer A, Neudeck H, Schlögl K (1976) Tetrahedron Lett 26:2233
127. Bitterwolf TE, Ceccon A, Gambaro A, Ganis P, Kuck D, Manoli F, Rheingold AL, Valle G, Venzo A (1997) J Chem Soc, Perkin Trans 2:735
128. Ceccon A, Gambaro A, Manoli F, Venzo A, Ganis P, Kuck D, Valle G (1992) J Chem Soc, Perkin Trans 2:1111
129. Trifonova OI, Galiullin RA, Ustynyuk YuA, Ustynyuk NA, Petrovskii PV, Kravtsov PN (1988) Proc Acad Sci USSR, Chem Ser 1218; Trifonova OI, Galiullin RA, Ustynyuk YUA, Ustynyuk NA, Petrovskii PV, Kravtsov PN (1987) J Organomet Chem 328:321
130. Ceccon A, Gambaro A, Manoli F, Venzo A, Ganis P, Valle G, Kuck D (1993) Chem Ber 126: 2053
131. Ceccon A, Gambaro A, Manoli F, Venzo A, Kuck D, Bitterwolf TE, Ganis P, Valle G (1991) J Chem Soc, Perkin Trans 2:233
132. Sun S, Dullaghan CA, Sweigart DA (1996) J Chem Soc, Dalton Trans 4493; Dullaghan CA, Curci R, Kuck D, Sweigart DA unpublished results
133. Eckrich R, Kuck D (1993) Synlett 344
134. Eckrich R (1993) Dissertation, Universität Bielefeld; Kuck D, Eckrich R (to be published)
135. Eckrich R, Neumann B, Stammler HG, Kuck D (1996) J Org Chem 61:3839
136. Tellenbröker J, Kuck D, to be published

Note added in proof: The tris(epoxide) **80** has been prepared by oxidation of **77** with dimethyldioxirane as a mixture of diastereomers. It proved to be extremely labile [80].

Author Index Volumes 151–196

The volume numbers are printed in italics

Adam W, Hadjiarapoglou L (1993) Dioxiranes: Oxidation Chemistry Made Easy. *164*:45–62

Alberto R (1996) High- and Low-Valency Organometallic Compounds of Technetium and Rhenium. *176*:149–188

Albini A, Fasani E, Mella M (1993) PET-Reactions of Aromatic Compounds. *168*:143–173

Allan NL, Cooper D (1995) Momentum-Space Electron Densities and Quantum Molecular Similarity. *173*:85–111

Allamandola LJ (1990) Benzenoid Hydrocarbons in Space: The Evidence and Implications. *153*:1–26

Alonso JA, Balbás LC (1996) Density Functional Theory of Clusters of Naontransition Metals Using Simple Models. *182*:119–171

Améduri B, Boutevin B (1997) Telomerisation Reactions of Fluorinated Alkenes. *192*:165–233

Anderson RC, McGall G, Lipshutz RJ (1998) Polynucleotide Arrays for Genetic Sequence Analysis. *194*:117–129

Anwander R (1996) Lanthanide Amides. *179*:33–112

Anwander R (1996) Routes to Monomeric Lanthanide Alkoxides. *179*:149–246

Anwander R, Herrmann WA (1996) Features of Organolanthanide Complexes. *179*:1–32

Artymiuk PJ, Poirette AR, Rice DW, Willett P (1995) The Use of Graph Theoretical Methods for the Comparison of the Structures of Biological Macromolecules. *174*:73–104

Astruc D (1991) The Use of p-Organoiron Sandwiches in Aromatic Chemistry. *160*:47–96

Azumi T, Miki H (1997) Spectroscopy of the Spin Sublevels of Transition Metal Complexes. *191*:1–40

Baerends EJ, see van Leeuwen R (1996) *180*:107–168

Baker BJ, Kerr RG (1993) Biosynthesis of Marine Sterols. *167*:1–32

Balbás LC, see Alonso JA (1996) *182*:119–171

Baldas J (1996) The Chemistry of Technetium Nitrido Complexes. *176*:37–76

Balzani V, Barigelletti F, De Cola L (1990) Metal Complexes as Light Absorption and Light Emission Sensitizers. *158*:31–71

Bardin VV, see Petrov VA (1997) *192*:39–95

Barigelletti F, see Balzani V (1990) *158*:31–71

Bassi R, see Jennings RC (1996) *177*:147–182

Battersby AR, Leeper FJ (1998) Biosynthesis of Vitamin B_{12}. *195*:143–193

Baumgarten M, Müllen K (1994) Radical Ions: Where Organic Chemistry Meets Materials Sciences. *169*:1–104

Beau J-M and Gallagher T (1997) Nucleophilic C-Glycosyl Donors for C-Glycoside Synthesis. *187*:1–54

Bechthold A F-W, see Kirschning A (1997) *188*:1–84

Begley TP, Kinsland C, Taylor S, Tandon M, Nicewonger R, Wu M, Chin H-J, Kelleher N, Campobasso N, Zhang Y (1998) Cofactor Biosynthesis: A Mechanistic Perspective. *195*:93–142

Berces A, Ziegler T (1996) Application of Density Functional Theory to the Calculation of Force Fields and Vibrational Frequencies of Transition Metal Complexes. *182*:41–85

Cimino G, Sodano G (1993) Biosynthesis of Secondary Metabolites in Marine Molluscs. *167*: 77–116.

Ciolowski J (1990) Scaling Properties of Topological Invariants. *153*:85–100

Clark T (1996) Ab Initio Calculations on Electron-Transfer Catalysis by Metal Ions. *177*:1–24

Cohen MH (1996) Strenghtening the Foundations of Chemical Reactivity Theory. *183*:143–173

Collet A, Dutasta J-P, Lozach B, Canceill J (1993) Cyclotriveratrylenes and Cryptophanes: Their Synthesis and Applications to Host-Guest Chemistry and to the Design of New Materials. *165*:103–129

Colombo M G, Hauser A, Güdel HU (1994) Competition Between Ligand Centered and Charge Transfer Lowest Excited States in bis Cyclometalated Rh^{3+} and Ir^{3+} Complexes. *171*:143–172

Cooper DL, Gerratt J, Raimondi M (1990) The Spin-Coupled Valence Bond Description of Benzenoid Aromatic Molecules. *153*:41–56

Cooper DL, see Allan NL (1995) *173*:85–111

Cordero FM, see Goti A (1996) *178*:1–99

Cyvin BN, see Chen RS (1990) *153*:227–254

Cyvin SJ, see Chen RS (1990) *153*:227–254

Cyvin BN, Brunvoll J, Cyvin SJ (1992) Enumeration of Benzenoid Systems and Other Polyhexes. *162*:65–180

Cyvin SJ, see Cyvin BN (1992) *162*:65–180

Cyvin BN, see Cyvin SJ (1993) *166*:65–119

Cyvin SJ, Cyvin BN, Brunvoll J (1993) Enumeration of Benzenoid Chemical Isomers with a Study of Constant-Isomer Series. *166*:65–119

Dartyge E, see Fontaine A (1989) *151*:179–203

De Cola L, see Balzani V (1990) *158*:31–71

Dear K (1993) Cleaning-up Oxidations with Hydrogen Peroxide. 16

de Meijere A, see Haag R (1998) *196*:137–165

de Mendoza J, see Seel C (1995) *175*:101–132

de Raadt A, Ekhart CW, Ebner M, Stütz AE (1997) Chemical and Chemo-Enzymatic Approaches to Glycosidase Inhibitors with Basic Nitrogen in the Sugar Ring. *187*:157–186

de Silva AP, see Bissell RA (1993) *168*:223–264

Descotes G (1990) Synthetic Saccharide Photochemistry. *154*:39–76

Dias JR (1990) A Periodic Table for Benzenoid Hydrocarbons. *153*:123–144

Dietrich-Buchecker Ch, see Chambron J-C (1993) *165*:131–162

Dobson JF (1996) Density Functional Theory of Time-Dependent Phenomena. *181*:81–172

Dohm J, Vögtle, F (1991) Synthesis of (Strained) Macrocycles by Sulfone Pyrolysis. *161*:69–106

Dolbier WR jr. (1997) Fluorinated Free Radicals. *192*:97–163

Drakesmith FG (1997) Electrofluorination of Organic Compounds. *193*:197–242

Dreizler RM (1996) Relativistic Density Functional Theory. *181*:1–80

Driguez H (1997) Thiooligosaccharides in Glycobiology. *187*:85–116

Dutasta J-P, see Collet A (1993) *165*:103–129

Eaton DF (1990) Electron Transfer Processes in Imaging. *156*:199–226

Ebner M, see de Raadt A (1997) *187*:157–186

Edelmann FT (1996) Rare Earth Complexes with Heteroallylic Ligands. *179*:113–148

Edelmann FT (1996) Lanthanide Metallocenes in Homogeneous Catalysis. *179*:247–276

Effenhauser CS (1998) Integrated Chip-Based Microcolumn Separation Systems. *194*:51–82

Ehrfeld W, Hessel V, Lehr H (1998) Microreactors for Chemical Synthesis and Biotechnology – Current Developments and Future Applications. *194*:233–252

Ekhart CW, see de Raadt A (1997) *187*:157–186

El-Basil S (1990) Caterpillar (Gutman) Trees in Chemical Graph Theory. *153*:273–290

Engel E (1996) Relativistic Density Functional Theory. *181*:1–80

Ernzerhof M, Perdew JP, Burke K (1996) Density Functionals: Where Do They Come From, Why Do They Work? *190*:1–30

Fasani A, see Albini A (1993) *168*:143–173

Fernández-Mayoralas A (1997) Synthesis and Modification of Carbohydrates using Glycosidases and Lipases. *186*:1–20

Printing: Saladruck, Berlin
Binding: Buchbinderei Lüderitz & Bauer, Berlin